Chemical Engineering Design
An Introduction

Chemical Engineering Design
An Introduction

Contributors :
Martha A. Grover,
Christine Y. He, *et al.*

AURIS REFERENCE LTD.
London, UK

Chemical Engineering Design: *An Introduction*
Contributors : Martha A. Grover *and* Christine Y. He, *et al.*

Auris Reference Ltd., UK

www.aurisreference.com

United Kingdom

Copyright 2016

Printed in 2017 for Sale in the Indian Subcontinent

Chemical Engineering Design: *An Introduction*

ISBN: 978-1-78154-509-6

British Library Cataloguing in Publication Data
A CIP record for this book is available from the British Library

PREFACE

Chemical engineering is a branch of science that applies physical sciences and life sciences together with applied mathematics and economics to produce, transform, transport, and properly use chemicals, materials and energy. Essentially, chemical engineers design large-scale processes that convert chemicals, raw materials, living cells, microorganisms and energy into useful forms and products.

Chemical engineering design concerns the creation of plans, specification, and economic analyses for pilot plants, new plants or plant modifications. Design engineers often work in a consulting role, designing plants to meet clients' needs. Design is limited by a number of factors, including funding, government regulations and safety standards. These constraints dictate a plant's choice of process, materials and equipment.

Process design requires the definition of equipment types and sizes as well as how they are connected together and the materials of construction. Details are often printed on a Process Flow Diagram which is used to control the capacity and reliability of a new or modified chemical factory.

This page left intentionally blank.

CONTENTS

This page left intentionally blank.

LIST OF CONTRIBUTORS

Martha A. Grover

School of Chemical & Biomolecular Engineering, Georgia Institute of Technology, 311 Ferst Dr. NW, Atlanta, GA 30032, USA; E-Mails: christine.he@chbe.gatech.edu (C.Y.H.); mhsieh34@gatech.edu (M.-C.H.); syu47@gatech.edu (S.-S.Y.)

Christine Y. He

School of Chemical & Biomolecular Engineering, Georgia Institute of Technology, 311 Ferst Dr. NW, Atlanta, GA 30032, USA; E-Mails: christine.he@chbe.gatech.edu (C.Y.H.); mhsieh34@gatech.edu (M.-C.H.); syu47@gatech.edu (S.-S.Y.)

Ming-Chien Hsieh

School of Chemical & Biomolecular Engineering, Georgia Institute of Technology, 311 Ferst Dr. NW, Atlanta, GA 30032, USA; E-Mails: christine.he@chbe.gatech.edu (C.Y.H.); mhsieh34@gatech.edu (M.-C.H.); syu47@gatech.edu (S.-S.Y.)

Sheng-Sheng Yu

School of Chemical & Biomolecular Engineering, Georgia Institute of Technology, 311 Ferst Dr. NW, Atlanta, GA 30032, USA; E-Mails: christine.he@chbe.gatech.edu (C.Y.H.); mhsieh34@gatech.edu (M.-C.H.); syu47@gatech.edu (S.-S.Y.)

This page left intentionally blank.

Chapter 1

CHEMICAL ENGINEERING

Chemical engineering is a discipline influencing numerous areas of technology. In broad terms, chemical engineers are responsible for the conception and design of processes for the purpose of production, transformation and transportation of materials. This activity begins with experimentation in the laboratory and is followed by implementation of the technology to full scale production.

The large number of industries which depend on the synthesis and processing of chemicals and materials place the chemical engineer in great demand.; In addition to traditional examples such as the chemical, energy and oil industries, opportunities in biotechnology, pharmaceuticals, electronic device fabrication, and environmental engineering are increasing. The unique training of the chemical engineer becomes essential in these areas whenever processes involve the chemical or physical transformation of matter. For example, chemical engineers working in the chemical industry investigate the creation of new polymeric materials with important electrical, optical or mechanical properties. This requires attention not only to the synthesis of the polymer, but also to the flow and forming processes necessary to create a final product. In biotechnology, chemical engineers have responsibilities in the design of production facilities to use microorganisms and enzymes to synthesize new drugs. Problems in environmental engineering that engage chemical engineers include the development of processes (catalytic converters, effluent treatment facilities) to minimize the release of or deactivate products harmful to the environment.

To carry out these activities, the chemical engineer requires a complete and quantitative understanding of both the engineering and scientific principles underlying these technological processes. This is reflected in the curriculum of the chemical engineering department which includes the study of applied mathematics, material and energy balances, thermodynamics, fluid mechanics, energy and mass transfer, separations technologies, chemical reaction kinetics and reactor design, and process design. These courses are built on a foundation in the sciences of chemistry, physics and biology.

Chemical engineering is a branch of engineering that applies the natural (or experimental) sciences (*e.g.* chemistry and physics) and life sciences (*e.g.* biology,microbiology and biochemistry) together with mathematics and economics to production, transformation, transportation and proper usage of chemicals, materials and energy. It essentially deals with the engineering of chemicals, energy and the processes that create and/or convert them. Modern chemical engineers are concerned with processes that convert raw-materials or (cheap) chemicals into more useful or valuable forms. In addition, they are also concerned with pioneering valuable materials and related techniques – which are often essential to related fields such as nanotechnology, fuel cells and bioengineering. Within chemical engineering, two broad subgroups include design, manufacture, and operation of plants and machinery in industrial chemical and related processes ("chemical process engineers") and development of new or adapted substances for products ranging from foods and beverages to cosmetics to cleaners to pharmaceutical ingredients, among many other products ("chemical product engineers").

ETYMOLOGY

A 1996 *British Journal for the History of Science* article cites James F. Donnelly for mentioning an 1839 reference to chemical engineering in relation to the production of sulfuric acid. In the same paper however, George E. Davis, an English consultant, was credited for having coined the term. The *History of Science in United States : An Encyclopedia* puts this at around 1890. "Chemical engineering", describing the use of mechanical equipment in the chemical industry, became common vocabulary in England after 1850. By 1910, the profession, "chemical engineer", was already in common use in Britain and the United States.

HISTORY

Chemical engineering emerged upon the development of unit operations, a fundamental concept of the discipline chemical engineering. Most authors agree that Davis invented unit operations if not substantially developed it. He gave a series of lectures on unit operations at the Manchester Technical School (University of Manchester today) in 1887, considered to be one of the earliest such about chemical engineering. Three years before Davis' lectures, Henry Edward Armstrong

Fig. Students inside an industrial chemistry laboratory at MIT.

taught a degree course in chemical engineering at the City and Guilds of London Institute. Armstrong's course "failed simply because its graduates ... were not especially attractive to employers." Employers of the time would have rather hired chemists and mechanical engineers. Courses in chemical engineering offered by Massachusetts Institute of Technology (MIT) in the United States, Owen's College in Manchester, England and University College London suffered under similar circumstances.

Starting from 1888, Lewis M. Norton taught at MIT the first chemical engineering course in the United States. Norton's course was contemporaneous and essentially similar with Armstrong's course. Both courses, however, simply merged chemistry and engineering subjects. "Its practitioners had difficulty convincing engineers that they were engineers and chemists that they were not simply chemists." Unit operations was introduced into the course by William Hultz Walker in 1905. By the early 1920s, unit operations became an important aspect of chemical engineering at MIT and other US universities, as well as at Imperial College London. The American Institute of Chemical Engineers (AIChE), established in 1908, played a key role in making chemical engineering considered an independent science, and unit operations central to chemical engineering. For instance, it defined chemical engineering to be a "science of itself, the basis of which is ... unit operations" in a 1922 report; and with which principle, it had published a list of academic institutions which offered "satisfactory" chemical engineering courses. Meanwhile, promoting chemical engineering as a distinct science in Britain lead to the establishment of the Institution of Chemical Engineers (IChemE) in 1922. IChemE likewise helped make unit operations considered essential to the discipline.

New Concepts and Innovations

By the 1940s, it became clear that unit operations alone was insufficient in developing chemical reactors. While the predominance of unit operations in chemical engineering courses in Britain and the United States continued until the 1960s, transport phenomena started to experience greater focus. Along with other novel concepts, such process systems engineering (PSE), a "second paradigm" was defined. Transport phenomena gave an analytical approach to chemical engineering while PSE focused on its synthetic elements, such as control system and process design. Developments in chemical engineering before and after World War II were mainly incited by the petrochemical industry, however, advances in other fields were made as well. Advancements in biochemical engineering in the 1940s, for example, found application in the pharmaceutical industry, and allowed for the mass production of various antibiotics, including penicillin and streptomycin. Meanwhile, progress in polymer science in the 1950s paved way for the "age of plastics".

Safety and Hazard Developments

Concerns regarding the safety and environmental impact of large-scale chemical manufacturing facilities were also raised during this period. *Silent Spring,* published in 1962, alerted its readers to the harmful effects of DDT, a potent insecticide. The 1974 Flixborough disaster in the United Kingdom resulted in 28 deaths, as well as damage to a chemical plant and three nearby villages. The 1984 Bhopal disaster in India resulted in almost 4,000 deaths. These incidents, along with other incidents, affected the reputation of the trade as industrial safety and environmental protection were given more focus. In response, the IChemE required safety to be part of every degree course that it accredited after 1982. By the 1970s, legislation and monitoring agencies were instituted in various countries, such as France, Germany, and the United States.

Recent Progress

Advancements in computer science found applications designing and managing plants, simplifying calculations and drawings that previously had to be done manually. The completion of the Human Genome Project is also seen as a major development, not only advancing chemical engineering but genetic engineering and genomics as well. Chemical engineering principles were used to produce DNA sequences in large quantities. While the application of chemical engineering principles to these fields only began in the 1990s, Rice University researchers see this as a trend towards biotechnology. The latest book 'Chemical Process Technology and Simulation' gives a very good comprehensive account of the various products and methods of its production and also deals with computer based process simulation of complex processes as practiced in industry.

Concepts

Chemical engineering involves the application of several principles. Key concepts are presented below.

Chemical Reaction Engineering

Chemical engineering involves managing plant processes and conditions to ensure optimal plant operation. Chemical reaction engineers construct models for reactor analysis and design using laboratory data and physical parameters, such as chemical thermodynamics, to solve problems and predict reactor performance.

Plant Design

Chemical engineering design concerns the creation of plans, specification, and economic analyses for new plants or plant modifications. Design engineers often work in a consulting role, designing plants to meet clients' needs. Design is limited by a number of factors, including funding, government regulations and safety standards. These constraints dictate a plant's choice of process, materials and equipment.

Process Design

A unit operation is a physical step in an individual chemical engineering process. Unit operations (such as crystallization, drying and evaporation) are used to prepare reactants, purifying and separating its products, recycling unspent reactants, and controlling energy transfer in reactors. On the other hand, a unit process is the chemical equivalent of a unit operation. Along with unit operations, unit processes constitute a process operation. Unit processes (such as nitration and oxidation) involve the conversion of material by biochemical, thermochemical and other means. Chemical engineers responsible for these are called process engineers.

Process Design is the most challenging field of chemical engineering. Overall process simulation is to be done using various software. The recent book "Chemical Process Technology and Simulation" by Srikumar Koyikkal gives many classical examples. It is also a text book of Chemical Process Technology.

Transport Phenomena

Transport phenomena occur frequently in industrial problems. These include fluid dynamics, heat transfer and mass transfer, which mainly concern momentum transfer, energy transfer and transport of chemical species respectively. Basic equations for describing the three transport phenomena in the macroscopic, microscopic and molecular levels are very similar. Thus, understanding transport phenomena requires thorough understanding of mathematics.

Applications and Practice

Fig. Chemical engineers use computers to manage automated systems in plants.

Chemical engineers "develop economic ways of using materials and energy". Chemical engineers use chemistry and engineering to turn raw materials into usable products, such as medicine, petrochemicals and plastics on a large-scale, industrial setting. They are also involved in waste management and research. Both applied and research facets could make extensive use of computers.

A chemical engineer may be involved in industry or university research where they are tasked in designing and performing experiments to create new and better ways of production, controlling pollution, conserving resources and making these processes safer. They may be involved in designing and constructing plants as a project engineer. In this field, the chemical engineer uses their knowledge in selecting plant equipment and the optimum method of production to minimize costs and increase profitability. After its construction, they may help in upgrading its equipment. They may also be involved in its daily operations. Chemical engineers may be permanently employed at chemical plants to manage operations. Alternatively, they may serve in a consultant role to troubleshoot problems, manage process changes and otherwise assist plant operators.

PROCESS DESIGN

In chemical engineering, process design is the design of processes for desired physical and/or chemical transformation of materials. Process design is central to chemical engineering, and it can be considered to be the summit of that field, bringing together all of the field's components.

Process design can be the design of new facilities or it can be the modification or expansion of existing facilities. The design starts at a conceptual level and ultimately ends in the form of fabrication and construction plans.

Process design is distinct from equipment design, which is closer in spirit to the design of unit operations. Processes often include many unit operations

Documentation

Process design documents serve to define the design and they ensure that the design components fit together. They are useful in communicating ideas and plans to other engineers involved with the design, to external regulatory agencies, to equipment vendors and to construction contractors.

In order of increasing detail, process design documents include :

- **Block Flow Diagrams (BFD)** : Very simple diagrams composed of rectangles and lines indicating major material or energy flows.
- **Process Flow Diagrams (PFD's)** : Typically more complex diagrams of major unit operations as well as flow lines. They usually include a material balance, and sometimes an energy balance, showing typical or design flowrates, stream compositions, and stream and equipment pressures and temperatures.
- **Piping and Instrumentation Diagrams (P&ID's)** : Diagrams showing each and every pipeline with piping class (carbon steel or stainless steel) and pipe size (diameter). They also show valving along with instrument locations and process control schemes.
- **Specifications** : Written design requirements of all major equipment items.

Process designers also typically write operating manuals on how to start-up, operate and shut-down the process.

Documents are maintained after construction of the process facility for the operating personnel to refer to. The documents also are useful when modifications to the facility are planned.

A primary method of developing the process documents is process flow-sheeting.

Design Considerations

Designs have objectives and constraints, and even a simple process requires a trade-off among such factors.

Objectives that a design may strive to include :
- Throughput rate
- Process yield
- Product purity

Constraints include :
- Capital cost
- Available space
- Safety concerns
- Environmental impact and projected effluents and emissions
- Waste production
- Operating and maintenance costs

Other factors that designers may include are :
- Reliability
- Redundancy
- Flexibility
- Anticipated variability in feedstock and allowable variability in product.

Sources of Design Information

Designers usually do not start from scratch, especially for complex projects. Often the engineers have pilot plant data available or data from full-scale operating facilities. Other sources of information include proprietary design criteria provided by process licensors, published scientific data, laboratory experiments, and input.

Computer Help

The advent of low cost powerful computers has aided complex mathematical simulation of processes, and simulation software is often used by design engineers. Simulations can identify weaknesses in designs and allow engineers to choose better alternatives.

However, engineers still rely on heuristics, intuition, and experience when designing a process. Human creativity is an element in complex designs.

Chapter 2

BASIC CONCEPT OF ANALYSIS

THE SYSTEM

A system is that portion of the universe which we have chosen to study.

A closed system is one in which no mass crosses the system boundaries.

An open system is one in which mass crosses the system boundaries. The system may gain or lose mass or simply have some mass pass through it.

An isolated system is one devoid of interactions of any kind with the surroundings (including mass exchange, heat, and work interactions).

System boundaries are established at the beginning of a problem, and simplification of balance equations depends on whether the system is open or closed. Therefore, the system boundaries should be clearly identified. If the system boundaries are changed, the simplification of the mass and energy balance equations should be performed again, because different balance terms are likely to be necessary. These guidelines become more apparent in Chapter. In many textbooks, especially those dealing with fluid mechanics, the system is called the control volume. The two terms are synonymous.

THE PLACEMENT OF SYSTEM BOUNDARIES IS A KEY STEP IN PROBLEM SOLVING

Equilibrium

A system is in equilibrium when there is no driving force for a change of intensive variables within the system. The system is "relaxed" relative to all forces and potentials.

An isolated system moves spontaneously to an equilibrium state. In the equilibrium state there are no longer any driving forces for spontaneous change of intensive variables.

The Mass Balance

Presumably, students in this course are familiar with mass balances from an introductory course in material and energy balances. The relevant relation is simply :

Equation 1.9

$$\begin{bmatrix} \text{rate of mass} \\ \text{accumulation within} \\ \text{system boundaries} \end{bmatrix} = \begin{bmatrix} \text{rate of mass flow} \\ \text{into system} \end{bmatrix} - \begin{bmatrix} \text{rate of mass flow} \\ \text{out of system} \end{bmatrix}$$

$$\dot{m} = \sum_{inlets} \dot{m}^{in} - \sum_{outlets} \dot{m}^{out}$$

The mass balance.

where $\dot{m} = \dfrac{dm}{dt}$ $\dot{m}^{in}$ and $\dot{m}^{out}$ are the *absolute values* of mass flow rates entering and leaving, respectively.

We may also write

Equation 1.10

$$dm = \sum_{inlets} dm^{in} - \sum_{outlets} dm^{out}$$

where mass differentials dm^{in} and dm^{out} are *always positive*. When all the flows of mass are analyzed in detail for several subsystems coupled together, this simple equation may not seem to fully portray the complexity of the application. The general study of thermodynamics is similar in that regard. A number of simple relations like this one are coupled together in a way that requires some training to understand. *In the absence of chemical reactions,* we may also write a mole balance by replacing mass with moles in the balance.

Heat – Sinks and Reservoirs

Heat is energy in transit between the source from which the energy is coming and a destination toward which the energy is going. When developing thermodynamic concepts, we frequently assume that our system transfers heat to/from a reservoir or sink. A heat reservoir is an infinitely large source or destination of heat transfer. The reservoir is assumed to be so large that the heat transfer does not affect the temperature of the reservoir. A sink is a special name sometimes used for a reservoir which can accept heat without a change in temperature. The assumption of constant temperature makes it easier to concentrate on the thermodynamic behavior of the system while making a reasonable assumption about the part of the universe assigned to be the reservoir.

A reservoir is an infinitely large source or destination for heat transfer.

The mechanics of heat transfer are also easy to picture conceptually from the molecular kinetics perspective. In heat conduction, faster-moving molecules collide with slower ones, exchanging kinetic energy and equilibrating the temperatures. In this manner, we can imagine heat being transferred from the hot surface to the center of a pizza in an oven until the center of the pizza is cooked. In heat convection, packets of hot mass are circulated and mixed, accelerating the equilibration process. Heat convection is important in getting the heat from the oven flame to the surface of the pizza. Heat radiation, the remaining mode of heat transfer, occurs by an entirely different mechanism having to do with waves of electromagnetic energy emitted from a hot body that are absorbed by a cooler body. Radiative heat transfer is typically discussed in detail during courses devoted to heat transfer.

Work

Work is a familiar term from physics. We know that work is a force acting through a distance. There are several ways forces may interact with the system which all fit under this category, including pumps, turbines, agitators, and pistons/cylinders. We will discuss the details of how we calculate work and determine its impact on the system in the next chapter.

Density

Density is a measure of the quantity per unit volume and may be expressed on a molar basis (molar density) or a mass basis (mass density). In some situations, it is expressed as the number of particles per unit volume (number density).

Intensive Properties

Intensive properties are those properties which are independent of the size of the system. For example, in a system at equilibrium without internal rigid/insulating walls, the temperature and pressure are uniform throughout the system and are therefore intensive properties. Specific properties are the total property divided by the mass and are intensive. For example, the molar volume ($[\equiv]$ length^3/mole), mass density ($[\equiv]$ mass/length^3), and specific internal energy ($[\equiv]$ energy/mass) are intensive properties. In this text, intensive properties are not underlined.

The distinction between intensive and extensive properties is key in selecting and using variables for problem solving.

Extensive Properties

Extensive properties depend on the size of the system, for example the volume ($[\equiv]$ length^3) and energy ($[\equiv]$ energy). Extensive properties are underlined; for example, $\underline{U} = n U$, where n is the number of moles and U is molar internal energy.

States and State Properties – The Phase Rule

Two state variables are necessary to specify the state of a *single-phase* pure fluid, that is, two from the set P, V, T, U. Other state variables to be defined later in the text which also fit in this category are molar enthalpy, molar entropy, molar Helmholtz energy, and molar Gibbs energy. *State variables must be intensive properties.* As an example, specifying P and T permits you to find the specific internal energy and specific volume of steam. Note, however, that you need to specify only one variable, the temperature or the pressure, if you want to find the properties of saturated vapor or liquid. This reduction in the needed specifications is referred to as a reduction in the "degrees of freedom." As another example in a ternary, two-phase system, the temperature and the mole fractions of two of the components of the lower phase are state variables (the third component is implicit in summing the mole fractions to unity), but the total number of moles of a certain component is not a state variable because it is extensive. In this example, the pressure and mole fractions of the upper phase may be calculated once the temperature and lower-phase mole fractions have been specified. The number of state variables needed to completely specify the state of a system is given by the Gibbs phase rule for a non-reactive system,[7] :

Equation 1.11

$$F = C - P + 2$$

where F is the number of state variables that can be varied while P phases exist in a system where C is the number of components (F is also known as the number of degrees of freedom). More details on the Gibbs phase rule are given in Chapters.

The Gibbs phase rule provides the number of state variables (intensive properties) to specify the state of the system.

Steady-State Open Systems

The term steady state is used to refer to open systems in which the inlet and outlet mass flow rates are invariant with time and there is no mass accumulation. In addition, steady state requires that state variables at all locations are invariant with respect to time. Note that state variables may vary with position. Steady state does not require the system to be at equilibrium. For example, in a heat exchanger operating at steady state with a hot and cold stream, each stream has a temperature gradient along its length, and there is always a driving force for heat transfer from the hotter stream to the colder stream. Section 2.13 describes this process in more detail.

Steady-state Flow is Very Common in the Process Industry

The Ideal Gas Law

The ideal gas is a "model" fluid where the molecules have no attractive potential energy and no size (and thus, no repulsive potential energy). Properties of the ideal gas are calculated from the ideal gas model :

Equation (ig) 1.12

$$P\underline{V} = nRT$$

The ideal gas law is a *model* that is not always valid, but gives an initial guess.

An equation of state relates the *P-V-T* properties of a fluid.

Note that scientists who first developed this formula empirically termed it a "law" and the name has persisted, but it should be more appropriately considered a "model." In the terminology we develop, it is also an equation of state, relating the *P-V-T* properties of the ideal gas law to one another as shown in Eqn. 1.12. We know that real molecules have potential energy of attraction and repulsion. Due to the lack of repulsive forces, ideal gas particles can "pass through" one another. Ideal gas molecules are sometimes called "point masses" to communicate this behavior. While the assumptions may seem extreme, we know experimentally that the ideal gas model represents many compounds, such as air, nitrogen, oxygen, and even water vapor at temperatures and pressures near ambient conditions. Use of this model simplifies calculations while the concepts of the energy and entropy balances are developed throughout Unit I. This does not imply that the ideal gas model is applicable to all vapors at all conditions, even for air, oxygen, and nitrogen. Analysis using more complex fluid models is delayed until Unit II. We rely on thermodynamic charts and tables until Unit II to obtain properties for gases that may not be considered ideal gases.

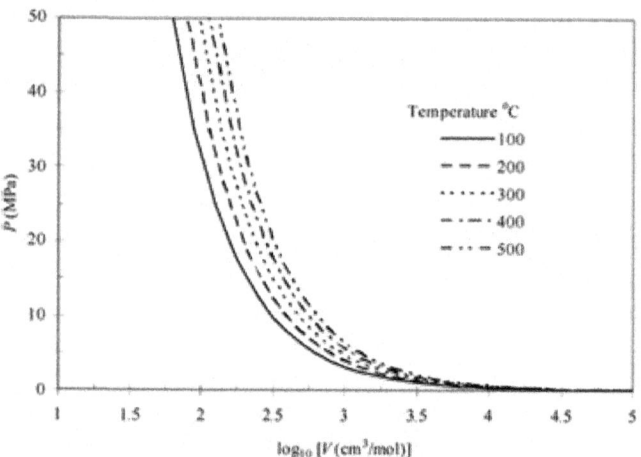

Fig. Ideal gas behavior at five temperatures.

Because kinetic energy is the only form of energy for an ideal gas, the internal energy of a monatomic ideal gas is given by summing the kinetic energy of the atoms and then relating this to temperature (*c.f.* Eqn. 1.1) :

Equation (ig) 1.13

$$\boxed{U^{ig} = \frac{Nm\langle v^2 \rangle}{2} = \frac{nN_A m\langle v^2 \rangle}{2} = \frac{nM_w \langle v^2 \rangle}{2} = \frac{3}{2}nRT = \frac{3}{2}NkT}\ \text{(monatomic ideal gas)}$$

The proportionality constant between temperature and internal energy is known as the ideal gas heat capacity at constant volume, denoted C_V. Eqn 1.13 shows that $C_V = 1.5R$ for a monatomic ideal gas. If you refer to the tables of constant pressure heat capacities (C_p) on the back flap of the text and note that $C_p = C_V + R$ for an ideal gas, you may be surprised by how accurate this ultrasimplified theory actually is for species like helium, neon, and argon at 298 K.[8]

While the equality in Eqn. 1.13 is valid for monatomic fluids only, the functionality $U^{ig} = U^{ig}(T)$ is universal for all ideal gases. For more multi-atom molecules, the heat capacity depends on temperature because vibrations hold some energy in a manner that depends on temperature. However, the observation that $U^{ig} = U^{ig}(T)$ is true for any ideal gas, not only for ultrasimplified, monatomic ideal gases. We build on this relation in Chapters, where we show how to compute changes in energy for any fluid at any temperature and density by systematically correcting the relatively simple ideal gas result. Let us explore more completely the assumptions of the ideal gas law by investigating the molecular origins of pressure.

Pressure

Pressure is the force exerted per unit area. We will be concerned primarily with the pressure exerted by the molecules of fluids upon the walls of their containers. Briefly, when molecules collide with the container walls, they must change momentum. The change in momentum creates a force on the wall. As temperature increases, the particles have more kinetic energy (and momentum) when they collide, so the pressure increases. We can understand this more fully with an ultra-simplified analysis of kinetic theory as it relates to the ideal gas law.

Suppose we have two hard spherical molecules in a container that are bouncing back and forth with 1D velocity in the x-direction only and not contacting one another. We wish to quantify the forces acting on each wall. Since the particles are colliding only with the walls at A_1 and A_2 in our idealized motion, these are the only walls we need to consider. Let us assume that particles bounce off the wall with the same speed which they had before striking the wall, but in the opposite direction (a perfectly elastic collision where the wall is perfectly rigid and absorbs no momentum).

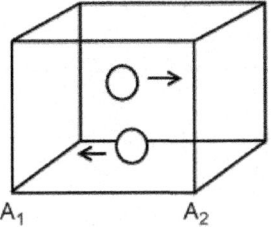

A_1 A_2

Thus, the kinetic energy of the particles is fixed. If $\vec{v}$ is the initial velocity of the particle (recall that $\vec{v}$ is a vector quantity and notation v represents a scalar) before it strikes a wall, the change in velocity due to striking the wall is $-2\vec{v}$. The

change in velocity of the particle indicates the presence of interacting forces between the wall and the particle. If we quantify the force necessary to change the
velocity of the particle, we will also quantify the forces of the particle on the wall
by Newton's third principle. To quantify the force, we may apply Newton's second
principle stated in terms of momentum : *The time rate of change of the momentum
of a particle is equal to the resultant force acting on the particle and is in the direction of
the resultant force.*

Equation 1.14

$$\frac{d\vec{p}}{dt} = \vec{F}$$

The application of this formula directly is somewhat problematic since the
change in direction is instantaneous, and it might seem that the time scale is important. This can be avoided by determining the time-averaged force, $\vec{F}_{avg}$ exerted on the wall during time Δt,

Equation 1.15

$$\int_{t^i}^{t^f} \vec{F} dt = \vec{F}_{avg} \Delta t = \int_{t^i}^{t^f} \frac{d\vec{p}}{dt} dt = \Delta \vec{p}$$

where $\Delta \vec{p}$ is the total change in momentum during time Δt. The momentum
change for each collision is $\Delta \vec{p} = -2m\vec{v}$ where m is the mass per particle. Collision
frequency can be related easily to the velocity. Each particle will collide with the
wall every Δt seconds, where $\Delta t = 2L/v$, where L is the distance between A_1 and A_2.
The average force is then

Equation 1.16

$$\vec{F}_{avg} = \frac{\Delta \vec{p}}{\Delta t} = -2m\vec{v} \frac{v}{2L}$$

where $\vec{v}$ is the velocity before the collision with the wall. Pressure is the force per
unit area, and the area of a wall is L^2, thus

Equation (*ig*) 1.17

$$P = \frac{m}{L^3}(v_1^2 + v_2^2) \text{ (1D ideal gas motion)}$$

where the subscripts denote the particles. If you are astute, you will recognize L^3 as
the volume of the box and the kinetic energy which we have shown earlier to
relate to the temperature.

P is proportional to the number of particles in a volume and to the kinetic
energy of the particles.

If the particle motions are generalized to motion in arbitrary directions,
collisions with additional walls in the analysis does not complicate the problem

dramatically because each component of the velocity may be evaluated independently. To illustrate, consider a particle bouncing around the centers of four walls in a horizontal plane. From the top view, the trajectory would appear as below :

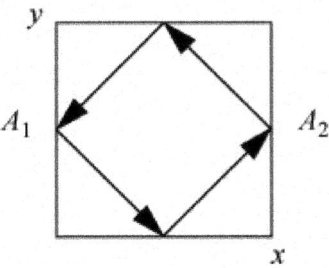

For the same velocity as the first case, the force of each collision would be reduced because the particle strikes merely a glancing blow. The time of collisions between walls is now dependent on the component of velocityperpendicular to the walls. We have chosen a special case to illustrate here, where the box is square and the particle impacts at a 45° angle in the center of each wall. The x-component of the force can be related to the magnitude of the velocity by noting that $v_x = v_y$, such that $v = (v_x^2 + v_y^2)^{1/2} = v_x 2^{1/2}$. The time between collisions with wall A_1 would be $4L/(v2^{1/2})$. The formula for the average force in two dimensions then becomes :

Equation 1.18

$$F_{avg,A_1} = -2mv_x \frac{v\sqrt{2}}{4L} = -2m\frac{v}{\sqrt{2}}\frac{v\sqrt{2}}{4L} = -2mv\frac{v}{2L} = -\frac{mv^2}{L}$$

and the pressure due to two particles that do not collide with one another in two dimensions becomes :

Equation (ig) 1.19

$$P = \frac{m}{2L^3}(v_1^2 + v_2^2)\,(\text{2D motion})$$

(More complicated impact angles and locations will provide the same results but require more tedious derivations.) The extension to three dimensions is more difficult to visualize, but comparing Eqn. 1.17 to Eqn. 1.19, you should not be surprised to learn that the pressure in three dimensions is :

Equation (ig) 1.20

$$P = \frac{m}{3L^3}(v_1^2 + v_2)^2) = \frac{m}{3\underline{V}}(v_1^2 + v_2^2)\,(\text{3D motion})$$

The problem gets more complicated when collisions between particles occur. We ignored that possibility here because the ideal gases being considered are point masses that do not collide with one another. Including molecular collisions is a straightforward implementation of "billiard ball" physics. This subject is discussed further in Section 7.10 on page 276 and with great interactive graphics in the discontinuous molecular dynamics (DMD) module at Etomica.org.

We see a relation developing between P and kinetic energy. When we insert the relation between temperature and kinetic energy (Eqn. 1.1) into Eqn. 1.20 we find that the ideal gas law results for a spherical (monatomic) molecule in 3D,

Equation (*ig*) 1.21

$$PV = \left(\frac{m}{3V} \sum_i^N v_i^2 \right) V = \frac{N}{3} m \langle v^2 \rangle = \frac{nN_A}{3} m \langle v^2 \rangle = \frac{nM_w}{3} \langle v^2 \rangle = nRT \text{ (3D motion)}$$

Check your units when using this equation. $1J = 1kg\text{-}m^2/s^2$.

where m is the mass per particle and M_w is the molecular weight. A similar derivation with Eqn. 1.19 gives the results for motions restricted to 2D,[10]

Equation (*ig*) 1.22

$$PV = \left(\frac{m}{2V} \sum_i v_i^2 \right) V = \frac{nN_A m}{2} \langle v^2 \rangle = \frac{nM_w}{3} \langle v^2 \rangle = nRT_{2D} \text{ (2D motion)}$$

Assumptions at the Splitting Point

The recombination point is relatively unpredictable because the composition of the stream leaving depends on both the composition of the feed and the composition of the recycle stream. However, the *spliitng point* is special because when a stream is split, it generally is split into two streams with equal composition. This is a piece of information that counts towards "additional information" when performing a degree of freedom analysis.

As an additional specification, it is common to know the *ratio* of splitting, *i.e.* how much of the exit stream from the process will be put into the outlet and how much will be recycled. This also counts as "additional information".

Assumptions at the Recombination Point

The recombination point is generally not specified like the splitting point, and also the recycle stream and feed stream are very likely to have different compositions. The important thing to remember is that you can generally use the properties of the stream coming from the splitting point for the stream entering the recombination point, unless it goes through another process in between (which is entirely possible).

Degree of Freedom Analysis of Recycle Systems

Degree of freedom analyses are similar for recycle systems to those for other systems, but with a couple important points that the engineer must keep in mind :

1. The recombination point and the splitting point must be counted in the degree of freedom analysis as "processes", since they can have unknowns that aren't counted anywhere else.

2. When doing the degree of freedom analysis on the splitting point, *you should not label the concentrations as the same but leave them as separate unknowns until after you complete the DOF analysis* in order to avoid confusion, since labeling the concentrations as identical "uses up" one of your pieces of information and then you can't count it.

As an example, let's do a degree of freedom analysis on the hypothetical system above, assuming that all streams have two components.

*   **Recombination Point** : 6 variables (3 concentrations and 3 total flow rates) - 2 mass balances = 4 DOF

*   **Process** : Assuming it's not a reactor and there's only 2 streams, there's 4 variables and 2 mass balances = 2 DOF

*   **Splitting Point** : 6 variables - 2 mass balances **- 1 knowing compositions are the same - 1 splitting ratio = 2 DOF**

So the total is 4 + 2 + 2 - 6 (in-between variables) = 2 DOF. Therefore, if the feed is specified then this entire system can be solved! Of course the results will be different if the process has more than 2 streams, if the splitting is 3-way, if there are more than two components, and so on.

Suggested Solving Method

The solving method for recycle systems is similar to those of other systems we have seen so far but as you've likely noticed, they are increasingly complicated. Therefore, the importance of making a plan becomes of the utmost importance. The way to make a plan is generally as follows :

1. Draw a completely labeled flow chart for the process.
2. Do a DOF analysis to make sure the problem is solvable.
3. If it is solvable, **a lot of the time, the best place to start with a recycle system is with a set of overall system balances, sometimes in combination with balances on processes on the border.** The reason for this is that the overall system balance cuts out the recycle stream entirely, since the recycle stream does not enter or leave the system as a whole but merely travels between two processes, like any other intermediate stream. Often, the composition of the recycle stream is unknown, so this simplifies the calculations a good deal.
4. Find a set of independent equations that will yield values for a certain set of unknowns (this is often most difficult the first time; sometimes, one of the unit operations in the system will have 0 DOF so start with that one. Otherwise it'll take some searching.)
5. Considering those variables as known, do a new DOF balance until something has 0 DOF. Calculate the variables on that process.

6. Repeat until all processes are specified completely.

Example Problem : Improving a Separation Process

It has been stated that recycle can help to This example helps to show that this is true and also show some limitations of the use of recycle on real processes.

Chapter 3

CHEMICAL PROCESS DESIGN

INTRODUCTION

Design in its most simplistic viewpoint is composed of the following steps :

1. Determine the problem and its constraints
2. Generate potential solutions
3. Develop sufficient detail that solutions can be compared and eliminated
4. Implement the preferred solution

Product Design and Chemical Process Design often share the objective of producing a product for a commercial purpose. In many ways the two are similar, but process design, as typically taught in the universities, has historically emphasised the manufacture of a known product more than the development of a completely new product. It should be noted that, to the chemical process engineer the word 'product' is meant to encompass not only chemicals, but also energy or other commercially useful "things".

The invention of new chemical products (*e.g.* Nylon® or Lycra®) to suit specific end needs or properties such as strength, weight, colour fastness, abrasion resistance, high temperature capability, or flexibility, *etc.* is typically the role of the molecular/chemical designers who are usually located in chemistry laboratories and research centers. However, the design of the modification steps to generate the chemical product in production quantities is more firmly in the domain of the process design engineer. In the chemical industry the product and its manufacturing process are so intrinsically linked that these two roles sometimes blur into one 'research and development' person. However, as we shall see below, process design engineers sometimes find themselves in different situations, such as :

1. The Output is known and a desirable route to its production is to be designed, including selecting appropriate inputs.
2. The Inputs are known and a desirable output is required along with.

3. The desirable properties are known and a product along with its manu-facturing process is to be designed.

4. The inputs, outputs and route are known, but an optimization of the route is to be performed.

In either Product or Process Engineering, the primary objective of the design engineer is usually to produce something at the lowest possible cost, with the most commercially desirable attributes, and to do so in a way that meets all applicable laws and standards and ensures safety and protection of society.

In the realm of the process designer, the objectives can be open ended, such as producing a more environmentally friendly or safer stream, or to meet a legislative requirement. But, more commonly, the end product is known and the route to its production is being designed, which tends to make the design problem less open ended. Another example of an open ended design problem would be when the objective of a process is to make use of a waste stream in any manner possible, which may or may not, require changing it's molecular properties (*i.e.* convert it to a useful product, burn it to recover energy, or digest it using biological means). In such a situation, the feeds are defined and the product of the process is unknown.

In either of these examples, there is a strong sense of the 'product' that will be produced and consumed, but this is not always the case.

Another potential situation for process design engineers is the optimization of a process. By its very nature, the objective of improving efficiency (less waste, less raw materials, less energy use, greater production rate, *etc.*), or improving safety, necessitates the use of creative thinking and engineering tools to satisfy the problem definition and constraints. In these situations, the final "product" may be performance criteria, and therefore the process itself becomes the product of the performance requirement. An analogy in "hard product" production would be improvement in the production process for higher throughput, lower scrap rate, higher quality, or lower cost.

Since the objectives of process and product design typically overlap, it follows that the engineering process to achieve the desired results should be common. In the next section some of the steps taken by the process engineer during the typical chemical plant process design will be discussed. The reader familiar with product design and/or manufacture is invited to draw comparisons with their own experience.

The Steps of Chemical Process Design

As with any engineering design problem, a goal is to eliminate non-optimal solutions with as little "effort cost" as possible. This usually leads to an iterative, or "bootstrapping" design methodology that begins with a low level of detail in the solutions and progressively creates more and more detail of fewer possible solutions until an optimal one is found. During the design development, one of the key decision making tools is economic viability, and producing design information that assists in making that analysis is principal to any methodology. The process of design has the following goals :

Goal 1. Eliminate solutions with as little effort as possible.

Goal 2. Produce a financial estimate

Goal 3. Understand the risk that the process poses to society and the environment

Goal 4. Produce the documentation required to build the process.

"Text book" process design is commonly broken down into the following stages :

1. Problem Definition

2. Process Synthesis

(Multiple solutions are generated and discarded as quickly as possible to produce a small number of favourable solutions that are taken to more detail)

3. Process Design

(One or two most favoured solutions are developed in enough detail that reasonable financial analysis can be performed, safety and environmental issues can be identified and their risk understood.)

4. Process Analysis

(In this stage optimization of the conditions or equipment will be performed.)

Experience suggests that the steps of process design are intrinsically linked to the phases of how projects in the chemical process industry are executed, and to a smaller degree, the opposite is true.

The following phases of a project are often used, with smaller projects combining some of the stages. The names may be different between companies but for the sake of simplicity we shall refer to them as :

Front End 1

Front End 2

Front End 3

Detailed Design

Construction

Start-up

As will be discussed, the design progressively provides not only more detail about the facility so that it may be built and operated, but equally importantly, it provides progressively more accurate estimates of its capital, operating costs, and business risk for business planning purposes. Some steps have a strong emphasis on technical design, some provide essential quality control, others prioritize costs and project management, and some focus more on societal impact. All are elements of the "process design process." At the end of each of the first four stages, an opportunity is provided for the business team to review the financial/project risk or the changed business environment, and a decision is made to either proceed, delay, or cancel the project.

Front End 1

Depending on the nature of a project, there is a wide range in the amount of available initial information. If the project is a duplication of, or similar to a previous project, then the core process is probably well defined but the interfaces (conditioning of inputs, provision of utility services, storage, (un)loading, *etc.*) are less so. If the project will be dealing with a new product, then much less information would be available. Regardless, this stage begins with a statement of business objectives and possibly some general information about the actual chemical or product that will be produced, and/or the feedstock that is available.

A thorough analysis of the opportunity and its constraints is usually documented in something called "Basic Data". The Basic Data remains a living document until the point where a process flow diagram and the heat and material balance (discussed later) are finalized. Once considered complete, the Basic Data is said to be 'frozen' and provides the guidance to all of the designers throughout the life of the project.

The first process design step is to generate a series of possible solutions to the problem. The most simplistic representation of a process begins with a block flow diagram, which is distinguished by the fact that no real equipment is required to be documented. In a chemical process, the inputs and outputs are usually chemicals, and the modification step is usually some molecular change or bulk property change that takes place (*e.g.* a separation, a reaction, or a change in heat, size, *etc.*). The above diagram does not provide much more information than allowing the comparison in costs between raw materials and products. It does allow for a first pass at eliminating some of the possible solutions, for instance, where the inputs are more costly than the outputs.

How the chemical engineer "invents" or synthesizes these alternatives (which are often called the process topography) is an interesting topic. Historically, creative thinking and experience are used,. but the opportunity to use the product designer's toolkit such as TRIZ, brainstorming, creative problem solving, and others are all possibilities deserving of some consideration.

THE AMMONIA SYNTHESIS

A brief summary of the Haber Process

The Haber Process combines nitrogen from the air with hydrogen derived mainly from natural gas (methane) into ammonia. The reaction is reversible and the production of ammonia is exothermic.

$$N_{2(g)} + 3H_{2(g)} \rightleftharpoons 2NH_{3(g)} \qquad \Delta H = -92\text{kJ mol}^{-1}$$

A flow scheme for the Haber Process looks like this :

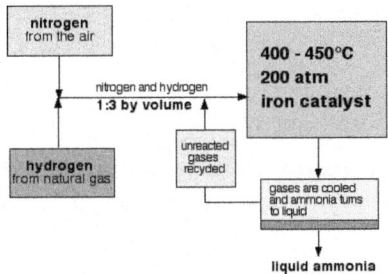

Some Notes on the Conditions

The catalyst

The catalyst is actually slightly more complicated than pure iron. It has potassium hydroxide added to it as a promoter - a substance that increases its efficiency.

The pressure

The pressure varies from one manufacturing plant to another, but is always high. You can't go far wrong in an exam quoting 200 atmospheres.

Recycling

At each pass of the gases through the reactor, only about 15% of the nitrogen and hydrogen converts to ammonia. (This figure also varies from plant to plant.) By continual recycling of the unreacted nitrogen and hydrogen, the overall conversion is about 98%.

Explaining the Conditions

The proportions of nitrogen and hydrogen.

The mixture of nitrogen and hydrogen going into the reactor is in the ratio of 1 volume of nitrogen to 3 volumes of hydrogen.

Avogadro's Law says that equal volumes of gases at the same temperature and pressure contain equal numbers of molecules. That means that the gases are going into the reactor in the ratio of 1 molecule of nitrogen to 3 of hydrogen.

That is the proportion demanded by the equation.

In some reactions you might choose to use an excess of one of the reactants. You would do this if it is particularly important to use up as much as possible of the other reactant - if, for example, it was much more expensive. That doesn't apply in this case.

There is always a down-side to using anything other than the equation proportions. If you have an excess of one reactant there will be molecules passing through the reactor which can't possibly react because there isn't anything for them to react with. This wastes reactor space - particularly space on the surface of the catalyst.

The Temperature

Equilibrium Considerations

You need to shift the position of the equilibrium as far as possible to the right in order to produce the maximum possible amount of ammonia in the equilibrium mixture.

The forward reaction (the production of ammonia) is exothermic.

$$N_{2(g)} + 3H_{2(g)} \rightleftharpoons 2NH_{3(g)} \qquad \Delta H = -92\text{kJ mol}^{-1}$$

According to Le Chatelier's Principle, this will be favoured if you lower the temperature. The system will respond by moving the position of equilibrium to counteract this - in other words by producing more heat.

In order to get as much ammonia as possible in the equilibrium mixture, you need as low a temperature as possible. However, 400 - 450°C isn't a low temperature!

Rate Considerations

The lower the temperature you use, the slower the reaction becomes. A manufacturer is trying to produce as much ammonia as possible per day. It makes no sense to try to achieve an equilibrium mixture which contains a very high proportion of ammonia if it takes several years for the reaction to reach that equilibrium.

You need the gases to reach equilibrium within the very short time that they will be in contact with the catalyst in the reactor.

The Compromise

400 - 450°C is a compromise temperature producing a reasonably high proportion of ammonia in the equilibrium mixture (even if it is only 15%), but in a very short time.

The Pressure

Equilibrium considerations

$$N_{2(g)} + 3H_{2(g)} \rightleftharpoons 2NH_{3(g)} \qquad \Delta H = -92\text{kJ mol}^{-1}$$

Notice that there are 4 molecules on the left-hand side of the equation, but only 2 on the right.

According to Le Chatelier's Principle, if you increase the pressure the system will respond by favouring the reaction which produces fewer molecules. That will cause the pressure to fall again.

In order to get as much ammonia as possible in the equilibrium mixture, you need as high a pressure as possible. 200 atmospheres is a high pressure, but not amazingly high.

Rate Considerations

Increasing the pressure brings the molecules closer together. In this particular instance, it will increase their chances of hitting and sticking to the surface of the catalyst where they can react. The higher the pressure the better in terms of the rate of a gas reaction.

Economic Considerations

Very high pressures are very expensive to produce on two counts.

You have to build extremely strong pipes and containment vessels to withstand the very high pressure. That increases your capital costs when the plant is built.

High pressures cost a lot to produce and maintain. That means that the running costs of your plant are very high.

The Compromise

200 atmospheres is a compromise pressure chosen on economic grounds. If the pressure used is too high, the cost of generating it exceeds the price you can get for the extra ammonia produced.

The catalyst

Equilibrium Considerations

The catalyst has no effect whatsoever on the position of the equilibrium. Adding a catalyst doesn't produce any greater percentage of ammonia in the equilibrium mixture. Its only function is to speed up the reaction.

Rate Considerations

In the absence of a catalyst the reaction is so slow that virtually no reaction happens in any sensible time. The catalyst ensures that the reaction is fast enough for a dynamic equilibrium to be set up within the very short time that the gases are actually in the reactor.

Separating the Ammonia

When the gases leave the reactor they are hot and at a very high pressure. Ammonia is easily liquefied under pressure as long as it isn't too hot, and so the temperature of the mixture is lowered enough for the ammonia to turn to a liquid. The nitrogen and hydrogen remain as gases even under these high pressures, and can be recycled.

Case Study : The Haber Process

The Haber process is the production of ammonia from a reaction between nitrogen and hydrogen, because of an iron substitute. This process is known for the commercial synthesis of ammonia. There is great abundance of nitrogen in the air when it is combined with hydrogen under extreme pressure and high temperature. This process is a great example of chemical equilibrium.

INTRODUCTION

The Haber process, also known as the Haber-Bosch Process, was founded by Fritz Haber and Carl Bosch, both who were German Chemists. Haber discovered the conditions for the formation of ammonia, and Bosch discovered the work of high-pressure on chemical reactions (developed into industrial process). Both were awarded the Nobel Prize. During the 1920's, there was a shortage of the world's supply for fixed nitrogen. Nitrogen was mainly used for fertilizer. Fertilizer was used in order to produce food, so that in WWI people could continue to fight. It only requires 1 percent of the world's energy to make 500 million tons of artificial fertilizer per year, which, in turn, helps feed 40 percent of the world's population.

THE PROCESS

The Haber process takes nitrogen gas from air and combines it with molecular hydrogen gas to form ammonia gas. This is anexothermic reaction, meaning it *releases* energy so that the sum of the enthalpies of N_2 and H_2 (the reactants) is greater than the enthalpy of NH_3 (the products).

$$N_2(g) + 3H_2(g) \rightarrow 2NH_3(g) \quad \Delta H = -92.4 \text{ kJ}$$

which is a reversible reaction :

$$2NH_3(g) \rightarrow N_2(g) + 3H_2(g) \quad \Delta H = +92.4 \text{ kJ mol-1}$$

Here's a visual to help convey the process :

From the flow chart above, we can see that methane and steam combine to form hydrogen and carbon monoxide, which in turn releases hydrogen. The hydrogen then combines with oxygen from the air to produce water. Finally, nitrogen gas is released which combines with hydrogen gas to form ammonia. This takes place under high pressure and temperature and with an **iron catalyst***** (mentioned later on).

Le Châtelier's Principle

The Haber process incorporates Le Chatlier's Principle, which is a good example of equilibrium principles. Uses of *Le Chatlier's Principle* are reversible reactions and reversible reactions involving gases. *Chemical equilibrium is when a reaction has no tendency to change the quantity of the products and reactants, so the reaction can go both ways.*

- Increasing the pressure and decreasing the temperature results in the higher yield of ammonia by causing a move of the reaction to the right.
- Because there are more molecules on the left side than the right side, when the pressure is increased, the system adapts to the change by moving the molecules left to right to decrease the overall pressure.
- For temperature, it moves from right to left when the temperature drops is because of the process being exothermic, where heat is released.
- The system adjusts to lessen the change, so it would make more heat to compensate for the energy lost, since that is the product of this. If more energy is made, then that would mean more ammonia is made, too. Even though decreasing the temperature is a slow reaction, if the temperature was increased to speed up the reaction, it would produce a smaller amount of ammonia yield.

Examples of Le Châtelier's Principle

$$N_2 (g) + 3H_2 (g) \leftrightarrow 2NH_3 (g)$$

- If the volume is decreased here, it has the same result as when the pressure is decreased.
- If the pressure is increased, in this equation, it will move right because there are fewer gas molecules are produced going to the right then the backwards one.

When you increase the pressure so that the least amount of molecules will be formed, there won't be an increase in collisions. However, if more gas molecules are formed, there will be an increase in collisions, thus moving the way that will produce the least amount of molecules.

$$PCl_3(g) + Cl_2(g) \leftrightarrow PCl_5(g) + energy$$

- If the temperature increased here, it would shift to the left because it would use the extra energy that is left over in the equation.
- In the equation it can go both ways, left and right, so one way is *endothermic* and the other is *exothermic*.
- If the temperature is increased, it would benefit the endothermic reactions, so there would be more energy for the reaction to take in. * If the temperature is decreased, it would be in the favor of an exothermic reaction, because then the reaction can release heat.

$$H_2 + I_2 \leftrightarrow 2HI$$

- Removing H_2 from the system will cause it to move towards the left, so more H_2 can be made.
- By lowering the concentration of one substance, the equation will shift in that direction so that it can produce more of that which was lowered.
- If one concentration increases, it will move in the direction that would help lower the concentration.

Catalyst

A catalyst is used to speed up a reaction by lowering the activation energy. So, in this reaction, the iron catalyst is used to lower the activation energy so that the N_2 and H_2 can be easier to break down.

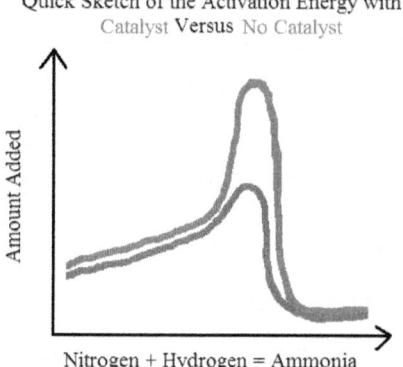

Quick Sketch of the Activation Energy with a
Catalyst Versus No Catalyst

Nitrogen + Hydrogen = Ammonia

From the above diagram the blue curve represents what the activation energy would look like with an iron catalyst. The purple curve shows what the activation energy would look like if a catalyst was *not* involved. Note that without the catalyst, the activation energy is much bigger. A catalyst is needed to break down the nitrogen and hydrogen gases.

Economic Effects

Nitric Acid

By mixing one part ammonia to nine parts air with the use of a catalyst, the ammonia will get oxidized to *nitric acid*.

$$4\,NH_3 + 5\,O_2 \rightarrow 4\,NO + 6\,H_2O$$
$$2\,NO + O_2 \rightarrow 2\,NO_2$$
$$2\,NO_2 + 2\,H_2O \rightarrow 2\,HNO_3 + H_2$$

Fertilizer

Ammonia is mixed with irrigation water to form a solution for fertilizer ingredients. Fertilizer is mainly used in fields that grow crops such as corn, barley, sorghum, rapeseed, soybean, and sunflower.

Ammonia Production with the Haber-Bosch Process

Nearly all modern commercial production of ammonia is based on the Haber-Bosch synthesis process. The process reaction is given by the equation:

$$N_2 + 3\,H_2 <==> 2\,NH_3 + Heat$$

This reaction happens in a special high pressure reactor in the presence of a special catalyst, usually a porous iron oxide. The reaction is exothermic which means that energy is released.

The equation shows that the nitrogen, hydrogen and the ammonia exist in equilirium which is determined by the conditions existing in the reactor. Typically for ammonia synthesis these condions are:

- Pressure - about 150 atmospheres
- Temperatures 370 - 500 °C

Definition: Under equilibrium conditions the proportion of reactants and the product of a chemcal reaction are balanced and detrmined by the existing physical contions such as pressure, temperature and concentrations.

Factors influencing the rate of ammonia production (rate of reaction r_x):

- Temperature - because the reaction is exothermic, lowering the temperature in the reactor will increase the yield of ammonia. But this also slows down the reaction therefore for the reason of efficiency in commercial production the temperature is kept as high as possible.
- Pressure - increasing the pressure will increase the yield of ammonia but there is a limit in pressure for safety reasons.

THE PROCESS (HABER-BOSCH)

In large scale commercial ammonia production plants the feed stock which makes up the reactants are water, methane and air. Through a series of chemical processes the feedstock is converted to nitrogen, hydrogen and carbon dioxide. After removal of the waste carbon dioxide, the remaining mixture of nitrogen and hydrogen is called "synthesis gas". The synthesis gas enters the ammonia reactor, where it undergoes the synthesis reaction shown above.

The synthesis reactor consists of a number of beds containing the catalyst through which the reactants pass and are converted to ammonia under equilibrium conditions. Because the reaction is exothermic, heat is removed by coolers (heat exchangers). This ensures that the maximum amount of ammonia is converted.

At the end of the reactor a stream of mixed gas (ammonia + nitrogen + hydrogen) is removed and cooled. The cooled ammonia condenses and is separated from the other gas which is then returned to the reactor for reprocessing. This is a continous process - at the one end ammonia is continuously removed and this is balanced by new "synthesis gas" which is continously added to the reactor.

Ammonia

Linde supplies ammonia plants using two different process concepts.

LAC™ (the Linde Ammonia Concept) makes possible the production of ammonia from natural gas or light hydrocarbons. High-pressure gasification makes use of refinery residue, heavy oils and coal with oxygen for the production of ammonia.

1. LAC™ Process

The Linde Ammonia Concept is a leading-edge process for the production of ammonia from natural gas or light hydrocarbons. It is based on a combination of proven process steps. An LAC plant primarily comprises a modern hydrogen plant, a standard nitrogen plant and high-efficiency ammonia synthesis.

In the hydrogen plant, the synthesis gas is purified by pressure swing adsorption in a single process step The pure nitrogen delivered from the nitrogen plant is first mixed in with the synthesis gas upstream of the synthesis gas compressor. With the exception of a third-party license for the NH3 synthesis of Ammonia Casale, all the other process steps are based on Linde's own technology.

The LAC process means a simplification vis-à-vis the classic process route and leads to savings in investment and operating costs, as well as simplified plant start-up and operation.

Additional savings in investment costs can also be brought about when the nitrogen is supplied by over-the-fence delivery. The specific operating costs can be further lowered by selling valuable by-products. Hydrogen and nitrogen are immediately available, *e.g.* as intermediate products of an LAC plant. Other by-products such as oxygen, argon and carbon dioxide can be produced if the plant is properly aligned.

Thus far, four plants based on the relatively new Linde Ammonia Concept have been constructed with capacities of between 230 to 1,350 mtd of NH3. The applied process steps have already frequently proven their reliability in large-scale operation and can also be partially installed in modularized form for cost savings. Linde has, for instance, built more than 2,700 air separators, 50 hydrogen plants and 300 PSA plants, while Ammonia Casale has more than 100 references for ammonia synthesis.

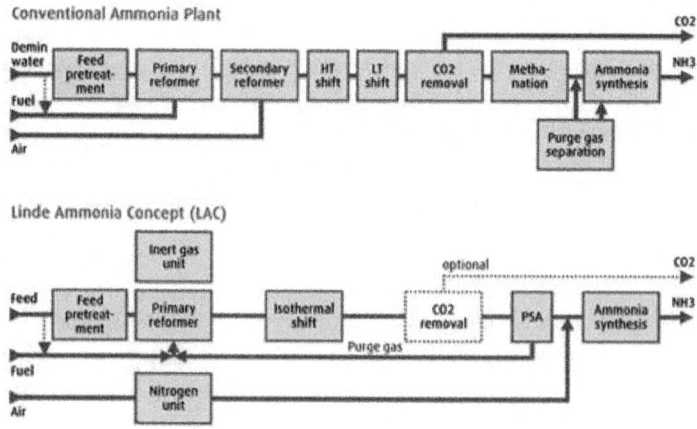

The first LAC plant was built for GSFC in Baroda, Gujarat, India and went into operation in 1998 with a capacity of 1,350 mtd. Urea is an final product of this plant. Further LAC plants have been constructed in Australia and China, and ammonium phosphate, ammonium nitrate and methylamine are among the final products produced in subsequent plants.

Chapter 4

THE BALANCE EQUATION

INTRODUCTION TO CHEMICAL ENGINEERING PROCESSES/UNITS

Consistency of Units

Most values that you'll run across as an engineer will consist of a number and a unit. Some do not have a unit because they are a pure number (like pi, π) or a ratio. In order to solve a problem effectively, all the types of units should be consistent with each other, or should be in the same *system*. A system of units defines each of the basic unit types with respect to some measurement that can be easily duplicated, so that, for example, 5 ft. is the same length in Australia as it is in the United States. There are five commonly-used base unit types or *dimensions* that one might encounter (shown with their abbreviated forms for the purpose of dimensional analysis) :

Length (L), or the physical distance between two positions with respect to some standard distance

Time (t), or how long something takes with respect to how long some natural phenomenon takes to occur

Mass (M), a measure of the inertia of a material relative to that of a standard

Temperature (T), a measure of the average kinetic energy of the molecules in a material relative to a standard

Electric Current (E), a measure of the total charge that moves in a certain amount of time

Note : It would make more commonsense to have **Electric Charge** as a base unit, since current is charge per time, and you may find it convenient to think of charge as the fundamental unit. However, current proved easier to measure very accurately and reproducibly, so the physicists decided it would be their reference.

There are several different consistent systems of units. In most of the world (apart from the US and to some extent the UK) the SI system is standard. It is also used in refereed scientific and engineering journals in these two countries. In practice, it is essential for a chemical engineer to be proficient in the SI system, but to be able to use data in units of other systems and to be able to specify designs in the preferred unit system for the job.

Units of Common Physical Properties

Every system of units has a large number of *derived* units which are, as the name implies, derived from the base units. These new units are based on the physical definitions of other quantities and involve the combination of different variables. Below is a list of several common derived system properties and the corresponding dimensions ($\doteq$ denotes unit equivalence). If you don't know what one of these properties is, you will learn it eventually :

Property	Dimensions	Property	Dimensions
Mass	M	Length	L
Time	t	Temperature	T
Area	L^2	Volume	L^3
Velocity	$\dfrac{L}{t}$	Acceleration	$\dfrac{L}{t^2}$
Force	$\dfrac{M*L}{t^2}$	Energy/Work/Heat	$\dfrac{M*L^2}{t^2}$
Power	$\dfrac{M*L^2}{t^3}$	Pressure	$\dfrac{M*L^2}{t^2}$
Density	$\dfrac{M}{L^3}$	Viscosity	$\dfrac{M}{L*t}$
Diffusivity	$\dfrac{L^2}{s}$	Thermal Conductivity	$\dfrac{M*L}{t^3*T}$
Specific Heat Capacity	$\dfrac{L^2}{t^2*T}$	Specific Enthalpy	$\dfrac{L^2}{t^2}$
Specific Gibbs Energy	$\dfrac{L^2}{t^2}$	Specific Energy	$\dfrac{L^2}{t^2*T}$

SI (kg-m-s) System

This is the most commonly-used system of units in the world, and is based heavily on factors of 10. It was originally based on the properties of water, though currently there are more precise standards in place. The major dimensions are :

Dimension name	SI unit	SI abbreviation
Length	meter	m
Time	second	s
Mass	kilogram	kg
Temperature	kelvin	K
Electric Current	ampere	A
Amount of substance	mole	mol

Note that the kilogram, not the gram, is a base unit.

The close relationship to water is that one m^3 of water weighs (approximately) 1000 kg.

A base unit that can be difficult to understand is the *mole*. A mole represents $6.022*10^{23}$ particles of *any substance*. (The number is known as Avogadro's Number, or the Avogadro constant.) This usually means the number of atoms or molecules of an element or compound. Chemical engineers commonly use kilomoles. The relative molecular mass (= molecular weight) of water H_2O is about 18, being made up of 2 H atoms (atomic mass = 1) and one O atom (atomic mass = 16). Thus 18 kilograms of water constitute 1 kilomole of H_2O and contain 2 kilomoles of H atoms and 1 kilomole of O atoms.

Each of these base units can be made smaller or larger in units of ten by adding the appropriate metric prefixes. The specific meanings are (from the SI page on Wikipedia) :

SI Prefixes									
Name	yotta	zetta	exa	peta	tera	giga	mega	kilo	hecto	deca
Symbol	Y	Z	E	P	T	G	M	k	h	da
Factor	10^{24}	10^{21}	10^{18}	10^{15}	10^{12}	10^9	10^6	10^3	10^2	10^1
Name	deci	centi	milli	micro	nano	pico	femto	atto	zepto	yocto
Symbol	d	c	m	μ	n	p	f	a	z	y
Factor	10^{-1}	10^{-2}	10^{-3}	10^{-6}	10^{-9}	10^{-12}	10^{-15}	10^{-18}	10^{-21}	10^{-24}

If you see a length of 1 km, according to the chart, the prefix "k" means there are 10^3 of something, and the following "m" means that it is meters. So 1 km = 10^3 meters. There should always be a space between the number and the unit and between different units which are multiplied together. There must not be a space between the multiplier and the unit. Thus 13 mA means 13 milliamps, but 13 m A means 13 meter-amps.

As noted above, the kilogram is a base unit, but the multipliers are added to the gram. 1000 kg = 1 Mg; 0.001 kg = 1 g.

In chemical engineering practice, we tend not to use the very large or small ends of the table, but you should know at least as large as mega (M), and as small as nano (n). The relationship between different sizes of metric units was deliberately made simple because you will have to do it all of the time. You may feel uncomfortable with it at first if you're from the U.S. but trust me, after working with the English system you'll learn to appreciate the simplicity of the Metric system.

Derived Units in the SI System

Imagine if every time you calculated a pressure, you would have to write the units in kg/(m s²). This would become cumbersome quickly, so the SI people set up derived units to use as shorthand for such combinations as these. Note that units named after a person do not start with a capital letter, but the abbreviation does! For example "a force of one newton" and " a force of 1.0 N". The most common ones used by chemical engineers are as follows :

Property name	Long SI Units	SI Name	SI Abbreviation	Equivalencies
Force	$\dfrac{kg * m}{s^2}$	newton	N	Mass * acceleration
Energy	$\dfrac{kg * m^2}{s^2}$	joule	J	$N * m,$ $Pa * m^3$
Power	$\dfrac{kg * m^2}{s^3}$	watt	W	$\dfrac{N * m}{s}$ or $\dfrac{J}{s}$
Pressure	$\dfrac{kg}{m * s^2}$	pascal	Pa	$\dfrac{N}{m^2}$

Allowed Units in the SI System

Some units are not simply derived from the base units or regular multiples, but are in common use and are therefore permitted. Thus, though periods of time can be expressed in kiloseconds or megaseconds, we are allowed to use minutes, hours and days. The term 'liter' (US) or 'litre' (European) is understood to be the same as 1 x 10⁻³ m³, and the term tonne (not ton) is understood to be the same as 1000 kg. The bar is a unit of pressure meaning 100 kPa, which is very close to the chemists' standard atmosphere (which is 101.325 kPa). The Celsius scale of temperature is understood to be the number of kelvin above 273.15 K. Thus we are allowed to write "the chemical reactor has a throughput of 4.3 tonnes per day at 5 bar and 200 °C" and we will be understood. However, it may be necessary to change to base or derived units in order to carry out calculations.

cgs (cm-g-s) System

This was the first metric system and may be found old publications (before 1960). There is no reason why a chemical engineer should work in it today, but you may have to convert data from old books. The base units of length and mass were the centimeter and gram. The unit of force was a *dyne*; the unit of energy was an *erg*. The value of *g*, the standard acceleration due to gravity was 981 cm/s/s. The viscosity units poise (especially centipoise, cP) and stokes (especially centistokes cSt) are a hangover from this system and may be found in relatively recent publications. You should convert them to SI.

Note that chemists often work with grams and cubic centimeters, but these are part of SI. Just because you work with cm, g, and s, does not mean you are using the cgs system. See w :cgs if you really want to know.

British, Imperial or American (gravitational) system

This system was established with the authority of the British Empire. It is known in Britain as the Imperial system, in America as the British (sometimes English) system, and in much of the world as the American system, since the USA is the only major market for chemical engineering which uses it. The engineering version uses a subset of this traditional or customary measure plus the *pound force* and the ampere.

Its peculiarity lies in the relationship between force and mass. According to Isaac Newton for a fixed mass accelerating under the influence of a force :

$$\text{force} = \text{mass} \times \text{acceleration or } f = m\,a$$

In the SI system a force of 1 newton acting on a mass of 1 kilogram produces an acceleration of 1 meter per second. Simple!

In the Imperial system a force of 1 pound-force acting on a mass of 1 pound produces an acceleration of 32 feet per second per second. This is because this is the natural acceleration under gravity. Older American books often include a *g* in the formula which do not appear in European versions of the same equation. The *g* represents the relationship between force and mass in the unit system (which is 1 in SI) : here it is 32. For a while, American (mainly) engineers used a version of the metric system including the kilogram-force and thus *g*, which had the value 9.81. Physicists call these both *gravitational* systems.

The common units are based on traditional measures which were practical in agriculture and shipping, and do not go in steps of 10, 100, 1000 *etc.* Instead of using prefixes you use names for larger units, and can use combinations of units for the same dimension, *e.g.* 6 yards 2 feet and 8 and a quarter inches (6 yd 2 ft 8¼ in) However, engineers tend to use one unit and a decimal, *e.g.* 20.7 ft, *e.g.* 13.47 in. The foot can also be denoted by a single mark and the inch by a double mark, *e.g.* 4 feet 7 inches was 4' 7". Note that the US gallon is smaller than the Imperial gallon (5/6 in fact), when you are doing conversions to SI.

The temperature scale is that of Fahrenheit, in which the melting point of ice is 32 °F. Absolute zero is -459.67 °F. For thermodynamic temperatures, the number of degrees Fahrenheit above absolute zero is the Rankine scale. Thus the melting point of ice is 459.67 °R.

The following are common units in this system.

Dimension name	Imperial unit	Imperial abbreviation
Length	foot, inch	ft, in
Time	second, minute, or hour	sec, min, and hr, respectively
Force	pound-force	lb_f
Temperature	degree Fahrenheit	°F
Electric current	ampere	A

A common derived unit is the pound(-force) per square inch, or psi. Note that psig or psi(g) means psi above atmospheric pressure. Energy is measured in British Thermal Units, generally BTU, sometimes B.Th U. Power is horsepower, hp.

"Parts-per" Notation

The "parts-per" notation is a unit that deals with very small traces of species within a mixture of gases or liquids. Parts-per million (ppm) and parts-per billion (ppb), as well as parts-per trillion (ppt) (American definition of trillion 10^{12}), refer to mass or mole ratios and communicate how many parts of the species are present-per million, billion, or trillion parts of the mixture. Generally mass ratios are used when dealing with liquids and mole ratios are used when dealing with gases, though either kind of ratio can be used for whichever phase a chemical is in (ratios are discussed in a later chapter).

Example :

Let's say the air around us contains 20 ppm He (Helium).

This means that, if one assumes that a molar basis is being used, for every million moles of air there are 20 moles of Helium. If the example was in terms of ppb, this would mean that for every billion moles of air there are 20 moles of Helium.

A Word About Conversions

It is generally safe to convert all data into SI then work your calculations out in that system, converting back if necessary at the end. If you are skilled enough in the American system, you may be able to carry some calculations within that system. It is best to consult a conversion table or program for the necessary changes and especially important to keep good track of the units.

However, do not make the mistake of just writing down the numbers you get from the calculator or program.

For example, if you have a pressure drop in a pipe of 16 psi, and the conversion factor 1 psi = 6.895 kPa, your calculator will give 16 x 6.895 = 110.32. However, your answer should be 110 kPa because your starting value was only given to a precision of two figures. The conversion factor cannot add accuracy!

If every value is written in terms of the same base units, and the equation that is used is correct, then the units of the answer will be consistent and in terms of the same base units.

How to Convert Between Units

Finding Equivalences

The first thing you need in order to convert between units is the equivalence between the units you want and the units you have. To do this use a conversion table. Seew : Conversion of units for a fairly extensive (but not exhaustive) list of common units and their equivalences.

Conversions within the metric system usually are not listed, because it is assumed that one can use the prefixes and the fact that 1 mL = 1 cm^3 to convert anything that is desired.

Conversions within the English system and especially between the English and metric system are sometimes (but not on Wikipedia) written in the form :

1 (unit 1) = (number) (unit 2) = (number) (unit 3) = ...

For example, you might recall the following conversion from chemistry class :

1 atm = 760 mmHg = 1.013 * 10^5 Pa = 1.013 bar = ...

The table on Wikipedia takes a slightly different approach : the column on the far left side is the unit we have 1 of, the middle is the definition of the unit on the left, and on the far right-hand column we have the metric equivalent. One listing is the conversion from feet to meters :

foot (International) ft = 1/3 yd = 0.3048 m

Both methods are common and one should be able to use either to look up conversions.

Using the equivalences

Once the equivalences are determined, use the general form :

$$\text{What you want} = \text{What you have} * \frac{\text{What you want}}{\text{What you have}}$$

The fraction on the right comes directly from the conversion tables.

Example :

Convert 800 mmHg into bars

Solution If you wanted to convert 800 mmHg to bars, using the horizontal list, you could do it directly :

$$bar\ s = 800\ \text{mmHg} * \frac{1.013\ \text{bar}}{760\ \text{mmHg}} = 1.066\ \text{bar}$$

Using the tables from Wikipedia, you need to convert to an intermediate (the metric unit) and then convert from the intermediate to the desired unit. We would find that

$$1 \text{ mmHg} = 133.322 \text{ Pa and } 1 \text{ bar} = 10^5 \text{ Pa}$$

Again, we have to set it up using the same general form, just we have to do it twice :

$$bar \ s = 800 \text{ mmHg} * \frac{133.322}{1 \text{ mmHg}} * \frac{1 \text{ bar}}{10^5 \text{Pa}} = 1.066 \text{ bar}$$

Setting these up takes practice, there will be some examples at the end of the section on this. It's a very important skill for any engineer.

One way to keep from avoiding "doing it backwards" is to write everything out and make sure your units cancel out as they should! If you try to do it backwards you'll end up with something like this :

$$bar \ s = 800 \text{ mmHg} * \frac{760 \text{ mmHg}}{1.013 \text{ bar}} = 6.0 * 10^5 \frac{mmHg^2}{bar}$$

If you write everything (even conversions within the metric system!) out, and make sure that everything cancels, you'll help mitigate unit-changing errors. About 30-40% of all mistakes I've seen have been unit-related, which is why there is such a long section in here about it. Remember them well.

Dimensional Analysis as a Check on Equations

Since we know what the units of velocity, pressure, energy, force, and so on should be in terms of the base units L, M, t, T, and E, we can use this knowledge to check the feasibility of equations that involve these quantities.

Example :

Analyze the following equation for dimensional consistency : $P = g * h$ where g is the gravitational acceleration and h is the height of the fluid

Solution We could check this equation by plugging in our units :

$$P \doteq M/(L*t^2), h \doteq L, g = L/t$$
$$g*h \doteq L^2/t^2 \neq M/(L*t^2)$$

Since $g*h$ doesn't have the same units as P, the equation must be wrong regardless of the system of units we are using! The correct equation, in fact, is :

$$P = \rho * g * h$$

where ρ is the density of the fluid. Density has base units of M/L^3 so

$$\rho * g * h \doteq M/L^3 * L^2/t^2 \doteq M/(L*t^2)$$ which are the units of pressure.

This does not tell us the equation is correct but it does tell us that the units are consistent, which is necessary though not sufficient to obtain a correct equation. This is a useful way to detect algebraic mistakes that would otherwise be hard to find. The ability to do this with an algebraic equation is a good argument against plugging in numbers too soon!

You may well be forced to do dimensional analysis in chemical engineering classes or if you do research. For much of the rest of the time, you will probably find it easier to check the units, particularly if you are using the SI system. In the above example, you think :

- Pressure = force / area
- Force = mass x acceleration
- Pressure = mass x acceleration / area

So 1 pascal (unit of pressure) = 1 kg x (m s^{-2}) / (m^2) = 1 kg m^{-1} s^{-2}

- Now g is 9.81 m s^{-2} and h is in meters
- So gh is in units m^2 s^{-2}

To make gh match pressure we need to multiply by something having the units kg m^{-3}, which we recognise as density.

Note dimensional analysis (or unit checking) does not tell you about numerical values that you might have to insert, such as 9.81 or π. Nor does it tell you if you should use the radius or the diameter of a pipe in fluid mechanics!

Importance of Significant Figures

Significant figures (also called significant digits) are an important part of scientific and mathematical calculations, and deals with the accuracy and precision of numbers. It is important to estimate uncertainty in the final result, and this is where significant figures become very important.

Precision and Accuracy

Before discussing how to deal with significant figures one should discuss what precision and accuracy in relation to chemical experiments and engineering. Precision refers to the reproducibility of results and measurements in an experiment, while accuracy refers to how close the value is to the actual or true value. Results can be both precise and accurate, neither precise nor accurate, precise and not accurate, or vice versa. The validity of the results increases as they are more accurate and precise.

An useful analogy that helps distinguish the difference between accuracy and precision is the use of a target. The bullseye of the target represents the true value, while the holes made by each shot (each trial) represents the validity.

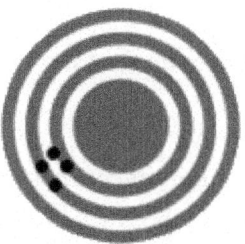

High precision, but unfortunately low accuracy

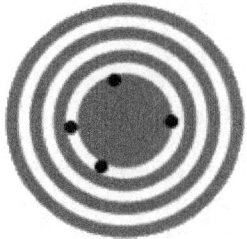

High accuracy, but not very good precision

As the above images show, the first has a lot of holes (black spots) covering a small area. The small area represents a precise experiment, yet it seems that there is a faultiness within the experiment, most likely due to systematic error, rather than random error. The second image represents an accurate though imprecise experiment. The holes are near the bullseye, even "touching" or within, though the problem is that they are spread out. This could be due to random error, systematic error, or not being careful in measuring.

Counting Significant Figures

There are three preliminary rules to counting significant. They deal with non-zero numbers, zeros, and exact numbers.

1. *Non-zero numbers* - all non-zero numbers are considered significant figures
2. *Zeros* - there are three different types of zeros
    *   *leading zeros* - zeros that precede digits - do not count as significant figures (example : .0002 has one significant figure)
    *   *captive zeros* - zeros that are "caught" between two digits - do count as significant figures (example : 101.205 has six significant figures)
    *   *trailing zeros* - zeros that are at the end of a string of numbers and zeros - only count if there is a decimal place (example : 100 has one significant figure, while 1.00 has three, as well as 100.)
3. *Exact numbers* - these are numbers not obtained by measurements, and are determined by counting. An example of this is if one counted the number of millimetres in a centimetre (10 - it is the definition of a millimetre), but another example would would be if you have 3 apples.

Example :

How many significant figures do the following numbers have? Assume none of them are exact numbers.

a) 4.2362 - all numbers, so five

b) 2.0 - zeros after a decimal point count, so two

c) 9900 - only two in this case, because there is no decimal point

d) .44205 - there is a "captive zero," which means it counts, so five

e) .05 - only the five counts, so one

f) 3.9400E9 - tricky one, but scientific notation helps make the zeros at the end noticeable; there are five

The Parable of the Cement Block

People new to the field often question the importance of significant figures, but they have great practical importance, for they are a quick way to tell how accurate a number is. Including too many can not only make your numbers harder to read, it can also have serious negative consequences.

As an anecdote, consider two engineers who work for a construction company. They need to order cement bricks for a certain project. They have to build a wall that is 10 feet wide, and plan to lay the base with 30 bricks. The first engineer does not consider the importance of significant figures and calculates that the bricks need to be 0.3333 feet wide and the second does and reports the number as 0.33, figuring that a precision of ± 0.01 ft (0.1 inches) would be precise enough for the work she was doing.

Now, when the cement company received the orders from the first engineer, they had a great deal of trouble. Their machines were precise but not so precise that they could consistently cut to within 0.0001 feet. However, after a good deal of trial and error and testing, and some waste from products that did not meet the specification, they finally machined all of the bricks that were needed. The other engineer's orders were much easier, and generated minimal waste.

When the engineers received the bills, they compared the bill for the services, and the first one was shocked at how expensive hers was. When they consulted with the company, the company explained the situation : they needed such a high precision for the first order that they required significant extra labor to meet the specification, as well as some extra material. Therefore it was much more costly to produce.

What is the point of this story? Significant figures matter. It is important to have a reasonable gauge of how accurate a number is so that you know not only what the number is but how much you can trust it and how limited it is. The engineer will have to make decisions about how precisely he or she needs to specify design specifications, and how precise measurement instruments (and control systems!) have to be. If you do not need 99.9999% purity then you probably don't need an expensive assay to detect generic impurities at a 0.0001% level (though the lab technicians will probably have to still test for heavy metals and such), and likewise you will not have to design nearly as large of a distillation column to achieve the separations necessary for such a high purity.

Mathematical Operations and Significant Figures

Most likely at one point, the numbers obtained in one's measurements will be used within mathematical operations. What does one do if each number has a different amount of significant figures? If one adds 2.0 litres of liquid with 1.000252 litres, how much does one have afterwards? What would 2.45 times 223.5 get?

For addition and subtraction, the result has the same number of decimal places as the least precise measurement use in the calculation. This means that 112.420020 + 5.2105231 + 1.4 would have have a single decimal place but there can be any amount of numbers to the left of the decimal point (in this case the answer is 119.0).

For multiplication and division, the number that is the least precise measurement, or the number of digits. This means that 2.499 is more precise than 2.7, since the former has four digits while the latter has two. This means that 5.000 divided by 2.5 (both being measurements of some kind) would lead to an answer of 2.0.

Rounding

So now you know how to pick which numbers to drop if there is a question about significant figures, but one also has to take into account rounding. Once one has decided which digit should be the last digit kept, one must decide whether to round up or down.

- If the number is greater than five (6 to 9), one rounds up - 1.36 becomes 1.4
- If the number is less than five (1 to 4), one rounds down - 1.34 becomes 1.3

What does one do when there is a five? There is a special case that deals with the number five, since, if you have not noticed, it is in the middle (between 1 and 9). Often in primary school one learns to just round up, but engineers tend to do something different, called unbiased rounding.

- If the number before the five is even, then one rounds down - 1.45 becomes 1.4
- If the number before the five is odd, then one rounds up - 1.55 becomes 1.6
- Another case is this : 1.4501, where the numbers after five are greater than zero, so one would round to 1.5

Note : Remember that rounding is generally done at the end of calculations, not before the calculations are made.

Why is this done? Engineers make many calculations that often matter, since time, money, *etc.* are being taken into account, it is best to make sure that the final results are not synthetic or untrue to what the actual value should be. This relates back to accuracy and precision.

Elementary mass balances

The "Black Box" Approach to Problem-solving

In this book, all the problems you'll solve will be "black-box" problems. This means that we take a look at a unit operation *from the outside*, looking at what goes into the system and what leaves, and extrapolating data about the properties of the entrance and exit streams from this. This type of analysis is important because it does not depend on the specific type of unit operation that is performed. *When doing a black-box analysis, we don't care about how the unit operation is designed, only what the net result is.* Let's look at an example :

Example :

Suppose that you pour 1L of water into the top end of a funnel, and that funnel leads into a large flask, and you measure that the entire liter of water enters the flask. If the funnel had no water in it to begin with, how much is left over after the process is completed?

Solution The answer, of course, is 0, because you only put 1L of water in, and 1L of water came out the other end. The answer to this does not depend on the how large the funnel is, the slope of the sides, or any other design aspect of the funnel, which is why it is a black-box problem.

Conservation Equations

The formal mathematical way of describing the black-box approach is with conservation equations which explicitly state that what goes into the system must either come out of the system somewhere else, get used up or generated by the system, or remain in the system and accumulate. The relationship between these is simple :

1. The streams entering the system cause an increase of the substance (mass, energy, momentum, *etc.*) in the system.
2. The streams leaving the system decrease the amount of the substance in the system.
3. Generating or consuming mechanisms (such as chemical reactions) can either increase or decrease the stuff in the system.
4. What's left over is the amount of stuff in the system

With these four statements we can state the following very important general principle :

$$Accumulation = In - Out + Generation - Consumption$$

Its so important, in fact, that you'll see it a million times or so, including a few in this book, and it is used to derive a variety of forms of conservation equations.

Common Assumptions on the Conservation Equation

The conservation equation is very general and applies to any property a system can have. However, it can also lead to complicated equations, and so in order to simplify calculations when appropriate, it is useful to apply assumptions to the problem.

- **Closed system** : A closed system is one which does not have flows in or out of the substance. Almost always, when one refers to a closed system, it is implied that the system is closed to *mass flow* but not to other flows such as energy or momentum. The equation for a closed system is :

$$Accumulation = Generation$$

The opposite of a closed system is an **open system** in which the substance is allowed to enter and/or leave the system. The funnel in the example was an open system because mass flowed in and out of it.

- **No generation** : Certain quantities are always **conserved** in the strict sense that they are never created or destroyed. These are the most useful quantities to do balances on because then the model does not need to include a generation term :

$$Accumulation = In - Out$$

The most commonly-used conserved quantities in this class are **mass** and **energy** (other conserved quantities include momentum and electric charge). However, it is important to note that though the *total* mass and *total* energy in a system are conserved, the mass of a single species is **not** (since it may be changed into something else in a reaction). Neither is the "heat" in a system if a so-called "heat-balance" is performed (since it may be transformed into other forms of energy. Therefore, one must be careful when deciding whether to discard the generation term).

- **Steady State** : A system which does not accumulate a substance is said to be at steady-state. Often times, this allows the engineer to avoid having to solve differential equations and instead use algebra.

$$In - Out + Generation - Consumption = 0$$

All problems in this text assume steady state but it is not always a valid assumption. It is mostly valid after a process has been running in a controlled manner for long enough that all the flow rates, temperatures, pressures, and other system parameters have reached reasonably constant values. It is not valid when a process is first warming up (or an operating condition is changed) and the system properties change significantly over time. How they change, and how long it takes to become close enough to steady state, is a subject for another course.

Conservation of Mass

TOTAL mass is a conserved quantity (except in nuclear reactions, let's not go there), as is the mass of any individual species if there is no chemical reaction occurring in the system. Let us write the conservation equation *at steady state* for such a case (with no reaction) :

$$In - Out = 0$$

Now, there are two major ways in which mass can enter or leave a system : diffusion and convection. However, if the velocity entering the unit operations is fairly large and the concentration gradient is fairly small, diffusion can be neglected and the only mass entering or leaving the system is due to convective flow :

A similar equation apply for the mass out.

In this book, we generally use the symbol $\dot{m}$ to signify a convective mass flow rate, in units of *mass/time*. Since the total flow in is the sum of individual flows, and the same with the flow out, the following steady state mass balance is obtained for the overall mass in the system :

$$\sum \dot{m}_{out} - \sum \dot{m}_{in} = 0$$

If it is a *batch* system, or if we're looking at how much has entered and left in a given period of time (rather than instantaneously), we can apply the same mass balance without the time component. In this book, a value without the dot signifies a value without a time component :

$$\sum m_{out} - \sum m_{in} = 0$$

Example :

Let's work out the previous example (the funnel), but explicitly state the mass balance. We're given the following information :

$$m_{in} = 1L$$

$$m_{out} = 1L$$

From the general balance equation,

$$In - Out = Accumulation$$

Therefore, $Accumulation = 1L - 1L = 0$.

Since the accumulation is 0, the system is at steady state.

This is a fairly trivial example, but it gets the concepts of "in", "out", and "accumulation" on a physical basis, which is important for setting up problems. In the next section, it will be shown how to apply the mass balance to solve more complex problems with only one component.

Converting Information into Mass Flows – Introduction

In any system there will be certain parameters that are easier (often considerably) to measure and/or control than others. When you are solving any problem and trying to use a mass balance or any other equation, *it is important to recognize what pieces of information can be interconverted.* The purpose of this section is to show some of the more common alternative ways that mass flow rates are expressed, mostly because it is easier to, for example, measure a velocity than it is to measure a mass flow rate directly.

Volumetric Flow Rates

A volumetric flow rate is a relation of how much volume of a gas or liquid solution passes through a fixed point in a system (typically the entrance or exit point of a process) in a given amount of time. It is denoted as :

$$\dot{V}_n \doteq \frac{Volume}{time} \text{ in stream n}$$

Volume in the metric system is typically expressed either in L (dm^3), mL (cm^3), or m^3. Note that a cubic meter is very large; a cubic meter of water weighs about 1000kg (2200 pounds) at room temperature!

Why They're Useful

Volumetric flow rates can be measured directly using flow meters. They are especially useful for gases since the volume of a gas is one of the four properties that are needed in order to use an *equation of state* (discussed later in the book) to calculate the molar flow rate. Of the other three, two (pressure, and temperature) can be specified by the reactor design and control systems, while one (compressibility) is strictly a function of temperature and pressure for any gas or gaseous mixture.

Limitations

Volumetric Flowrates are Not Conserved. We can write a balance on volume like anything else, but the "volume generation" term would be a complex function of system properties. Therefore if we are given a volumetric flow rate we should change it into a mass (or mole) flow rate before applying the balance equations.

Volumetric flowrates also do not lend themselves to splitting into components, since when we speak of volumes in practical terms we generally think of the total solution volume, not the partial volume of each component (the latter is a useful tool for thermodynamics, but that's another course entirely). There are some things that are measured in volume fractions, but this is relatively uncommon.

How to convert volumetric flow rates to mass flow rates

Volumetric flowrates are related to mass flow rates by a relatively easy-to-measure physical property. Since $\dot{m} \doteq mass / time$ and $\dot{V} \doteq volume / time$, we need a property with units of $mass / volume$ in order to convert them. The density serves this purpose nicely!

$$\dot{V}_n * \rho_n = \dot{m}_n \text{ in stream n}$$

The "i" indicates that we're talking about one particular flow stream here, since each flow may have a different density, mass flow rate, or volumetric flow rate.

Velocities

The velocity of a bulk fluid is *how much lateral distance along the system (usually a pipe) it passes per unit time*. The velocity of a bulk fluid, like any other, has units of :

$$v_n = \frac{dis\,tan\,ce}{time} \text{ in stream n}$$

By definition, the bulk velocity of a fluid is related to the volumetric flow rate by :

$$v_n = \frac{\dot{V}_n}{A_n} \text{ in stream n}$$

This distinguishes it from the velocity of the fluid at a certain point (since fluids flow faster in the center of a pipe). The bulk velocity is about the same as the instantaneous velocity for relatively fast flow, or especially for flow of gasses.

For purposes of this class, all velocities given will be bulk velocities, not instantaneous velocities.

Why They're Useful

(Bulk) Velocities are useful because, like volumetric flow rates, they are relatively easy to measure. They are especially useful for liquids since they have constant density (and therefore a constant pressure drop at steady state) as they pass through the orifice or other similar instruments. This is a necessary prerequisite to use the design equations for these instruments.

Limitations

Like volumetric flowrates, velocity is not conserved. Like volumetric flowrate, velocity changes with temperature and pressure of a gas, though for a liquid velocity is generally constant along the length of a pipe.

Also, velocities can't be split into the flows of individual components, since all of the components will generally flow at the same speed. They need to be converted into something that can be split (mass flow rate, molar flow rate, or pressure for a gas) before concentrations can be applied.

How to convert velocity into mass flow rate

In order to convert the velocity of a fluid stream into a mass flow rate, you need two pieces of information :

1. The **cross sectional area** of the pipe.

2. The **density** of the fluid.

In order to convert, first use the definition of bulk velocity to convert it into a volumetric flow rate :

$$\dot{V}_n = v_n * A_n$$

Then use the density to convert the volumetric flow rate into a mass flow rate.

$$\dot{m}_n = \dot{V}_n * \rho_n$$

The combination of these two equations is useful :

$$\dot{m}_n = v_n * \rho_n * A_n \text{ in stream n}$$

Molar Flow Rates

The concept of a molar flow rate is similar to that of a mass flow rate, it is the number of moles of a solution (or mixture) that pass a fixed point per unit time :

$$\dot{n}_n \doteq \frac{moles}{time} \text{ in stream n}$$

Why They're Useful

Molar flow rates are mostly useful because *using moles instead of mass allows you to write material balances in terms of reaction conversion and stoichiometry*. In other words, there are a lot fewer unknowns when you use a mole balance, since the stoichiometry allows you to consolidate all of the changes in the reactant and product concentrations in terms of one variable. This will be discussed more in a later chapter.

Limitations

Unlike mass, total moles are not conserved. Total mass flow rate is conserved whether there is a reaction or not, but the same is not true for the number of moles. For example, consider the reaction between hydrogen and oxygen gasses to form water :

$$H_2 + \frac{1}{2}O_2 \rightarrow H_2O$$

This reaction consumes 1.5 moles of reactants for every mole of products produced, and therefore the total number of moles entering the reactor will be more than the number leaving it.

However, since neither mass nor moles of individual components is conserved in a reacting system, it's better to use moles so that the stoichiometry can be exploited, as described later.

The molar flows are also somewhat less practical than mass flow rates, since you can't measure moles directly but you can measure the mass of something, and then convert it to moles using the molar flow rate.

How to Change from Molar Flow Rate to Mass Flow Rate

Molar flow rates and mass flow rates are related by the molecular weight (also known as the molar mass) of the solution. In order to convert the mass and molar flow rates of the *entire solution*, we need to know the average molecular weight of the solution. This can be calculated from the molecular weights and mole fractions of the components using the formula :

$$\overline{MW}_n = [(MW_i * y_i)]_n$$

where i is an index of *components* and n is the *stream* number. signifies *mole fraction* of each component (this will all be defined and derived later).

Once this is known it can be used as you would use a molar mass for a single component to find the total molar flow rate.

A Typical Type of Problem

Most problems you will face are significantly more complicated than the previous problem and the following one. In the engineering world, problems are presented as so-called "word problems", in which a system is described and the problem must be set up and solved (if possible) from the description. This section will attempt to illustrate through example, step by step, some common techniques and pitfalls in setting up mass balances. Some of the steps may seem somewhat excessive at this point, but if you follow them carefully on this relatively simple problem, you will certainly have an easier time following later steps.

Single Component in Multiple Processes : A Steam Process

Example :

A feed stream of pure liquid water enters an evaporator at a rate of 0.5 kg/s. Three streams come from the evaporator : a vapor stream and two liquid streams. The flowrate of the vapor stream was measured to be $4*10^6$ L/min and its density was 4 g/m^3. The vapor stream enters a turbine, where it loses enough energy to condense fully and leave as a single stream. One of the liquid streams is discharged as waste, the other is fed into a heat exchanger, where it is cooled. This stream leaves the heat exchanger at a rate of 1500 pounds per hour. Calculate the flow rate of the discharge and the efficiency of the evaporator.

Note that one way to define efficiency is in terms of conversion, which is intended here :

$$efficiency = \frac{\dot{m}_{vapor}}{\dot{m}_{feed}}$$

Step 1 : Draw a Flowchart

The problem as it stands contains an awful lot of text, but it won't mean much until you *draw what is given to you*. First, ask yourself, what processes are in use in this problem? Make a list of the processes in the problem :

1. Evaporator (A)
2. Heat Exchanger (B)
3. Turbine (C)

Once you have a list of all the processes, you need to find out how they are connected (it'll tell you something like "the vapor stream enters a turbine"). Draw a basic sketch of the processes and their connections, and label the processes. It should look something like this :

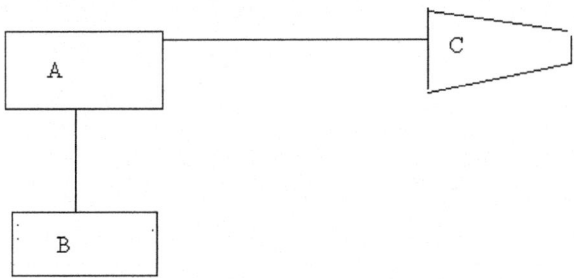

Remember, we don't care what the actual processes look like, or how they're designed. At this point, we only really label what they are so that we can go back to the problem and know which process they're talking about.

Once all your processes are connected, find any streams that are not yet accounted for. In this case, we have not drawn the feed stream into the evaporator, the waste stream from the evaporator, or the exit streams from the turbine and heat exchanger.

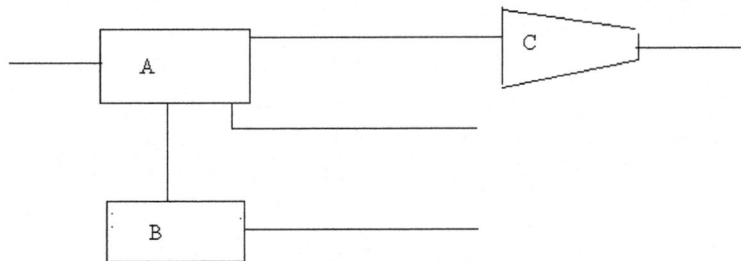

The third step is to Label all your flows. Label them with any information you are given. Any information you are not given, and even information you are given should be given a different variable. It is usually easiest to give them the same variable as is found in the equation you will be using (for example, if you have an unknown flow rate, call it $\dot{m}$ so it remains clear what the unknown value is physically. Give each a different subscript corresponding to the number of the feed stream (such as $\dot{m}_1$ for the feed stream that you call "stream 1"). Make sure you include all units on the given values!

In the example problem, the flowchart I drew with all flows labeled looked like this :

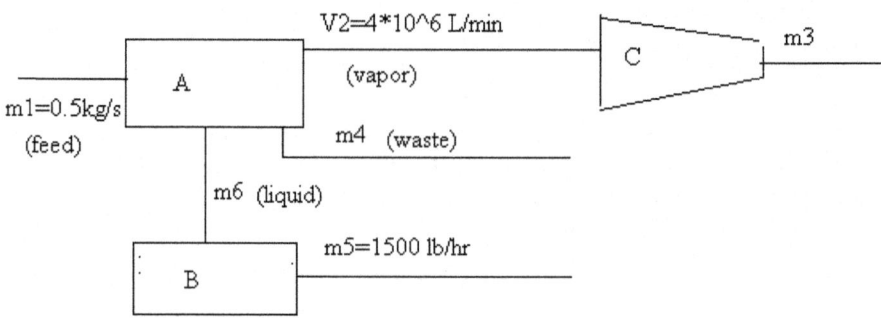

Notice that for one of the streams, a *volume* flow rate is given rather than a *mass* flow rate, so it is labeled as such. This is very important, so that you avoid using a value in an equation that isn't valid (for example, there's no such thing as "conservation of volume" for most cases)!

The final step in drawing the flowchart is to write down any additional given information in terms of the variables you have defined. In this problem, the density of the water in the vapor stream is given, so write this on the side for future reference.

Carefully drawn flowcharts and diagrams are half of the key to solving any mass balance, or really a lot of other types of engineering problems. They are just as important as having the right units to getting the right answer.

Step 2 : Make sure your units are consistent

The second step is to make sure all your units are consistent and, if not, to convert everything so that it is. In this case, since the principle that we'll need to use to solve for the flow rate of the waste stream ($\dot{m}_3$) is conservation of mass, everything will need to be on a mass-flow basis, and also in the *same* mass-flow units.

In this problem, since two of our flow rates are given in metric units (though one is a volumetric flow rate rather than a mass flow rate, so we'll need to change that) and only one in English units, it would save time and minimize mistakes to convert $\dot{V}_2$ and $\dot{m}_5$ to kg/s.

From the previous section, the equation relating volumetric flow rate to mass flow rate is :

$$\dot{V}_i * \rho_i = \dot{m}_i$$

Therefore, we need the density of water vapor in order to calculate the mass flow rate from the volumetric flow rate. Since the density is provided in the problem statement (if it wasn't, we'd need to calculate it with methods described later), the mass flow rate can be calculated :

$$\dot{V}_2 = \frac{4 * 10^6 L}{1\ min} * \frac{1\ m^3}{1000\ L} * \frac{1min}{60\ s} = 66.67\ \frac{m^3}{s}$$

$$\rho_2 = 4\frac{g}{m^3} * \frac{1\,kg}{1000\,g} = 0.004\,\frac{kg}{m^3}$$

$$\dot{m}_2 = 66.67\frac{m^3}{s} * 0.004\frac{kg}{m^3} = 0.2666\frac{kg}{s}$$

Note that since the density of a gas is so small, a huge volumetric flow rate is necessary to achieve any significant mass flow rate. This is fairly typical and is a practical problem when dealing with gas-phase processes.

The mass flow rate $\dot{m}_5$ can be changed in a similar manner, but since it is already in terms of mass (or weight technically), we don't need to apply a density :

$$\dot{m}_5 = 1500\frac{lb}{hr} * \frac{1\,kg}{2.2lb} * \frac{1hr}{3600s} = 0.1893\frac{kg}{s}$$

Now that everything is in the same system of units, we can proceed to the next step.

Step 3 : Relate your variables

Since we have the mass flow rate of the vapor stream we can calculate the efficiency of the evaporator directly :

$$efficiency = \frac{\dot{m}_2}{\dot{m}_1} = \frac{0.2666\frac{kg}{s}}{0.5\frac{kg}{s}} = 53.3\%$$

Finding $\dot{m}_4$, as asked for in the problem, will be somewhat more difficult. One place to start is to write the mass balance on the evaporator, since that will certainly contain the unknown we seek. Assuming that the process is steady state we can write :

$$In - Out = 0$$
$$\dot{m}_1 - \dot{m}_2 = \dot{m}_4 - \dot{m}_6 = 0$$

Problem : we don't know $\dot{m}_6$ so with only this equation we cannot solve for $\dot{m}_4$. Have no fear, however, because there is another way to figure out what $\dot{m}_6$ is... can you figure it out? Try to do so before you move on.

So you want to check your guess? Alright then read on.

The way to find $\dot{m}_6$ is to do a mass balance on the heat exchanger, because the mass balance for the heat exchanger is simply :

$$\dot{m}_6 - \dot{m}_5 = 0$$

Since we know $\dot{m}_5$ we can calculate $\dot{m}_6$ and thus the waste stream flowrate $\dot{m}_4$.

Note : Notice the strategy here : we first start with a balance on the operation containing the stream we need information about. Then we move to balances on other operations in order to garner additional information about the unknowns in the process. This takes practice to figure out when you have enough information to solve the problem or you need to do more balances or look up information.

It is also of note that any process has a limited number of independent balances you can perform. This is not as much of an issue with a relatively simple problem like this, but will become an issue with more complex problems. Therefore, a step-by-step method exists to tell you exactly how many independent mass balances you can write on any given process, and therefore how many total independent equations you can use to help you solve problems.

Step 4 : Calculate your unknowns.

Carrying out the plan on this problem :

$$\dot{m}_6 - 0.1893 \frac{kg}{s} = 0$$

$$\dot{m}_6 = 0.1893 \frac{kg}{s}$$

Hence, from the mass balance on the evaporator :

$$\dot{m}_4 = \dot{m}_1 - \dot{m}_2 - \dot{m}_6 = (0.5 - 0.2666 - 0.1893) \frac{kg}{s} = 0.0441 \frac{kg}{s}$$

So the final answers are :

Evaporator Efficiency = 53.3%

Waste stream rate = $0.0441 \frac{kg}{s}$

Step 5 : Check your work.

Ask : Do these answers make sense? Check for things like negative flow rates, efficiencies higher than 100%, or other physically impossible happenings; if something like this happens (and it will), you did something wrong. Is your exit rate higher than the total inlet rate (since no water is created in the processes, it is impossible for this to occur)?

In this case, the values make physical sense, so they may be right. It's always good to go back and check the math *and* the setup to make sure you didn't forget to convert any units or anything like that.

Mass balances on multicomponent systems

Component Mass Balance

Most processes, of course, involve more than one input and/or output, and therefore it must be learned how to perform mass balances on . The basic idea remains the same though. We can write a mass balance in the same form as the overall balance for each component :

In − Out + Generation = Accumulation

For steady state processes, this becomes :

$In - Out + Generation = 0$

The overall mass balance at steady state, recall, is :

$$\Sigma \dot{m}_{in} - \Sigma \dot{m}_{out} + m_{gen} = 0$$

The mass of each component can be described by a similar balance.

$$\Sigma \dot{m}_{A,in} - \Sigma \dot{m}_{A,out} + m_{A,gen} = 0$$

The biggest difference between these two equations is that The total generation of mass m_{gen} is zero due to conservation of mass, but since individual species can be consumed in a reaction, $m_{A,gen} \neq 0$ for a reacting system

Concentration Measurements

You may recall from general chemistry that a concentration is a measure of the amount of some species in a mixture relative to the total amount of material, or relative to the amount of another species. Several different measurements of concentration come up over and over, so they were given special names.

Molarity

The first major concentration unit is the molarity which relates the moles of one particular species to the total volume of the solution.

$$Molarity\ (A) = [A] = \frac{n_A}{V_{sln}} \text{ where } n \doteq mol, V \doteq L$$

A more useful definition for flow systems that is equally valid is :

$$[A] = \frac{\dot{n}_A}{\dot{V}_n} \text{ where } \dot{n}_A \doteq mol/s, \dot{V}_s \doteq L/s$$

Molarity is a useful measure of concentration because it takes into account the volumetric changes that can occur when one creates a mixture from pure substances. Thus it is a very practical unit of concentration. However, since it involves volume, it can change with temperature so *molarity should always be given at a specific temperature*. Molarity of a gaseous mixture can also change with pressure, so it is not usually used for gasses.

Mole Fraction

The mole fraction is one of the most useful units of concentration, since it allows one to directly determine the molar flow rate of any component from the total flow rate. It also conveniently is *always* between 0 and 1, which is a good check on your work as well as an additional equation that you can always use to help you solve problems.

The mole fraction of a component A in a mixture is defined as :

$$x_A = \frac{n_A}{n_n}$$

where n_A signifies moles of A. Like molarity, a definition in terms of flowrates is also possible :

Mole Fraction Definition

$$x_A = \frac{\dot{n}_A}{\dot{n}_n}$$

If you add up all mole fractions in a mixture, you should always obtain 1 (within calculation and measurement error), because sum of individual component flow rates equals the total flow rate :

$$\Sigma x_i = 1$$

Note that each stream has its own independent set of concentrations. This fact will become important when you are performing mass balances.

Mass Fraction

Since mass is a more practical property to measure than moles, flowrates are often given as *mass* flowrates rather than *molar* flowrates. When this occurs, it is convenient to express concentrations in terms of mass fractions defined similarly to mole fractions.

In most texts mass fraction is given the same notation as mole fraction, and which one is meant is explicitly stated in the equations that are used or the data given.

Note : In this book, assume that a percent concentration has the same units as the total flowrate unless stated otherwise. So if a flowrate is given in kg/s, and a composition is given as "30%", assume that it is 30% by mass.

The definition of a mass fraction is similar to that of moles :

$$x_A = \frac{m_A}{m_n} \quad \text{for batch system}$$

Mass Fraction of Continuous Systems

$$x_A = \frac{\dot{m}_A}{\dot{m}_n}$$

where m_A is the mass of A. It doesn't matter what the units of the mass are as long as they are the same as the units of the total mass of solution.

Like the mole fraction, the total mass fraction in any stream should always add up to 1.

$$\Sigma x_i = 1$$

Calculations on Multi-component Streams

Various conversions must be done with multiple-component streams just as they must for single-component streams. This section shows some methods to combine the properties of single-component streams into something usable for multiple-component streams(with some assumptions).

Average Molecular Weight

The *average molecular weight* of a mixture (gas or liquid) is the multicomponent equivalent to the molecular weight of a pure species. It allows you to convert between the mass of a mixture and the number of moles, which is important for reacting systems especially because balances must usually be done in moles, but measurements are generally in grams.

To find the value of $\overline{MW}_n = \dfrac{g\,s\ln}{mole\,s\ln}$, we split the solution up into its com-

ponents as follows, for k components :

$$\frac{g\,s\ln}{mole\,s\ln} = \frac{\Sigma m_i}{n_n} = \Sigma \frac{m_i}{n_n}$$

$$= \Sigma(\frac{m_i}{n_i} * \frac{n_i}{n_n}) = \Sigma(MW_i * x_i)$$

where x_i is the mole fraction of component i in the mixture. Therefore, we have the following formula :

$$\overline{MW}_n = \Sigma(MW_i * x_i)$$

where x_i is the mole fraction of component i in the mixture.

This derivation only assumes that mass is additive, which it is, so this equation is valid for *any* mixture.

Density of Liquid Mixtures

Let us attempt to calculate the density of a liquid mixture from the density of its components, similar to how we calculated the average molecular weight. This time, however, we will notice one critical difference in the assumptions we have to make. We'll also notice that there are two different equations we could come up with, depending on the assumptions we make.

First Equation

By definition, the density of a single component i is: $\rho_i = \dfrac{m_i}{V_i}$ The corresponding definition for a solution is $\rho = \dfrac{m \, sln}{V \, sln}$. Following a similar derivation to the
above for average molecular weight:

$$\frac{m \, sln}{V \, sln} = \frac{\Sigma m_i}{V_n} = \Sigma \frac{m_i}{V_n}$$

$$= \Sigma \frac{m_i}{V_n} * \frac{V_i}{V_n} = \Sigma(\rho i * \frac{V_i}{V_n})$$

Now we make the assumption that The volume of the solution is proportional
to the mass. This is true for any pure substance (the proportionality constant is the
density), but it is further assumed that the proportionality constant is the same for
both pure k and the solution. This equation is therefore useful for two substances
with similar pure densities. If this is true then:

$$\frac{V_i}{V} = \frac{m_i}{m_n} = x_i \, , \text{ where } x_i \text{ is the mass fraction of component i. Thus :}$$

$$\rho_n = \Sigma(x_i * \rho_i)_n$$

where x_i is the mass fraction (not the mole fraction) of component i in the mixture.

Second Equation

This equation is easier to derive if we assume the equation will have a form
similar to that of average molar mass. Since density is given in terms of mass, it
makes sense to start by using the definition of mass fractions:

$$x_i = \frac{m_i}{m_n}$$

To get this in terms of only *solution* properties (and not *component* properties),
we need to get rid of m_i. We do this first by dividing by the density:

$$\frac{x_i}{\rho_i} = \frac{m_i}{m_n} * \frac{V_i}{m_i}$$

$$= \frac{V_i}{m_n}$$

Now if we add all of these up we obtain:

$$\Sigma(\frac{x_i}{\rho_i}) = \frac{\Sigma V_i}{m_n}$$

Now we have to make an assumption, and it's different from that in the first
case. This time we assume that the Volumes are additive. This is true in two cases:

1. In an ideal solution. The idea of an ideal solution will be explained more later, but for now you need to know that ideal solutions :
    *   Tend to involve similar compounds in solution with each other, or when one component is so dilute that it doesn't effect the solution properties much.
    *   Include Ideal Gas mixtures at constant temperature and pressure.
2. In a Completely immiscible non-reacting mixture. In other words, if two substances don't mix at all (like oil and water, or if you throw a rock into a puddle), the total volume will not change when you mix them. And the total volume in this case will be sum of volume of individual components.

    If the solution is ideal, then we can write :

$$\frac{\Sigma \dot{V}_i}{\dot{m}_n} = \frac{\dot{V}_n}{\dot{m}_n} = \frac{1}{\rho_n}$$

Hence, for an ideal solution,

$\dfrac{1}{\rho_n} = \Sigma(\dfrac{x_i}{\rho_i})n$ here x_i is the mass fraction of component i in the mixture.

Note that this is significantly different from the previous equation! This equation is more accurate for most cases. In all cases, however, it is most accurate to look up the value in a handbook such as Perry's Chemical Engineers Handbook if data is available on the solution of interest.

General Strategies for Multiple-Component Operations

The most important thing to remember about doing mass balances with multiple components is that *for each component, you can write one independent mass balance*. What do I mean by independent? Well, remember we can write the general, overall mass balance for any steady-state system :

$$\Sigma \dot{m}_{in} - \Sigma \dot{m}_{out} = 0$$

And we can write a similar mass balance for any *component* of a stream :

$$\Sigma \dot{m}_{a,in} - \Sigma \dot{m}_{a,out} + m_{a,gen} = 0$$

This looks like we have three equations here, but in reality only two of them are independent because :

1. The sum of the masses of the components equals the total mass
2. The total mass generation due to reaction is always zero (by the law of mass conservation)

Therefore, if we add up all of the mass balances for the *components* we obtain the *overall* mass balance. Therefore, we can choose any set of *n* equations we want, where *n* is the number of components, but if we choose the overall mass balance as one of them we cannot use the mass balance on one of the components.

The choice of which balances to use depends on two particular criteria :

1. Which component(s) you have the most information on; if you don't have enough information you won't be able to solve the equations you write.

2. Which component(s) you can make the most reasonable assumptions about. For example, if you have a process involving oxygen and water at low temperatures and pressures, you may say that there is no oxygen dissolved in a liquid flow stream, so it all leaves by another path. This will simplify the algebra a good deal if you write the mass balance on that component.

Multiple Components in a Single Operation : Separation of Ethanol and Water

Example : Suppose a stream containing ethanol and water (two fully miscible compounds) flows into a distillation column at 100 kg/s. Two streams leave the column : the vapor stream contains 80% ethanol by mass and the liquid bottoms has an ethanol concentration of 4M. The total liquid stream flowrate is 20 kg/s. Calculate the composition of the entrance stream.

Following the step-by-step method makes things easier.

Step 1 : Draw a Flowchart

The first step as always is to draw the flowchart, as described previously. If you do that for this system, you may end up with something like this, where x signifies mass fraction, [A] signifies molarity of A, and numbers signify stream numbers.

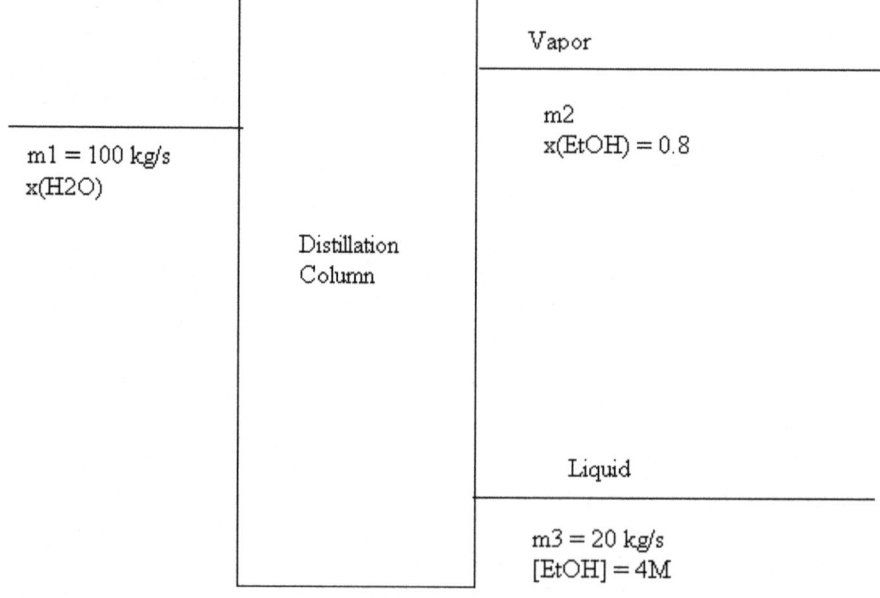

Step 2 : Convert Units

Now, we need to turn to converting the concentrations into appropriate units. Since the total flowrates are given in terms of mass, a unit that expresses the concentration in terms of mass of the components would be most useful. The vapor stream compositions are given as mass percents, which works well with the units of flow. However, the liquid phase concentration given in terms of a molarity is not useful for finding a mass flow rate of ethanol (or of water). Hence we must convert the concentration to something more useful.

Note : Converting between Concentration Measurements

The easiest way to convert between concentrations is to take a careful look at the units of both what you want and what you have, and ask what physical properties (*i.e.* molar mass, density) you could use to interchange them. In this example, we want to convert a molarity into a mass fraction. We have from the definitions that :

$$[A] = \frac{mol_A}{L_{sln}}$$

$$x_A = \frac{m_A}{m_{sln}}$$

To convert the numerators, we need to convert moles of A to mass of A, so we can use the molar mass for this purpose. Similarly, to convert the denominators we need to change Liters to Mass, which means we'll use a density. Hence, the conversion from molarity to mass fraction is :

$$x_A = [A] * \frac{(MW)_A}{\rho_{SLN}}$$

Since we have ways to estimate ρ_{SLN} (remember them?), we can inter-convert the conversions.

In order to convert the molarity into a mass fraction, then, we need the molecular weight of ethanol and the density of a 4M ethanol solution. The former is easy if you know the chemical formula of ethanol : CH_3CH_2OH. Calculating the molecular weight (as you did in chem class) you should come up with about $46\frac{g}{mol}$.

Calculating the density involves plugging in mass fractions in and of itself, so you'll end up with an implicit equation. Recall that one method of estimating a solution density is to assume that the solution is ideal (which it probably is not in this case, but if no data are available or we just want an estimate, assumptions like these are all we have, as long as we realize the values will not be exact) :

$$\frac{1}{\rho_{SLN}} = \Sigma(\frac{x_k}{\rho_k})$$

In this case, then,

$$\frac{1}{\rho_{SLN}} = \frac{x_{EtOH}}{\rho_{EtOH}} + \frac{x_{H2O}}{\rho_{H2O}}$$

We can look up the densities of pure water and pure ethanol, they are as follows (from Wikipedia's articles w : Ethanol and w : Water) :

$$\rho_{EtOH} = 0.789\frac{g}{cm^3} = 789\frac{g}{L}$$

$$\rho_{H2O} = 1.00\frac{g}{cm^3} = 1000\frac{g}{L}$$

Therefore, since the mass fractions add to one, our equation for density becomes :

$$\frac{1}{\rho_{sln}} = \frac{x_{EtOH}}{789\frac{g}{L}} + \frac{1 - x_{EtOH}}{1000\frac{g}{L}}$$

From the NOTE above, we can now finally convert the molarity into a mass fraction as :

$$x_{EtOH} = [EtOH] * \frac{(MW)_{EtOH}}{\rho_{SLN}} = 4\frac{mol}{L} * 46\frac{g}{mol} * \left(\frac{x_{EtOH}}{789\frac{g}{L}} + \frac{1 - x_{EtOH}}{1000\frac{g}{L}}\right)$$

Solving this equation yields :

$x_{EtOH} = 0.194$ (unitless)

Step 3 : Relate your Variables

Since we are seeking properties related to mass flow rates, we will need to relate our variables with mass balances.

Remember that we can do a mass balance on any of the N independent species and one on the overall mass, but since the sum of the individual masses equals the overall only $N - 1$ of these equations will be independent. It is often easiest mathematically to choose the overall mass balance and $N - 1$ individual species balances, since you don't need to deal with concentrations for the overall measurements.

Since our concentrations are now in appropriate units, we can do any two mass balances we want. Lets choose the overall first :

$$\dot{m}_1 - \dot{m}_2 - \dot{m}_3 = 0$$

Plugging in known values :

$$\dot{m}_2 = 100\frac{kg}{s} - 20\frac{kg}{s}$$

$$\dot{m}_2 = 80\frac{kg}{s}$$

Now that we know $\dot{m}_2$ we can do a mass balance on either ethanol or water to find the composition of the input stream. Lets choose ethanol (A) :

$$\dot{m}_{A1} = \dot{m}_{A2} + \dot{m}_{A3}$$

Written in terms of mass fractions this becomes :

$$\dot{m}_{A1} * \dot{m}_1 = x_{A2} * \dot{m}_2 + x_{A3} * \dot{m}_3$$

Plugging in what we know :

$$x_{A1} * 100\frac{kg}{s} = 0.8 * 80\frac{kg}{s} + 0.194 * 20\frac{kg}{s}$$

$$x_{A1} = 0.68$$

Hence, the feed is 68% Ethanol and 34% Water.

Introduction to Problem Solving with Multiple Components and Processes

In the vast majority of chemical processes, in which some raw materials are processed to yield a desired end product or set of end products, there will be more than one raw material entering the system and more than one unit operation through which the product must pass in order to achieve the desired result. The calculations for such processes, as you can probably guess, are considerably more complicated than those either for only a single component, or for a single-operation process. Therefore, several techniques have been developed to aid engineers in their analyses. This section describes these techniques and how to apply them to an example problem.

Degree of Freedom Analysis

For more complex problems than the single-component or single-operation problems that have been explored, it is essential that you have a method of determining if a problem is even solvable given the information that you have. There are three ways to describe a problem in terms of its solvability :

1. If the problem has a finite (not necessarily unique!) set of solutions then it is called **well-defined**.

2. The problem can be **over determined** (also known as **over specified**), which means that you have too much information and it is either redundant or inconsistent. This could possibly be fixed by consolidating multiple data into a single function or, in extreme cases, a single value (such as a slope of a linear correlation), or it could be fixed by removing an assumption about the system that one had made.

3. The problem can be **under determined** (or **under specified**), which means that you don't have enough information to solve for all your unknowns. There are several ways of dealing with this. The most obvious is to gather additional information, such as measuring additional temperatures, flow

rates, and so on until you have a well-defined problem. Another way is to use additional equations or information about what we want out of a process, such as how much conversion you obtain in a reaction, how efficient a separation process is, and so on. Finally, we can **make assumptions** in order to simplify the equations, and perhaps they will simplify enough that they become solvable.

The method of analyzing systems to see whether they are over or under-specified, or if they are well-defined, is called a degree of freedom analysis. It works as follows for mass balances on a single process :

1. From your flowchart, determine the number of **unknowns** in the process. What qualifies as an unknown depends on what you're looking for, but in a material balance calculation, masses and concentrations are the most common. In equilibrium and energy balance calculations, temperature and pressure also become important unknowns. In a reactor, you should include the conversion as an unknown unless it is given OR you are doing an atom balance.

2. Subtract the number of **Equations** you can write on the process. This can include mass balances, energy balances, equilibrium relationships, relations between concentrations, and any equations derived from additional information about the process.

3. The number you are left with is the degrees of freedom of the process.

If the degrees of freedom are negative that means the unit operation is over-specified. If it is positive, the operation is under specified. If it is zero then the unit operation iswell-defined, meaning that it is theoretically possible to solve for the unknowns with a finite set of solutions.

Degrees of Freedom in Multiple-Process Systems

Multiple-process systems are tougher but not undoable. Here is how to analyze them to *see if a problem is uniquely solvable* :

1. Label a flowchart completely with all the relevant unknowns.
2. Perform a degree of freedom analysis on each unit operation, as described above.
3. Add the degrees of freedom for each of the operations.
4. Subtract the number of variables in *intermediate streams, i.e.* streams between two unit operations. This is because each of these was counted twice, once for the operation it leaves and once for the one it enters.

The number you are left with is the process degrees of freedom, and this is what will tell you if the process as a *whole* is overspecified, underspecified, or well-defined.

Note : If any single process is overspecified, and is found to be inconsistent, then the problem as a whole cannot be solved, regardless of whether the process as a whole is well-defined or not.

Using Degrees of Freedom to Make a Plan

Once you have determined that your problem is solvable, you still need to figure out how you'll solve for your variables. This is the suggested method.

1. Find a unit operation or combination of unit operations for which the degrees of freedom are zero.

2. Calculate all of the unknowns involved in this combination.

3. **Recalculate** the degrees of freedom for each process, treating the calculated values as known rather than as variables.

4. Repeat these steps until everything is calculated (or at least that which you seek)

Note : You must be careful when recalculating the degrees of freedom in a process. You have to be aware of the sandwich effect, in which calculations from one unit operation can trivialize balances on another operation. For example, suppose you have three processes lined up like this :

$$-> A -> B -> C ->$$

Suppose also that through mass balances on operations A and C, you calculate the exit composition of A and the inlet composition of C. Once these are performed,the mass balances on B are already completely defined. The moral of the story is that before you claim that you can write an equation to solve an unknown, write the equation and make sure that it contains an unknown. Do not count equations that have no unknowns in your degree of freedom analysis.

Multiple Components and Multiple Processes : Orange Juice Production

Example :

Consider a process in which raw oranges are processed into orange juice. A possible process description follows :

The oranges enter a crusher, in which all of the water contained within the oranges is released.

The now-crushed oranges enter a strainer. The strainer is able to capture 90% of the solids, the remainder exit with the orange juice as pulp.

The velocity of the orange juice stream was measured to be $30\frac{m}{s}$ and the radius of the piping was 8 inches. Calculate :

a) The mass flow rate of the orange juice product.

b) The number of oranges per year that can be processed with this process if it is run 8 hours a day and 360 days a year. Ignore changes due to unsteady state at startup.

Use the following data : Mass of an orange : 0.4 kg Water content of an orange : 80% Density of the solids : Since its mostly sugars, its about the density of glucose, which is $1.540\frac{g}{cm^3}$

Step 1 : Draw a Flowchart

This time we have multiple processes so it's especially important to label each one as its given in the problem.

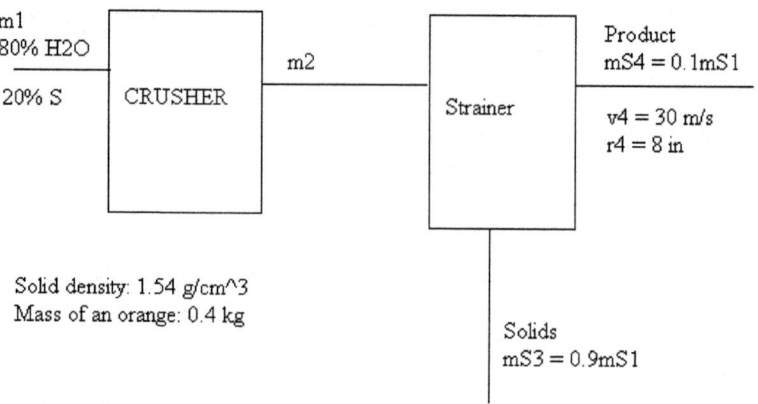

Notice how I changed the 90% capture of solids into an algebraic equation relating the mass of solids in the solid waste to the mass in the feed. This will be important later, because *it is an additional piece of information that is necessary to solve the problem.*

Also note that from here in, "solids" are referred to as S and "water" as W.

Step 2 : Degree of Freedom Analysis

Recall that for each stream there are C independent unknowns, where C is the number of components in the individual stream. These generally are concentrations of C-1 species and the total mass flow rate, since with C-1 concentrations we can find the last one, but we cannot obtain the total mass flow rate from only concentration.

Let us apply the previously described algorithm to determining if the problem is well-defined.

On the strainer :

- There are **6** unknowns : m2, xS3, m3, xS3, m4, and xS4
- We can write **2** independent mass balances on the overall system (one for each component).
- We are given a conversion and enough information to write the mass flow rate in the product in terms of only concentration of one component (which eliminates one unknown). Thus we have **2** additional pieces of information.
- Thus the degrees of freedom of the strainer are 6-2-2 = **2 DOF**

Note : We are given the mass of an individual orange, but since we cannot use that information alone to find a total mass flow rate of oranges in the feed, and we already have used up our allotment of C-1 independent concentrations, we cannot count this as "given information". If, however, we were told the number of oranges produced per year, then we could use the two pieces of information in tandem to eliminate a single unknown (because then we can find the mass flow rate)

On the crusher :

- There are **3** unknowns (m1, m2, and xS2).
- We can write **2** independent mass balances.
- Thus the crusher has 3-2 = **1 DOF**

Therefore for the system as a whole :

- Sum of DOF for unit operations = 2 + 1 = 3 DOF
- Number of intermediate variables = 2 (m2 and xS2)
- Total DOF = 3 - 2 = **1 DOF**.

Hence the problem is underspecified.

So how to we solve it?

In order to solve an underspecified problem, one way we can obtain an additional specification is to make an assumption. What assumptions could we make that would reduce the number of unknowns (or equivalently, increase the number of variables we do know)?

The most common type of assumption is to assume that something that is relatively insignificant is zero.

In this case, one could ask : will the solid stream from the strainer contain any water? It might, of course, but this amount is probably very small compared to both the amount of solids that are captured and how much is strained, provided that it is cleaned regularly and designed well. If we make this assumption, then *this specifies that the mass fraction of water in the waste stream is zero* (or equivalently, that the mass fraction of solids is one). Therefore, we know one additional piece of information and the degrees of freedom for the overall system become zero.

Step 3 : Convert Units

This step should be done *after* the degree of freedom analysis, because that analysis is independent of your unit system, and if you don't have enough information to solve a problem (or worse, you have too much), you shouldn't waste time converting units and should instead spend your time defining the problem more precisely and/or seeking out appropriate assumptions to make.

Here, the most sensible choice is either to convert everything to the cgs system or to the m-kg-s system, since most values are already in metric. Here, the latter route is taken.

$$r_4 = 8 \text{ in} * \frac{2.54 \text{ cm}}{\text{in}} * \frac{1 \text{ m}}{100 \text{ cm}} = 0.2032 \text{ m}$$

$$\rho_S = 1.54 \frac{g}{cm^3} = 1540 \frac{kg}{m^3}$$

Now that everything is in the same system, we can move on to the next step.

Step 4 : Relate Your Variables

First we have to relate the velocity and area given to us to the mass flowrate of stream 4, so that we can actually use that information in a mass balance. From chapter, we can start with the equation :

$$\rho_n * v_n * A_n = \dot{m}_n$$

Since the pipe is circular and the area of a circle is $\pi * r^2$, we have :

$$A_4 = \pi * 0.2032^2 = 0.1297 \text{ m}^2$$

So we have that :

$$\rho_4 * 30 * 0.1297 = 3.8915 * \rho_4 = \dot{m}_4$$

Now to find the density of stream 4 we assume that volumes are additive, since the solids and water are essentially immiscible (does an orange dissolve when you wash it?). Hence we can use the ideal-fluid model for density :

$$\frac{1}{\rho_4} = \frac{x_{S4}}{\rho_S} + \frac{x_{W4}}{\rho_W} = \frac{x_{S4}}{\rho_S} = \frac{1 - x_{S4}}{\rho_W}$$

$$= \frac{x_{S4}}{1540} + \frac{1 - x_{S4}}{1000}$$

Hence, we have the equation we need with only concentrations and mass flowrates :

$$\text{EQUATION 1}: \frac{x_{S4}}{1540} + \frac{1 - x_{S4}}{1000} = \frac{3.8915}{\dot{m}_4}$$

Now we have an equation but we haven't used either of our two (why two?) independent mass balances yet. We of course have a choice on which two to use.

In this particular problem, since we are directly given information concerning the amount of *solid* in stream 4 (the product stream), it seems to make more sense to do the balance on this component. Since we don't have information on stream 2, and finding it would be pointless in this case (all parts of it are the same as those of stream 1), lets do an overall-system balance on the solids :

$$\Sigma \dot{m}_{S,in} - \Sigma \dot{m}_{S,out} = 0$$

Note : Since there is no reaction, the generation term is 0 even for individual-species balances.

Expanding the mass balance in terms of mass fractions gives :

$$\dot{m}_1 * x_{S1} = \dot{m}_3 * x_{S3} + \dot{m}_4 * x_{S4}$$

Plugging in the known values, with the assumption that stream 3 is pure solids (no water) and hence $x_{s3} = 1$:

EQUATION 2 : $0.2 * \dot{m}_1 = (0.9 * 0.2 * \dot{m}_1) * 1 + x_{S4} * \dot{m}_4$

Finally, we can utilize one further mass balance, so let's use the easiest one : the overall mass balance. This one again assumes that the total flowrate of stream 3 is equal to the solids flowrate.

EQUATION 3 : $\dot{m}_1 = 0.9 * 0.2 * \dot{m}_1 + \dot{m}_4$

We now have three equations in three unknowns $(\dot{m}_1, \dot{m}_4, x_{S4})$ so the problem is solvable. This is where all those system-solving skills will come in handy.

If you don't like solving by hand, there are numerous computer programs out there to help you solve equations like this, such as MATLAB, POLYMATH, and many others. You'll probably want to learn how to use the one your school prefers eventually so why not now?

Using either method, the results are :

$$\dot{m}_1 = 4786 \frac{kg}{s}$$

$$\dot{m}_4 = 3925.07$$

$$x_{S4} = 0.0244$$

We're almost done here, now we just have to calculate the number of oranges per year.

$$4786 \frac{kg}{s} * 1 \frac{orange}{0.4\ kg} * 3600 \frac{s}{hr} * 8 \frac{wk\ hr}{day} * 360 \frac{wk\ day}{year}$$

Yearly Production : $1.24 * 10^{11} \frac{oranges}{year}$

Mass balances with recycle

What is Recycle?

Recycling is the act of taking one stream in a process and reusing it in an earlier part of the process rather than discarding it. It is used in a wide variety of processes.

Uses and Benefit of Recycle

The use of recycle makes a great deal of environmental *and* economic sense, for the following reasons among others :

• Using recycle lets a company achieve a wider range of separations

This will be demonstrated in the next section. However, there is a tradeoff : the more dilute or concentrated you want your product to be, the lower the flowrate you can achieve in the concentrated or dilute stream.

- By using recycle, in combination with some sort of separation process, a company can increase the overall conversion of an **equilibrium reaction**.

You may recall from general chemistry that many reactions do not go to completion but only up to a certain point, because they are *reversible*. How far the reaction goes depends on the concentrations (or partial pressures for a gas) of the products and the reactants, which are related by the reaction stoichiometry and the equilibrium constant K. If we want to increase the amount of conversion, one way we can do this is to separate out the products from the product mixture and re-feed the purified reactants in to the reactor. By Le Chatlier's Principle, this will cause the reaction to continue moving towards the products.

- By using recycle, it is possible to recover expensive **catalysts and reagents**.

Catalysts aren't cheap, and if we don't try to recycle them into the reactor, they may be lost in the product stream. This not only gives us a contaminated product but also wastes a lot of catalyst.

- Because of the previous three uses, recycle can decrease the amount of equipment needed to get a process meet specifications and consumer demand.

For example, it may improve reaction conversion enough to eliminate the need for a second reactor to achieve an economical conversion.

- Recycle reduces the amount of waste that a company generates.

Not only is this the most environmentally sounds way to go about it, it also saves the company money in disposal costs.

- Most importantly, all of these things can save a company money.

By using less equipment, the company saves maintainence as well as capital costs, and probably gets the product faster too, if the proper analysis is made.it is use to stop the wastage of the material

Differences between Recycle and Non-recycle Systems

The biggest difference between recycle and non-recycle systems is that *the extra splitting and recombination points must be taken into account, and the properties of the streams change from before to after these points.* To see what is meant by this, consider any arbitrary process in which a change occurs between two streams :

Feed -> Process -> Outlet

If we wish to implement a recycle system on this process, we often will do something like this :

The "extra" stream between the splitting and recombination point must be taken into account, but the way to do this is *not* to do a mass balance on the process, since *the recycle stream itself does not go into the process, only the recombined stream does.*

Instead, we take it into account by performing a mass balance on the *recombination point* and one on the *splitting point*.

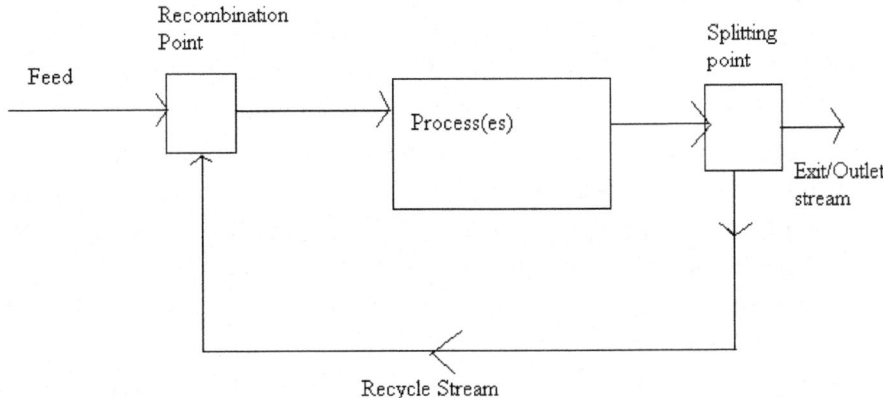

Assumptions at the Splitting Point

The recombination point is relatively unpredictable because the composition of the stream leaving depends on both the composition of the feed and the composition of the recycle stream. However, the *spliitng point* is special because when a stream is split, it generally is split into two streams with equal composition. This is a piece of information that counts towards "additional information" when performing a degree of freedom analysis.

As an additional specification, it is common to know the *ratio* of splitting, *i.e.* how much of the exit stream from the process will be put into the outlet and how much will be recycled. This also counts as "additional information".

Assumptions at the Recombination Point

The recombination point is generally not specified like the splitting point, and also the recycle stream and feed stream are very likely to have different compositions. The important thing to remember is that you can generally use the properties of the stream coming from the splitting point for the stream entering the recombination point, unless it goes through another process in between (which is entirely possible).

Degree of Freedom Analysis of Recycle Systems

Degree of freedom analyses are similar for recycle systems to those for other systems, but with a couple important points that the engineer must keep in mind :

1. The recombination point and the splitting point must be counted in the degree of freedom analysis as "processes", since they can have unknowns that aren't counted anywhere else.

2. When doing the degree of freedom analysis on the splitting point, *you should not label the concentrations as the same but leave them as separate unknowns until after you complete the DOF analysis* in order to avoid confusion,

since labeling the concentrations as identical "uses up" one of your pieces of information and then you can't count it.

As an example, let's do a degree of freedom analysis on the hypothetical system above, assuming that all streams have two components.

- **Recombination Point** : 6 variables (3 concentrations and 3 total flow rates) - 2 mass balances = 4 DOF

- **Process** : Assuming it's not a reactor and there's only 2 streams, there's 4 variables and 2 mass balances = 2 DOF

- **Splitting Point** : 6 variables - 2 mass balances - **1 knowing compositions are the same - 1 splitting ratio = 2 DOF**

So the total is 4 + 2 + 2 - 6 (in-between variables) = 2 DOF. Therefore, if the feed is specified then this entire system can be solved! Of course the results will be different if the process has more than 2 streams, if the splitting is 3-way, if there are more than two components, and so on.

Suggested Solving Method

The solving method for recycle systems is similar to those of other systems we have seen so far but as you've likely noticed, they are increasingly complicated. Therefore, the importance of making a plan becomes of the utmost importance. The way to make a plan is generally as follows :

1. Draw a completely labeled flow chart for the process.

2. Do a DOF analysis to make sure the problem is solvable.

3. If it is solvable, **a lot of the time, the best place to start with a recycle system is with a set of overall system balances, sometimes in combination with balances on processes on the border**. The reason for this is that the overall system balance cuts out the recycle stream entirely, since the recycle stream does not enter or leave the system as a whole but merely travels between two processes, like any other intermediate stream. Often, the composition of the recycle stream is unknown, so this simplifies the calculations a good deal.

4. Find a set of independent equations that will yield values for a certain set of unknowns (this is often most difficult the first time; sometimes, one of the unit operations in the system will have 0 DOF so start with that one. Otherwise it'll take some searching.)

5. Considering those variables as known, do a new DOF balance until something has 0 DOF. Calculate the variables on that process.

6. Repeat until all processes are specified completely.

Example Problem : Improving a Separation Process

It has been stated that recycle can help to This example helps to show that this is true and also show some limitations of the use of recycle on real processes.

Consider the following proposed system without recycle.

Example :

A mixture of 50% A and 50% B enters a separation process that is capable of splitting the two components into two streams : one containing 60% of the entering A and half the B, and one with 40% of the A and half the B (all by mass) :

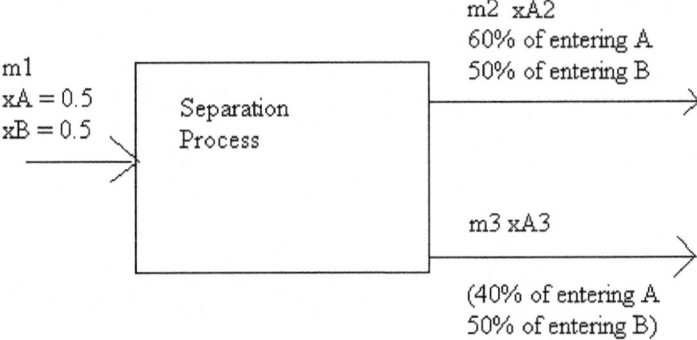

If 100 kg/hr of feed containing 50% A by mass enters the separator, what are the concentrations of A in the exit streams?

A degree of freedom analysis on this process :

4 unknowns ($\dot{m}_2, x_{A2}, \dot{m}_3$, and x_{A3}), 2 mass balances, and 2 pieces of information (knowing that 40% of A and half of B leaves in stream 3 is not independent from knowing that 60% of A and half of B leaves in stream 2) = 0 DOF.

Methods of previous chapters can be used to determine that $\dot{m}_2 = 55\frac{kg}{hr}$, $x_{A2} = 0.545$, $\dot{m}_3 = 45\frac{kg}{hr}$ and $x_{A3} = 0.444$. This is good practice for the interested reader.

If we want to obtain a greater separation than this, one thing that we can do is use a *recycle system*, in which a portion of one of the streams is siphoned off and remixed with the feed stream in order for it to be re-separated. The choice of which stream should be re-siphoned depends on the desired properties of the exit streams. The effects of each choice will now be assessed.

Implementing Recycle on the Separation Process

Example :

Suppose that in the previous example, a recycle system is set up in which half of stream 3 is siphoned off and recombined with the feed (which is still the same composition as before). Recalculate the concentrations of A in streams 2 and 3. Is the separation more or less effective than that without recycle? Can you see a major limitation of this method? How might this be overcome?

This is a rather involved problem, and must be taken one step at a time. The analyses of the cases for recycling each stream are similar, so the first case will be considered in detail and the second will be left for the reader.

Step 1 : Draw a Flowchart

You must be careful when drawing the flowchart because the separator separates 60% of all the A that enters it into stream 2, not 60% of the fresh feed stream.

Note : there is a mistake in the flow scheme. m6 and xA6 before the process is actually m4 and xA4

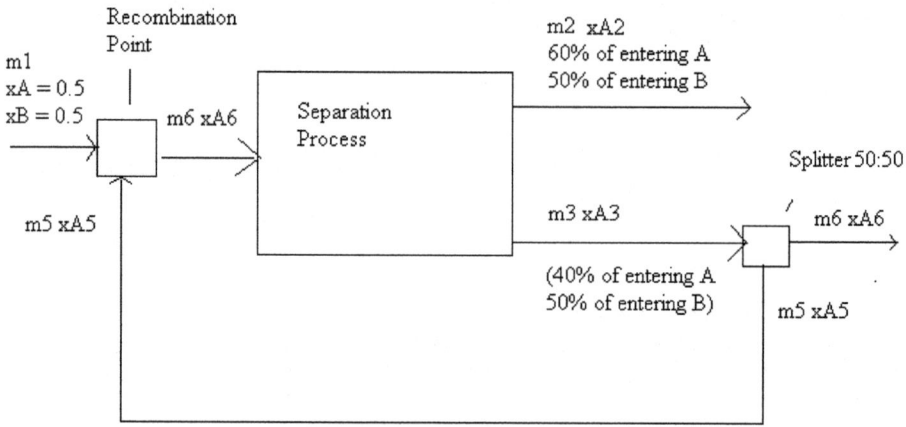

Step 2 : Do a Degree of Freedom Analysis

Recall that you must include the recombination and splitting points in your analysis.

• **Recombination point** : 4 unknowns - 2 mass balances = **2 degrees of freedom**

• **Separator** : 6 unknowns (nothing is specified) - 2 independent pieces of information - 2 mass balances = **2 DOF**

• **Splitting point** : 6 unknowns (again, nothing is specified) - 2 mass balances - 1 assumption that concentration remains constant - 1 splitting ratio = **2 DOF**

• **Total** = 2 + 2 + 2 - 6 = 0. Thus the problem is completely specified.

Step 3 : Devise a Plan and Carry it Out

First, look at the entire system, since none of the original processes individually had 0 DOF.

• **Overall mass balance on A** : $0.5 * 100 \frac{kg}{h} = \dot{m}_2 * x_{A2} + \dot{m}_6 * x_{A6}$

- **Overall mass balance on B** : $50\dfrac{kg}{h} = \dot{m}_2 * (1 - x_{A2}) + \dot{m}_6 * (1 - x_{A6})$

We have 4 unknowns and 2 equations at this point. This is where the problem solving requires some ingenuity. First, lets see what happens when we combine this information with the splitting ratio and constant concentration at the splitter :

- **Splitting Ratio** : $\dot{m}_6 = \dfrac{\dot{m}_3}{2}$

- **Constant concentration** : $x_{A6} = x_{A3}$

Plugging these into the overall balances we have :

- **On A** : $50 = \dot{m}_2 * x_{A2} + \dfrac{\dot{m}_3}{2} * x_{A3}$

- **Total** : $50 = \dot{m}_2 * (1 - x_{A2}) + \dfrac{\dot{m}_3}{2} * (1 - x_{A3})$

Again we have more equations than unknowns *but we know how to relate everything in these two equations to the inlet concentrations in the separator*. This is due to the conversions we are given :

- **60% of entering A goes into stream 2** means $\dot{m}_2 * x_{A2} = 0.6 * x_{A4} * \dot{m}_4$
- **40% of entering A goes into stream 3** means $\dot{m}_3 * x_{A3} = 0.4 * x_{A4} * \dot{m}_4$
- **50% of entering B goes into stream 2** means
  $\dot{m}_2 * (1 - x_{A2}) = 0.5 * (1 - x_{A4}) * \dot{m}_4$

- **50% of entering B goes into stream 3** means $\dot{m}_3 * x_{A3} = 0.5 * (1 - x_{A4}) * \dot{m}_4$

Spend some time trying to figure out where these equations come from, it's all definition of mass fraction and translating words into algebraic equations.

Plugging in all of these into the existing balances, we finally obtain 2 equations in 2 unknowns :

On A : $50 = 0.6\dot{m}_4 * x_{A4} + \dfrac{0.4}{2}\dot{m}_4 * x_{A4}$

$\dot{m}_4 = 129.17\dfrac{kg}{h}, x_{A4} = 0.484$

On B : $50 = 0.5\dot{m}_4 * (1 - x_{A4}) + \dfrac{0.5}{2}\dot{m}_4 * (1 - x_{A4})$

Solving these equations gives :

Note : Notice that two things happened as expected : the concentration of the stream entering the evaporator went down (because the feed is mixing with a more dilute recycle stream), and the total flowrate went up (again due to contribution from the recycle stream). This is always a good rough check to see if your answer makes sense, for example if the flowrate was lower than the feed rate you'd know something went wrong.

Once these values are known, you can choose to do a balance either on the separator or on the recombination point, since both now have 0 degrees of freedom. We choose the separator because that leads directly to what we're looking for.

The mass balances on the separator can be solved using the same method as that without a recycle system, the results are :

$$\dot{m}_2 = 70.83\frac{kg}{hr}, x_{A2} = 0.530, \dot{m}_3 = 58.33\frac{kg}{hr}, x_{A3} = 0.429$$

Now since we know the flowrate of stream 3 and the splitting ratio we can find the rate of stream 6 :

$$\dot{m}_6 = \frac{\dot{m}_3}{2} = 29.165\frac{kg}{hr}, x_{A6} = x_{A3} = 0.429$$

Note : You should check to make sure that m2 and m6 add up to the total feed rate, otherwise you made a mistake.

Now we can assess how effective the recycle is. The concentration of A in the liquid stream *was* reduced, by a small margin of 0.015 mole fraction. However, this extra reduction came at a pair of costs : the flow rate of dilute stream was significantly reduced : from 45 to 29.165 kg/hr! This limitation is important to keep in mind and also explains why we bother trying to make very efficient separation processes.

Mass/Mole Balances in Reacting Systems

Review of Reaction Stoichiometry

Up until now, all of the balances we have done on systems have been in terms of mass. However, mass is inconvenient for a reacting system because it does not allow us to take advantage of the *stoichiometry* of the reaction in relating the relative amounts of reactants and of products.

Stoichiometry is the relationship between reactants and products in a balanced reaction as given by the ratio of their coefficients. For example, in the reaction :

$$C_2H_2 + 2H_2 \rightarrow C_2H_6$$

the reaction stoichiometry would dictate that for every one molecule of C_2H_2 (acetylene) that reacts, two molecules of H_2 (hydrogen) are consumed and one molecule of C_2H_6 are formed. However, this does not hold for grams of products and reactants.

Even though the number of molecules in *single substance* is proportional to the mass of that substance, the constant of proportionality (the molecular mass) is not the same for every molecule. Hence, it is necessary to use the molecular weight of each molecule to convert from grams to *moles* in order to use the reaction's coefficients.

Molecular Mole Balances

We can write balances on moles like we can on anything else. We'll start with our ubiquitous general balance equation :

Input − Output = Accumulation − Generation

As usual we assume that accumulation = 0 in this book so that :

Input − Output + Generation = 0

Let us denote molar flow rates by $\dot{n}$ to distinguish them from mass flow rates. We then have a similar equation to the mass balance equation :

$$\Sigma \dot{n}_{in} - \Sigma \dot{n}_{out} + n_{gen} = 0$$

The same equation can be written in terms of each individual species.

There are a couple of important things to note about this type of balance as opposed to a mass balance :

1. Just like with the mass balance, in a mole balance, a non-reactive system has $n_{gen} = 0$ for all species.

2. **Unlike the mass balance**, the TOTAL generation of moles isn't necessarily 0 even for the overall mole balance! To see this, consider how the total number of moles changes in the above reaction; the final number of moles will not equal the initial number because 3 total moles of molecules are reacting to form 1 mole of products.

Why would we use it if the generation isn't necessarily 0? We use the molecular mole balance because if we know how much of any one substance is consumed or created in the reaction, we can find all of the others from the reaction stoichiometry. This is a very powerful tool because *each reaction only creates one new unknown* if you use this method! The following section is merely a formalization of this concept, which can be used to solve problems involving reactors.

Extent of Reaction

In order to formalize the previous analysis of reactions in terms of a single variable, let us consider the generic reaction :

$aA + bB \rightarrow cC + dD$

The Molar Extent of Reaction X is defined as :

$$X = -\frac{\Delta n_A}{a} = -\frac{\Delta n_B}{b} = \frac{\Delta n_C}{c} = \frac{\Delta n_D}{d}$$

Since all of these are equivalent, it is possible to find the change in moles of any species in a given reaction if the extent of reaction X is known.

Note : Though they won't be discussed here, there are other ways in which the extent of reaction can be defined. Some other definitions are dependent on the percentachange of a particular substrate, and the stoichiometry is used in a different way to determine the change in the others. This definition makes X independent of the substrate you choose.

The following example illustrates the use of the extent of reaction.

Example :

Consider the reaction $H_2O_2 + O_3 \rightarrow H_2O + 2O_2$. If you start with 50 g of H_2O_2 and 25 grams of O_3, and 25% of the moles of O_3 are consumed, find the molar extent of reaction and the changes in the other components.

Solution : First we need to convert to moles, since stoichiometry is not valid when units are in terms of mass.

$$50 \ g \ H_2O_2 * \frac{1 \ mol}{34 \ g} = 1.471 \ moles \ H_2O_2$$

$$25g \ O_3 * \frac{1 \ mol}{48 \ g} = 0.5208 \ moles \ O_3$$

Clearly ozone is the limiting reactant here. Since 25% is consumed, we have that :

$$\Delta n(O_3) = -0.25 * 05208 = -0.1302 \ moles \ O_3$$

Hence, by definition, $X = \dfrac{-0.1302}{1} = 0.1302$

And then we have $\Delta n \ (H_2O_2) = -1302$, $\Delta n \ (H_2O) = 0.1302$, $\Delta n \ (O_2) = 2 * 0.1302 = 0.2604$, all in moles of the appropriate substrate.

Mole Balances and Extents of Reaction

The mole balance written above can be written in terms of extent of reaction *if we notice that the Δn (A) term defined above is exactly the number of moles of a generated or consumed by the reaction.*

Note : This is only useful for individual species balances, not the overall mole balance. When doing balances on reactive systems, unlike with non-reactive systems, it is generally easier to use all individual species balances possible, rather than the total mole balance and then all but one of the individual species. This is because the total generation of moles in a reaction is generally not 0, so no algebraic advantage is gained by using the total material balance on the system.

Therefore we can write that :

$$n_{A,gen} \ \Delta n \ (A) = -X * a$$

where X is the molar extent of reaction and a is the stoichiometric coefficient of A. Plugging this into the mole balance derived earlier, we arrive at the molecular mole balance equation :

$$\Sigma \dot{n}_{A,out} - \Sigma n_{A,in} - X * a = 0$$ if A is consumed, or +Xa if it is generated in the reaction

Problem Considerations with Molecular Balances

Degree of Freedom Analysis on Reacting Systems

If we have N different molecules in a system, we can write N mass balances *or* N mole balances, whether a reaction occurs in the system or not. The only difference is that in a reacting system, we have one additional unknown, the molar extent of reaction, for each reaction taking place in the system. Therefore each reaction taking place in a process will add one degree of freedom to the process.

Complications

Unfortunately, life is not ideal, and even if we want a single reaction to occur to give us only the desired product, this is either impossible or uneconomical compared to dealing with byproducts, side reactions, equilibrium limitations, and other non-idealities.

Independent and Dependent Reactions

When you have more than one reaction in a system, you need to make sure that they are independent. The idea of independent reactions is similar to the idea of linear independence in mathematics, $-b, c, d, 0]$

Lets consider the following two general parallel competing reactions :

- $aA + aB \rightarrow cC + dD$
- $a_2A + b_2B \rightarrow e_2E$

We can represent each of the reactions by a *vector* of the coefficients :

- $V = $ [A coeff, B coeff, C coeff, D coeff, E coeff]
- $v_1 = [-a, -b, c, d, 0]$
- $v_2 = [-a_2, -b_2, 0, 0, e_2]$

This site

Note : The site above gives a nice tool to tell whether any number of vectors are linearly dependent or not. Lacking such a tool, it is necessary to assess by hand whether the equations are independent. Only independent equations should be used in your analysis of multiple reactions, so if you have dependent equations, you can eliminate reactions from consideration until you've obtained an independent set.

By definition a set of vectors is only linearly independent if the equation :

$$K_1 * v_1 + K_2 * v_2 = 0$$

where K_1 and K_2 are constants only has one solution : $K_1 = K_2 = 0$.

Lets plug in our vectors :

$$K_1 * [-a, -b, c, d, 0] + K_2 * [-a_2, -b_2, 0, 0, e_2] = 0$$

Since *all components* must add up to 0, the following system follows :

- $K_1 * a_1 - K_2 * a_2 = 0$
- $- K_1 * b_1 - K_2 * b_2 = 0$
- $K_1 * c + 0 = 0$
- $K_1 * d + 0 = 0$
- $0 + K_2 * e_2 = 0$

Obviously, the last three equations imply that unless $c = d = 0$ and $e_2 = 0$, $K_1 = K_2 = 0$ and thus the reactions are independent.

Linearly Dependent Reactions

There is one rule to keep in mind whenever you are checking for reaction dependence or independence, which is summarized in the following box.

If any non-zero multiple of one reaction can be added to a multiple of a second reaction to yield a third reaction, then the three reactions are not independent.

Therefore, if the following reaction could occur in the same system as the two above :

$$(a + a_2) A + (b + b_2) B \rightarrow cC + dD + e_2E$$

then it would not be possible to analyze all three reactions at once, since this reaction is the sum of the first two. Only two can legitimately be analyzed at the same time.

All degree of freedom analyses in this book assume that the reactions are independent. You should check this by inspection or, for a large number of reactions, with numerical methods.

Extent of Reaction for Multiple Independent Reactions

When you are setting up extents of reaction in a molecular species balance, you *must* make sure that you set up one for *each* reaction, and include both in your mole balance. So really, your mole balance will look like this :

$$\Sigma n_{A,in} - \Sigma n_A, out + \Sigma a_k X_k = 0$$

for all k reactions. In such cases it is generally easier, if possible, to use an atom balance instead due to the difficulty of solving such equations.

Equilibrium Reactions

In many cases (actually, the majority of them), a given reaction will be reversible, meaning that instead of reacting to completion, it will stop at a certain point and not go any farther. How far the reaction goes is dictated by the value of the *equilibrium coefficient*. Recall from general chemistry that the equilibrium coefficient for the reaction $aA + bB \rightarrow cC + dD$ is defined as follows :

$$K = \frac{C_{C,eq}^c * C_{D,eq}^d}{C_{A,eq}^a * C_{B,eq}^b}$$ with concentration C_i expressed as molarity for liquid

solutes or partial pressure for gasses

Here [A] is the equilibrium concentration of A, usually expressed in molarity for an aqueous solution or partial pressure for a gas. *This equation can be remembered as "products over reactants"*.

Usually solids and solvents are omitted by convention, since their concentrations stay approximately constant throughout a reaction. For example, in an aqueous solution, if water reacts, it is left out of the equilibrium expression.

Often, we are interested in obtaining the extent of reaction of an equilibrium reaction when it is in equilibrium. In order to do this, first recall that :

$$X = \frac{-\Delta n_A}{a}$$

and similar for the other species.

Liquid-phase Analysis

Rewriting this in terms of molarity (moles per volume) by dividing by volume, we have :

$$\frac{X}{V} = \frac{[A]_0 - [A]_f}{a}$$

Or, since the final state we're interested in is the equilibrium state,

$$\frac{X}{V} = \frac{[A]_0 - [A]_{eq}}{a}$$

Solving for the desired equilibrium concentration, we obtain the equation for equilibrium concentration of A in terms of conversion :

$$[A]_{eq} = [A]_0 - \frac{aX}{V}$$

Similar equations can be written for B, C, and D using the definition of extent of reaction. Plugging in all the equations into the expression for K, we obtain :

$$K = \frac{([C]_0 + \frac{cX}{V})^c ([D]_0 + \frac{dX}{V})^d}{([A]_0 - \frac{aX}{V})^a ([B]_0 - \frac{bX}{V})^b}$$

At equilibrium for liquid-phase reactions only

Using this equation, knowing the value of K, the reaction stoichiometry, the initial concentrations, and the volume of the system, the equilibrium extent of reaction can be determined.

Note : If you know the reaction reaches equilibrium in the reactor, this counts as an additional piece of information in the DOF analysis because it allows you to find X. This is the same idea as the idea that, if you have an irreversible reaction and know it goes to completion, you can calculate the extent of reaction from that.

Gas-phase Analysis

By convention, gas-phase equilibrium constants are given in terms of partial pressures which, for ideal gasses, are related to the mole fraction by the equation :

$P_A = Y_A P$ for ideal gasses only

If A, B, C, and D were all gases, then, the equilibrium constant would look like this :

$$\frac{P_C^c P_D^d}{P_A^a P_B^b}$$

Gas-Phase Equilibrium Constant

In order to write the gas equilibrium constant in terms of extent of reaction, let us assume for the moment that we are dealing with ideal gases. You may recall from general chemistry that for an ideal gas, we can write the ideal gas law for *each species* just as validly as we can on the *whole gas* (for a non-ideal gas, this is in general not true). Since this is true, we can say that :

$$\frac{n_A}{V} = [A]\frac{P_A}{RT}$$

Plugging this into the equation for $\frac{X}{V}$ above, we obtain :

$$\frac{aX}{V} = [A] - [A]_{eq} = \frac{P_{A0}}{RT} - \frac{P_{A,eq}}{RT}$$

Therefore,

$$P_{a,eq} = P_{A0} - \frac{aX\,RT}{V}$$

Similar equations can be written for the other components. Plugging these into the equilibrium constant expression :

$$K = \frac{(P_{C0} + \frac{xX\,RT}{V})^c\,(P_{D0} + \frac{dX\,RT}{V})^d}{(P_{A0} + \frac{aX\,RT}{V})^a\,(P_{B0} + \frac{bX\,RT}{V})^b}$$

Gas Phase Ideal-Gas Equilibrium Reaction at Equilibrium

Again, if we know we are at equilibrium and we know the equilibrium coefficient (which can often be found in standard tables) we can calculate the extent of reaction.

Special Notes about Gas Reactions

You need to remember that *In a constant-volume, isothermal gas reaction, the total pressure will change as the reaction goes on*, unless the same number of moles are created as produced. In order to show that this is true, you only need to write the ideal gas law for the total amount of gas, and realize that the total number of moles in the system changes.

This is why we don't want to use *total* pressure in the above equations for K, we want to use *partial* pressures, which we can conveniently write in terms of extent of reaction.

Inert Species

Notice that all of the above equilibrium equations depend on *concentration* of the substance, in one form or another. Therefore, if there are species present that don't react, they may still have an effect on the equilibrium because they will decrease the concentrations of the reactants and products. Just make sure you take them into account when you're calculating the concentrations or partial pressures of each species in preparation for plugging into the equilibrium constant.

Example Reactor Solution using Extent of Reaction and the DOF

Example :

Consider the reaction of Phosphene with oxygen : $4PH_3 + 8O_2 \; P_4O_{10} + 6H_2O$

Suppose a 100-kg mixture of 50% PH_3 and 50% O_2 by mass enters a reactor in a single stream, and the single exit stream contains 25% O_2 by mass. Assume that all the reduction in oxygen occurs due to the reaction. How many degrees of freedom does this problem have? If possible, determine mass composition of all the products.

It always helps to draw a flowchart :

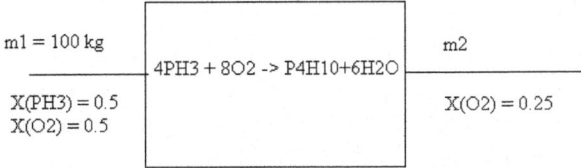

There are four independent unknowns : the total mass (mole) flowrate out of the reactor, the concentrations of two of the exiting species (once they are known, the forth can be calculated), and the extent of reaction.

Additionally, we can write four independent equations, one on each reacting substance. Hence, there are 0 DOF and this problem can be solved.

Let's illustrate how to do it for this relatively simple system, which illustrates some very important things to keep in mind.

First, recall that total mass is conserved even in a reacting system. Therefore, we can write that :

$$\dot{m}_{out} = \dot{m}_{in} = 100 \text{ kg}$$

Now, since component masses aren't conserved, we need to convert as much as we can into moles so we can apply the extent of reaction.

$$\dot{n}_{PH3,in} = 0.5 * (100 \text{ kg}) * \frac{1 \text{ mol}}{0.034 \text{ kg}} = 1470.6 \text{ moles } PH_3 \text{ in}$$

$$\dot{n}_{O2,in} = 0.5 * (100 \text{ kg}) * \frac{1 \text{ mol}}{0.032 \text{ kg}} = 1562.5 \text{ moles } O_3 \text{ in}$$

$$\dot{n}_{O2,out} = 0.25 * (100 \text{ kg}) * \frac{1 \text{ mol}}{0.032 \text{ kg}} = 781.25 \text{ mole } O_2 \text{ out}$$

Let's use the mole balance on oxygen to find the extent of reaction, since we know how much enters and how much leaves. Recall that :

$$\Sigma \dot{n}_{A,in} - \Sigma \dot{n}_{A,out} - a * X = 0$$

where a is the stoichiometric coefficient for A. Plugging in known values, including a = 8 (from the reaction written above), we have :

$$1562.5 - 781.25 - 8X = 0$$

Solving gives :

$$X = 97.66 \text{ moles}$$

Now let's apply the mole balances to the other species to find how much of them is present :

- $PH_3 : 1470.6 - \dot{n}_{PH3,out} - 4(97.66) = 0 \rightarrow \dot{n}_{PH3,out} = 1080.0 \text{ moles } PH_3$

- $P_4H_{1}0:0 - \dot{n}_{P4H10,out} + 1(97.66) = 0 \rightarrow \dot{n}_{P4H10,out} = 97.66 \text{ moles } P_4O_{10}$

  (note it's + instead of - because it's being generated rather than consumed by the reaction)

- $H_2O:0 - \dot{n}_{H2O,out} + 6(97.66) = 0 \rightarrow \dot{n}_{H2O,out} = 586.0 \text{ moles } H_2O$

Finally, the last step we need to do is find the mass of all of these, and divide by the total mass to obtain the mass percents. As a sanity check, all of these plus 25 kg of oxygen should yield 100 kg total.

- Mass PH_3 out = 1080 moles * 0.034 kg\mole = 36.72 kg

- Mass P_4O_{10} out = 97.66 moles * .284 kg\mole = 27.74 kg

- Mass H_2O out = 586 moles * 0.018 kg\mole = 10.55 kg

Sanity check : 36,72 + 27.74 + 10.55 + 25 (oxygen) = 100 kg (total), so we're still sane.

Hence, we get : 36.72% PH_3, 27.74% P_4H_{10}, 10.55% H_2O, 25% O_2 by mass

Example Reactor with Equilibrium

Example :

Suppose that you are working in an organic chemistry lab in which 10 kg of compound A is added to 100 kg of a 16% aqueous solution of B (which has a density of 57 lb/ft^3) The following reaction occurs :

$$A + 2B \longleftrightarrow 3C + D$$

A has a molar mass of 25 g/mol and B has a molar mass of 47 g/mol. If the equilibrium constant for this reaction is 187 at 298K, how much of compound C could you obtain from this reaction? Assume that all products and reactants are soluble in water at the design conditions. Adding 10 kg of A to the solution causes the volume to increase by 5 L. Assume that the volume does not change over the course of the reaction.

Solution : First, draw a flowchart of what we're given.

Since all of the species are dissolved in water, we should write the equilibrium constant in terms of molarity :

$$K = 187 = \frac{[C]^3[D]}{[A][B]^2}$$

We use initial *molarities* of A and B, while we are given mass percents, so we need to convert.

Let's first find the number of *moles* of A and B we have initially :

$$n_{A0} = 10 \text{ kg A} * \frac{1 \text{ mol A}}{0.025 \text{ mol A}} = 400 \text{ mol A}$$

$$n_{B0} = 100 \text{ kg solution} * \frac{0.16 \text{ mol B}}{\text{kg sln}} = 16 \text{ kg B} * \frac{1 \text{ mol B}}{0.047 \text{ kg B}} = 340.43 \text{ mol B}$$

Now, the volume contributed by the 100kg of 16% B solution is :

$$V = \frac{m}{\rho} = \frac{100 \text{ kg}}{57 \frac{lb}{ft^3} * \frac{1 \text{ kg}}{2.2 \text{ lb}} * \frac{1 \text{ } ft^3}{28.317 \text{ L}}} = 109.3 \text{ L}$$

Since adding the A contributes 5L to the volume, the volume after the two are mixed is 109.3 L + 5 L = 114.3 L.

By definition then, the molarities of A and B before the reaction occurs are :

- $[A]_0 = \dfrac{400 \text{ moles A}}{114.3 \text{ L}} = 3.500M$

- $[B]_0 = \dfrac{340.42 \text{ moles B}}{114.3 \text{ L}} = 2.978M$

In addition, there is no C or D in the solution initially :

- $[C]_0 = [D]_0 = 0$

According to the stoichiometry of the reaction, $a = 1$, $b = 2$, $c = 3$, $d = 1$. Therefore we now have enough information to solve for the conversion. Plugging all the known values into the equilibrium equation for liquids, the following equation is obtained :

$$187 = \frac{\left(\dfrac{3X}{114.3}\right)^3 \left(\dfrac{X}{114.3}\right)}{\left(3.5 - \dfrac{X}{114.3}\right)\left(2.978 - \dfrac{2X}{114.3}\right)^2}$$

This equation can be solved using Goalseek or one of the numerical methods in appendix 1 to give :

    $X = 146.31$ moles

Since we seek the amount of compound C that is produced, we have :

- $X = \dfrac{\Delta n_C}{c}$

- Since $c = 3$, $n_{C0} = 0$, and $X = 146.31$ this yields $n_C = 3 * 146.31 = 438.93$ moles C 438.93 moles of C can be produced by this reaction.

Introduction to Reactions with Recycle

Reactions with recycle are very useful for a number of reasons, most notably because they can be used to improve the selectivity of multiple reactions, push a reaction beyond its equilibrium conversion, or speed up a catalytic reaction by removing products. A recycle loop coupled with a reactor will generally contain a separation process in which unused reactants are (partially) separated from products. These reactants are then fed back into the reactor along with the fresh feed.

Example Reactor with Recycle

Example :

Consider a system designed for the hydrogenation of ethylene into ethane :

$$2H_2 + C_2H_2 \rightarrow C_2H_6$$
$$(2\,A + B \rightarrow C)$$

The reaction takes too long to go to completion (and releases too much heat) so the designers decided to implement a recycle system in which, after only part of the reaction had finished, the mixture was sent into a membrane separator. There, most of the ethylene was separated out, with little hydrogen or ethylene contamination. After this separation, the cleaned stream entered a splitter, where some of the remaining mixture was returned to the reactor and the remainder discarded.

The system specifications for this process were as follows :

Feed : 584 kg/h ethylene, 200 kg/h hydrogen gas

Outlet stream from reactor contains 15% hydrogen by mass

Mass flows from membrane separator : 100 kg/h, 5% Hydrogen and 93% ethane

Splitter : 30% reject and 70% reflux

What was the extent of reaction for this system? What would the extent of reaction be if there was no separation/recycle process after (assume that the mass percent of hydrogen leaving the reactor is the same)? What limits how effective this process can be?

Solution :

Let's first draw our flowchart as usual :

DOF Analysis

- On reactor : 6 unknowns ($\dot{m}_5, x_{A5}, x_{B5}, \dot{m}_3, x_{B3}, X$) - 3 equations = 3 DOF
- On separator : 5 unknowns ($\dot{m}_3, x_{B3}, \dot{m}_5, x_{A5}, x_{B5}$) - 3 equations = 2 DOF
- On splitter : 3 unknowns - 0 equations (we used all of them in labeling the chart) -> 3 DOF
- Duplicate variables : 8 ($\dot{m}_5, x_{A5}, x_{B5}$ twice each and $\dot{m}_3, x_{B3}$ once)
- Total DOF = 8 - 8 = 0 DOF

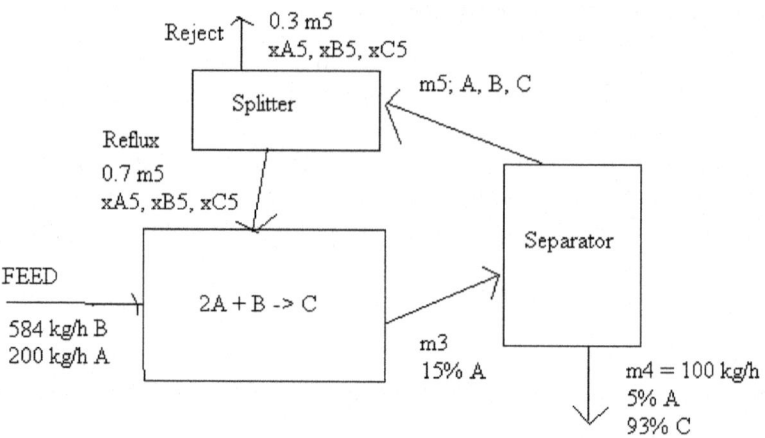

Plan and Solution

Generally, though not always, it is easiest to deal with the reactor itself *last* because it usually has the most unknowns. Lets begin by looking at the overall system because we can often get some valuable information from that.

Overall System DOF(overall system) = 4 unknowns ($\dot{m}_5, x_{A5}, X, x_{B5}$) - 3 equations = 1 DOF.

Note : We CANNOT say that total mass of A and B is conserved because we have a reaction here! Therefore we must include the conversion X in our list of unknowns for both the reactor and the overall system. However, the total mass in the system is conserved so we can solve for $\dot{m}_5$.

Let's go ahead and solve for m_5 though because that'll be useful later.

$784 = 100 + 0.3\,(\dot{m}_5)$

$\dot{m}_5 = 2280$ kg/h

We can't do anything else with the overall system without knowing the conversion so lets look elsewhere.

DOF(separator) = 4 unknowns ($\dot{m}_3, x_{B3}, x_{A5}, x_{B5}$) - 3 equations = 1 DOF.

Let's solve for those variables we can though.

We can solve for m_3 because from the overall material balance on the separator :

- $\dot{m}_3 = \dot{m}_4 + \dot{m}_5$

- $\dot{m}_3 = 100 + 2280$

$\dot{m}_3 = 2380$ kg/h

Then we can do a mass balance on A to solve for x_{A5} :

- $\dot{m}_3 x_{A3} = \dot{m}_4 x_{A4} + \dot{m}_5 x_{A5}$

- $2380(0.15) = 100\,(0.05) + 2380\,(x_{A5})$

$x_{A5} = .1544$

Since we don't know x_{A5} or x_{B3}, we cannot use the mass balance on B or C for the separator, so lets move on. Let's now turn to the reactor :

Reactor Analysis

DOF : 3 unknowns remaining (x_{B3}, x_{B5}, and X) - 2 equations (because the overall balance is already solved!) = 1 DOF. Therefore we still cannot solve the reactor completely. However, we can solve for the conversion and generation terms given what we know at this point. Lets start by writing a *mole* balance on A in the reactor.

$$\dot{n}_{A1} + \dot{n}_{A,recycle} - X * a = \dot{n}_{A3}$$

To find the three n_A terms we need to convert from mass to moles (since A is hydrogen, H_2, the molecular weight is $\dfrac{1\ \text{mol}}{0.002016\ \text{g}}$) :

- $\dot{n}_{A1} = 200\,\dfrac{kg}{h} * \dfrac{1\ \text{mol}}{0.002016\ kg} = 99206\,\dfrac{\text{mol A}}{h}$

- $\dot{n}_{A,recycle} = 0.7 * \dfrac{m_5 * x_{A5}}{MW_A} = \dfrac{0.7(2280)\,(0.1544)}{0.002016} = 122000\,\dfrac{\text{mol A}}{h}$

Thus the *total* amount of A entering the reactor is :

- $\dot{n}_{A,in} = 99206 + 122000 = 221428\,\dfrac{\text{mol A}}{h}$

The amount exiting is :

- $\dot{n}_{A,out} = \dfrac{\dot{m}_3 * x_{A3}}{M\,M_A} = \dfrac{2380 * 0.15}{0.002016} = 177083\,\dfrac{\text{mol A}}{h}$

Therefore we have the following from the mole balance :

- $221428 - 2X = 177083$

$X = 22173\,\dfrac{\text{mol A}}{h}$

Now that we have this we can calculate the mass of B and C generated :

- $m_{B,gen} = -Xb * MW_B = 22173\,\dfrac{\text{mol B}}{h} * 0.026\,\dfrac{kg}{\text{mol B}} = -576.5\,\dfrac{\text{mol B}}{h}$

- $m_{C,gen} = +Xc * MW_C = 22173\,\dfrac{\text{mol C}}{h} * 0.030\,\dfrac{kg}{\text{mol C}} = +665.2\,\dfrac{kg\ C}{h}$

At this point you may want to calculate the amount of B and C leaving the reactor with the mass balances on B and C :

- $584 + 0.7 * x_{B5} * 2280 - 576.5 = * x_{B3} * 2380$
- (1) $0.7 * (1 - 0.1544 - x_{B5}) * 2280 + 665.2 = (1 - 0.15 - * x_{B3}) * 2380$

However, *these equations are exactly the same!* Therefore, we have proven our assertion that there is still 1 DOF in the reactor. So we need to look elsewhere for something to calculate x_{B5}. That place is the separator balance on B :

- $\dot{m}_3 * x_{B3} = \dot{m}_4 * x_{B4} + \dot{m}_5 * x_{B5}$
- (2) $2380 \, x_{B3} = 0.2 \, (100) + 2280 \, x_{B5}$

Solving these two equations (1) and (2) yields the final two variables in the system :

$x_{B3} = 0.00856, \; x_{B5} = 0.008058$

Note that this means the predominant species in stream 5 is also C ($x_{C5} = 0.838$). However, the separator/recycle setup does make a big difference, as we'll see next.

Comparison to the Situation without the Separator/Recycle System

Now that we know how much ethane we can obtain from the reactor after separating, let's compare to what would happen without any of the recycle systems in place. With the same data as in the first part of this problem, the new flowchart looks like this :

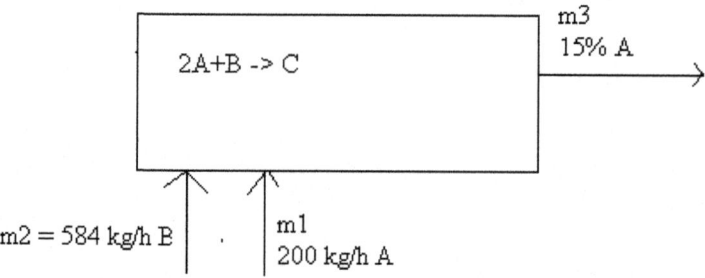

There are three unknowns ($\dot{m}_3, x_{B3}, X$) and three independent material balances, so the problem can be solved. Starting with an overall mass balance because total mass is conserved :

- $\dot{m}_1 + \dot{m}_2 = \dot{m}_3$

- $\dot{m}_3 = 789 \dfrac{kg}{h}$

We can carry out the same sort of analysis on the reactor as we did in the previous section to find the conversion and mass percent of C in the exit stream, which is left as an exercise to the reader. The result is that :

- $X = 20250$ moles, $x_{C3} = 0.77$

Compare this to the two exit streams in the recycle setup. Both of the streams were richer in C than 77%, even the reject stream. This occurred because the unreacted A and B was allowed to re-enter the reactor and form more C, and the separator was able to separate almost all the C that formed from the unreacted A and B.

Introduction to Reactions with Recycle

Reactions with recycle are very useful for a number of reasons, most notably because they can be used to improve the selectivity of multiple reactions, push a reaction beyond its equilibrium conversion, or speed up a catalytic reaction by removing products. A recycle loop coupled with a reactor will generally contain a separation process in which unused reactants are (partially) separated from products. These reactants are then fed back into the reactor along with the fresh feed.

The Concept of Atom Balances

Let's begin this section by looking at the reaction of hydrogen with oxygen to form water :

$$H_2 + O_2 \rightarrow H_2O$$

We may attempt to do our calculations with this reaction, but there is something seriously wrong with this equation! It is not *balanced*; as written, it implies that an atom of oxygen is somehow "lost" in the reaction, but this is in general impossible. Therefore, we must compensate by writing :

$$H_2 + \frac{1}{2}O_2 \rightarrow H_2O$$

or some multiple thereof.

Notice that in doing this we have made use of the following conservation law, which is actually the basis of the conservation of mass :

The number of atoms of any given element does not change in any reaction (assuming that it is not a nuclear reaction).

Since by definition the number of moles of an element is proportional to the number of atoms, this implies that $\dot{n}_{A,gen} = 0$ where A represents any *element* in atomic form.

Mathematical Formulation of the Atom Balance

Now recall the general balance equation :

In — Out + Generation = Accumulation

In this course we're assuming *Accumulation* = 0. Since the *moles of atoms* of any element are conserved, *generation* = 0. So we have the following balance on a given element A :

For a given element A,

$$\Sigma \dot{n}_{A,in} - \Sigma \dot{n}_{A,out} = 0$$

Note : When analyzing a reacting system you must choose *either* an atom balance *or* a molecular species balance but not both. Each has advantages; an atom balance often yields simpler algebra (especially for multiple reactions; the actual reaction that takes place is irrelevant!) but also will not directly tell you the extent(s) of reaction, and will not tell you if the system specifications are actually impossible to achieve for a given set of equilibrium reactions.

Degree of Freedom Analysis for the Atom Balance

As before, to do a degree of freedom analysis, it is necessary to count the number of unknowns and the number of equations one can write, and then subtract them. However, there are a couple of important things to be aware of with these balances.

- **When doing atom balances, the extent of reaction does not count as an unknown, while with a molecular species balance it does.** This is the primary advantage of this method : the extent of reaction does not matter since atoms of elements are conserved regardless of how far the reaction has proceeded.

- You need to make sure each atom balance will be **independent**. This is difficult to tell unless you write out the equations and look to see if any two are identical.

- In reactions with **inert species**, each molecular balance on the inert species counts as an *additional equation*. This is because of the following important note :

Note : When you're doing an atom balance you should only include reactive species, not inerts.

Example :

Suppose a mixture of nitrous oxide (N_2O) and oxygen is used in a natural gas burner. The reaction $CH_4 + 2O_2 \rightarrow 2H_2O + CO_2$ occurs in it.

There would be four equations that you could write : 3 atom balances (C, H, and O) and a molecular balance on nitrous oxide. You would not include the moles of nitrous oxide in the atom balance on oxygen.

Example of the Use of the Atom Balance

Let's re-examine a problem from the previous section. In that section it was solved using a molecular species balance, while here it will be solved using atom balances.

Example :

Consider the reaction of Phosphene with oxygen :

$$4PH_3 + 8O_2 \rightarrow P_4O_{10} + 6H_2O$$

Suppose a 100-kg mixture of 50% PH_3 and 50% O_2 by mass enters a reactor in a single stream, and the single exit stream contains 25% O_2 by mass. Assume that all the reduction in oxygen occurs due to the reaction. How many degrees of freedom does this problem have? If possible, determine mass composition of all the products.

For purposes of examination, the flowchart is re-displayed here :

```
                    ┌─────────────────────────────────┐
 m1 = 100 kg        │                                 │         m2
                    │   4PH3 + 8O2 -> P4H10+6H2O      │
 ─────────────────  │                                 │  ─────────────────
                    │                                 │
 X(PH3) = 0.5       │                                 │       X(O2) = 0.25
 X(O2) = 0.5        │                                 │
                    │                                 │
                    └─────────────────────────────────┘
```

Degree of Freedom Analysis

There are three elements involved in the system (P, H, and O) so we can write three atom balances on the system.

There are likewise three unknowns (since the extent of reaction is NOT an unknown when using the atom balance) : the outlet concentrations of PH_3, P_4O_{10}, H_2O

Therefore, there are 3 - 3 = 0 unknowns.

Problem Solution

Let's start the same as we did in the previous section : by finding converting the given information into moles. The calculations of the previous section are repeated here :

- $\dot{m}_{out} = \dot{m}_{in} = 100$ kg

- $\dot{n}_{PH3,in} = 0.5 * (100 \text{ kg}) * \dfrac{1 \text{ mol}}{0.034 \text{ kg}} = 1470.6$ moles PH_3 in

- $\dot{n}_{O2,in} = 0.5 * (100 \text{ kg}) * \dfrac{1 \text{ mol}}{0.032 \text{ kg}} = 1562.5$ moles O_2 in

- $\dot{n}_{O2,out} = 0.25 * (100 \text{ kg}) * \dfrac{1 \text{ mol}}{0.032 \text{ kg}} = 781.25$ moles O_2 out

Now we start to *diverge* from the path of molecular balances and instead write atom balances on each of the elements in the reaction. Let's start with Phosphorus. How many moles of Phosphorus *atoms* are entering?

- Inlet : Only PH_3 provides P, so the inlet moles of P are just $1 * 1470.6 = 1470.6$ moles P in

- Outlet: There are *two* ways phosphorus leaves : as unused PH_3 or as the product P_4H_{10}. Therefore, the moles of PH_3 out are $1 * n_{PH3,\,out} + 4 * nP_4O_{10,\,out}$. Note that the 4 in this equation comes from the fact that there are 4 Phosphorus atoms in every mole of P_4O_{10}.

Therefore the atom balance on Phosphorus becomes :

Phosphorus

$$1 * n_{PH3,out} + 4 * n_{P4O10,out} = 1470.6$$

Similarly, on Oxygen we have :

- Inlet : $2 * n_{O2,in} = 2 * 1562.5 = 3125$ moles O_2
- Outlet: $2 * n_{O2,out} + 10 * n_{P4O10,out} + 1 * n_{H2O,out} = 1562.5 + 10 * n_{P4O10,out} + 1 * n_{H2O,out}$

Oxygen

$$1562.5 + 10 * n_{P4O10,out} + 1 * n_{H2O,out} = 3125$$

Finally, check to see if you can get the following Hydrogen balance as a practice problem :

Hydrogen

$$2 * n_{H2O,out} + 3 * n_{PH3,out} = 4411.8$$

Solving these three linear equations, the solutions are :

$$n_{PH3,out} = 1080, n_{O2,out} = 586, n_{P4O10,out} = 97.66 \text{ moles}$$

All of these answers are identical to those obtained using extents of reaction. Since the remainder of the solution to that problem is identical to that in the previous section, the reader is referred there for its completion.

Example of Balances with Inert Species

Sometimes it's more difficult to choose which type of balance you want, because both are possible but one is significantly easier than the other. As an example, lets consider a basic pollution control system.

Example :

Suppose that you are running a power plant and your burner releases a lot of pollutants into the air. The flue gas has been analyzed to contain 5% SO_2, 3% NO_2, 7% O_2 and 15% CO_2 by moles. The remainder was determined to be inert.

Local regulations require that the emissions of sulfur dioxide be less than 200 ppm (by moles) from your plant. They also require you to reduce nitrogen dioxide emissions to less than 50 ppm. You decide that the most economical method for control of these for your plant is to utilize ammonia-based processes. The proposed system is as follows :

Put the flue gas through a denitrification system, into which (pure) ammonia is pumped. The amount of ammonia pumped in is three times as much as would theoretically be needed to use all of the nitrogen dioxide in the flue gas.

Allow it to react a specified amount of time.

Pump it into a desulfurization system. Nothing new is injected here, it just has a different catalyst than the denitrification, and the substrates are at a different temperature and pressure.

The reactions that occur are :

$$2NO_2 + 4NH_3 + O_2 \rightarrow 3N_2 + 6H_2O$$
$$H_2O + 2NH_3 + SO_2 \rightarrow (NH_4)_2SO_3$$

If your plant makes $130\dfrac{ft^3}{s}$ of flue gas at $T = 900K$ and $P = 2$ atm, how must ammonia do you need to purchase for each 8-hour shift? How much of it remains unused? Why do we want to have a significant amount of excess ammonia?

Assume that the flue gas is an ideal gas. Recall the ideal gas law, $PV = nRT$, where

$$R = 0.0821\dfrac{L * atm}{mol * K}$$

Step 1 : Flowchart

Flowcharts are becoming especially important now as means of organizing all of that information!

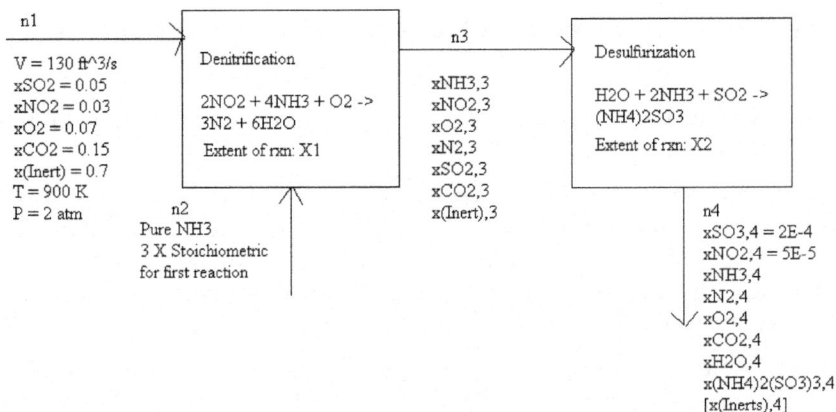

Step 2 : Degrees of Freedom

Let's consider an atomic balance on each reactor.

- Denitrification system : 9 unknowns (all concentrations in stream 3, and $\dot{n}_2$.) - 3 atom balances (N, H, and O) - 3 inert species (CO_2, SO_2, inerts) - 1 additional info (3X stoichiometric feed) = **2 DOF**
- Desulfurization system : 15 unknowns - 4 atom balances (N, H, O, and S) - 5 inerts (CO_2, O_2, NO_2, N_2, inerts) = **6 DOF**
- Total = 2 + 6 - 8 shared = **0 DOF**, hence the problem has a unique solution.

We can also perform the same type of analysis on molecular balances.

- Denitrification system : 10 unknowns (now the conversion X_1 is also unknown) - 8 molecular species balances - 1 additional info = **1 DOF**
- Desulfurization system : 16 unknowns (now the conversion X_2 is unknown) - 9 balances = **7 DOF**.
- Total = 1 + 7 - 8 shared = **0 DOF**.

Therefore the problem is theoretically solvable by both methods.

Step 3 : Units

The only weird units in this problem (everything is given in moles already so no need to convert) are in the volumetric flowrate, which is given in $\frac{ft^3}{s}$. Lets convert this to $\frac{moles}{s}$ using the ideal gas law. To use the law with the given value of R is is necessary to change the flowrate to units of $\frac{L}{S}$:

$$130\frac{ft^3}{s} * \frac{28.317\ L}{ft^3} = 3681.2\frac{L}{s}$$

$$P\dot{V} = \dot{n}RT \rightarrow 2 * 3681.2 = \dot{n}_1(0.0821)(900)$$

$$\dot{n}_1 = 99.64\frac{moles}{s}$$

Now that everything is in good units we can move on to the next step.

Step 4 : Devise a plan

We can first determine the value of $\dot{n}_2$ using the additional information. Then, we should look to an *overall system balance*.

Since none of the individual reactors is completely solvable by itself, it is necessary to look to combinations of processes to solve the problem. The best way to do an overall system balance with multiple reactions is to treat the entire system as if it was a single reactor in which multiple reactions were occurring. In this case, the flowchart will be revised to look like this :

Before we try solving anything, we should check to make sure that we still have no degrees of freedom.

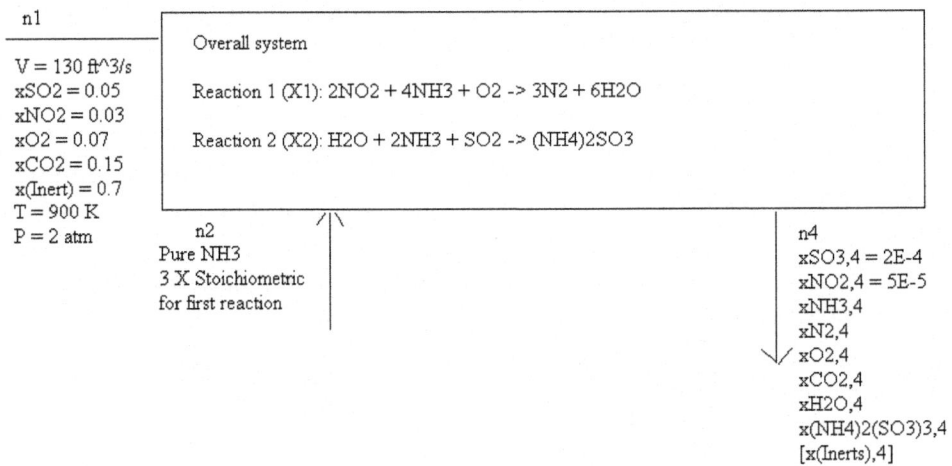

n1
V = 130 ft^3/s
xSO2 = 0.05
xNO2 = 0.03
xO2 = 0.07
xCO2 = 0.15
x(Inert) = 0.7
T = 900 K
P = 2 atm

Overall system

Reaction 1 (X1): 2NO2 + 4NH3 + O2 -> 3N2 + 6H2O

Reaction 2 (X2): H2O + 2NH3 + SO2 -> (NH4)2SO3

n2
Pure NH3
3 X Stoichiometric
for first reaction

n4
xSO3,4 = 2E-4
xNO2,4 = 5E-5
xNH3,4
xN2,4
xO2,4
xCO2,4
xH2O,4
x(NH4)2(SO3)3,4
[x(Inerts),4]

ATOM BALANCE

There are 8 unknowns (don't count conversions when doing atom balances), 4 types of atoms (H, N, O, and S), 2 species that never react, and 1 additional piece of information (3X stoichiometric), so there is 1 DOF. This is obviously a problem, which occurs because when performing atom balances you cannot distinguish between species that react in only ONE reaction and those that take part in more than one.

In this case, then, it is necessary to look to molecular-species balances.

Molecular-species Balance

In this case, there are 10 unknowns, but we can do molecular species balances on 9 species (SO_2, NO_2, NH_3, N_2, O_2, CO_2, H_2O, $(NH_4)_2 SO_3$, inerts) and have the additional information, so there are 0 DOF when using this method.

Once we have all this information, getting the information about stream 3 is trivial from the definition of extent of reaction.

Step 5 : Carry Out the Plan

First off we can determine $\dot{n}_2$ by using the definition of a stoichiometric feed.

$$\dot{n}_{NO_2,in} = 0.03 * 99.64 = 2.9892 \frac{mol}{s}$$

The stoichiometric amount of ammonia needed to react with this is, from the reaction,

$$\frac{4 \text{ moles } NH_3}{2 \text{ moles } NO_2} * 2.9892 = 5.96 \frac{\text{moles } NH_3}{s}$$

Since the problem states that three times this amount is injected into the denitrification system, we have :

$$\dot{n}_2 = 17.88 \frac{moles}{s}$$

Now, we are going to have a very complex system of equations with the 9 molecular balances. This may be a good time to invest in some equation-solving software.

See if you can derive the following system of equations from the overall-system flowchart above.

$$NH_3 : \dot{n}_4 * x_{NH3,4} = 17.88 - 4 * X_1 - 2 * X_2$$

$$SO_2 : \dot{n}_4 * 2 * 10^{-4} = 0.05 * 99.64 - X_2$$

$$NO_2 : \dot{n}_4 * 5 * 10^{-5} = 0.03 * 99.64 - 2 * X_1$$

$$N_2 : \dot{n}_4 * x_{N2,4} = 3 * X_1$$

$$O_2 : \dot{n}_4 * x_{O2,4} = 0.07 * 99.64 - X_1$$

$$H_2O : \dot{n}_4 * x_{H2,4} = 6 * X_1 - X_2$$

$$CO_2 : \dot{n}_4 * x_{CO2,4} = 0.15 * 99.64$$

$$(NH_4)_2(SO_3) : \dot{n}_4 * x_{(NH4)2SO3,4} = X_2$$

$$Inerts : \dot{n}_4 * (1 - 2 * 10^{-4} - 5 * 10^{-5} - x_{NH3,4} - x_{N2,4} - x_{O2,4} - x_{H2O,4}$$
$$-x_{CO2,4} - x_{(NH4)2SO3,4}) = 0.7 * 99.64$$

Using an equation-solving package, the following results were obtained :

$X_1 = 1.492$ moles

$X_2 = 4.961$ moles

$$\dot{n}_3 = 105.62 \frac{mol}{s}$$

$x_{NH3,4} = 0.01884$

$x_{N2,4} = 0.04238$

$x_{O2,4} = 0.05191$

$x_{H2O,4} = 0.03778$

$x_{CO2,4} = 0.1415$

$x_{CO2,4} = 0.1415$

$x_{(NH4)2SO3,4} = 0.04697$

$x_1 = 1 - \Sigma(\text{other components}) = 0.6606$

Stream 3

Now that we have completely specified the composition of stream 4, it is possible to go back and find the compositions of stream 3 using the extents of reaction and feed composition. Although this is not necessary to answer the problem statement, it should be done, so that we can then test to make sure that all of the numbers we have obtained are consistent.

Chapter 5

MULTIPLE-PHASE SYSTEMS, INTRODUCTION TO PHASE EQUILIBRIUM

NON-IDEAL GAS-PHASE ANALYSIS

What is an Ideal Gas?

Recall from general chemistry that the volume, pressure, temperature, and moles of a gas in a closed system can be related by the following equation, which is referred to as the ideal gas law :

$PV = nRT$

R is referred to as the Universal Gas Constant, and it has the following values for different units of P, V, n, and T :

$$R = 0.0821 \frac{L * atm}{mol * K} = 8.31 \frac{J}{mol * K} = 8.31 \frac{Pa * m^3}{mol * K}$$

One thing that may have been de-emphasized in an introductory chemistry course is the fact that *gasses do not always follow this law*. In fact, they only do under very special circumstances.

Theoretical Background on the Ideal Gas Law

The ideal gas law relies on several rather strong assumptions about the nature of gasses, which make up the classical Kinetic Theory of Gases. These assumptions are :

1. That gas molecules do not interact with each other whatsoever.
2. That all collisions between molecules and each other or with the walls of the container are completely **elastic**, meaning the mean kinetic energy of the molecules stays the same.

3. That gas molecules are very small compared to the distance between them.

There are several other assumptions as well, but these will suffice to explain why deviations occur.

Important Facts about Ideal Gasses

Ideal gasses are nice because they have several properties that make them relatively easy and useful to work with :

1. A mixture of ideal gasses is also an ideal gas. Therefore, you can use the ideal gas law on the entire mixture or on any of its components without loss of validity.

2. The **Partial Pressure** of a component in an ideal gas mixture is related to the total pressure by the equation $P_A = x_A * P$ where x_A is the mole fraction of component A in the mixture.

3. The ideal gas equation, if it is valid, is independent of the properties of the gas. Therefore, there is no need to look up or measure gas-specific parameters, all you need to know is 3 of the unknowns (P, V, T, and n) and you can solve for the fourth one.

4. The enthalpy and entropy of an ideal gas only depend on temperature (not pressure or volume).

5. Many gasses are close to ideal at low pressures and high temperatures. Therefore it can be used as a realistic **reference state** to which a real gas can be compared.

There are many other useful properties of ideal gasses that will be discussed in thermodynamics.

Deviations at High Pressure

Suppose that you have water vapor in a small water bottle. What happens when you apply pressure to that vapor by shrinking the volume? If you apply more pressure to the bottle, the gas molecules within become closer and closer together. However, the closer molecules of a substance are to one another, the larger the dispersion forces (and polar forces, if applicable) are between them. Eventually, the dispersion forces become significant, and the kinetic theory of gasses as stated above is no longer valid. Therefore, the ideal gas law no longer applies, or becomes a rough estimate at best.

At very high pressures, a vapor might even condense into a liquid, which would also invalidate the use of the law.

Deviations at Low Temperature

By definition of temperature, when the temperature of a substance decreases, this indicates a lower average kinetic energy of the molecules of that substance.

Molecules with less kinetic energy also have less momentum, and therefore are more easily swayed by dispersion forces from other molecules and by gravity. Eventually, the temperature might become low enough that the forces from other molecules cause significant deviations in the path of a molecule, and in this case the ideal gas law becomes less valid.

As with pressure, a very low temperature can cause a gas to condense.

Rule of Thumb for Use of the Ideal Gas Law

Due to the above two discussions, we can claim that the ideal gas law is a decent assumption at high temperatures and low pressures, but should not be used outside that realm.

The Idea of Equations of State

The ideal gas law is not the only way to express relationships between system properties. There are an infinite number of possible ways that you *could* propose to correlate system variables, although thermodynamics provides some guidelines regarding how many variables to correlate and what types of correlations we should look for. In particular, it is most useful to correlate relationships between state variables, for which changes in the properties do not depend on how the change occurred.

Any equation that relates state variables is called an equation of state. The most commonly-used equations of state relate the variables P, T, V, and n (pressure, temperature, volume, and number of moles) since they are all measurable variables whereas many other possible variables are not directly measurable (such as enthalpy, which will be discussed later).

Compressibility

There is a highly useful quantity for describing how much a gaseous system deviates from ideality, which is called the compressibility of the gas. The compressibility Z is defined as :

$$Z = \frac{P * V}{n * R * T}$$

For an ideal gas, since $PV = nRT$, $Z = 1$. Therefore, any deviation of the compressibility from 1 is a non-ideality.

The compressibility of a liquid is very small, due to a small volume per mole of substance compared to that of a gas.

Ideal Gas Law Extension

One use of the compressibility is that it allows a simple extension of the ideal gas law, which is completely general.

REAL-GAS EXTENSION TO THE IDEAL GAS LAW

$PV = nRT\,Z$

Unfortunately, Z is not a constant for any material, but changes with pressure and temperature. However, in a later section you will learn a technique called the generalized compressibility method with which you can estimate the value of Z for *any* substance, given certain data. Once it is known, this extension can be used to calculate an unknown system property.

Alternatives to the Ideal Gas Law 1 : Van der Waals Equation

One of the oldest equations that was developed to take non-ideality into account is called the Van der Waals Equation, which has two substance-dependent parameters. One takes into account the interactions between particles, and the other the fact that particles have volume, sometimes substantial.

The Van der Waals equation is as follows :

$\left(p + \dfrac{n^2 a}{V^2}\right)(V - nb) = nRT$ where a and b depend on the gas being analyzed

A list of values for a and b can be found on Wikipedia at this page.

The Van der Waals equation is significantly more accurate than the ideal gas law and can be used to crudely predict when a gas will condense. However, it is not sufficiently accurate for many industrial purposes, and therefore other methods have been sought since then.

Alternatives to the Ideal Gas Law 2 : Virial Equation

The Virial equation is an equation which has a potentially infinite number of parameters that depend on the properties of the substances involved. It is important because it can be shown to be a valid extension to the ideal gas law using statistical theories (whereas the other equations of state have been derived semi-empirically).

The virial equation can take several forms, depending on the data one has available. The one in terms of molar volume is :

$$Z = \frac{PV_m}{RT} = 1 + \frac{B}{V_m} + \frac{C}{V_m^2} + \frac{D}{V_m^3} + \cdots$$

where V_m is the molar volume, which is the same as $\dfrac{V}{n}$.

Alternatives to the Ideal Gas Law 3 : Peng-Robinson Equations

One of the more modern equations of state is the Peng-Robinson equation, which is most useful for describing nonpolar molecules such as hydrocarbons or nitrogen. The Peng-Robinson equation has two parameters like the Van der Waals equation, but unlike the latter, one of the parameters is not constant; it depends on the temperature of the system as well as the properties of the substance inside.

VAPOR-LIQUID EQUILIBRIUM

Phase Equlibrium

Many processes in chemical engineering do not only involve a single phase but a combination of two immiscible liquids, or a stream containing both gas and liquid. It is very important to recognize and be able to calculate when these phases are in equilibrium with each other, and how much is in each phase. This knowledge will be especially useful when you study separation processes, for many of these processes work by somehow distorting the equilibrium so that one phase is especially rich in one component, and the other is rich in the other component.

More specifically, there are three important criteria for different phases to be in equilibrium with each other :

1. The **temperature** of the two phases is the same at equilibrium.
2. The **partial pressure** of every component in the two phases is the same at equilibrium.
3. The **Gibbs free energy**' of every component in the two phases is the same at equilibrium.

The third criteria will be explored in more depth in another course; it is a consequence of the first two criteria and the second law of thermodynamics.

Single-Component Phase Equilibrium

If there is only a single component in a mixture, there is only a single possible temperature (at a given pressure) for which phase equilibrium is possible. For example, water at standard pressure (1 atm) can *only* remain in equilibrium at 100°C. Below this temperature, all of the water condenses, and above it, all of the water vaporizes into steam.

At a given temperature, the unique *atmospheric* pressure at which a pure liquid boils is called its vapor pressure. If the atmospheric pressure is higher than the vapor pressure, the liquid will not boil. Vapor pressure is strongly temperature-dependent. Water at 100°C has a vapor pressure of 1 atmosphere, which explains why water on Earth (which has an atmosphere of about 1 atm) boils at 100°C. Water at a temperature of 20°C(a typical room temperature) will only boil at pressures under 0.023 atm, which is its vapor pressure at that temperature.

Multiple-Component Phase Equilibrium : Phase Diagrams

In general, chemical engineers are not dealing with single components; instead they deal with equilibrium of mixtures. When a mixture begins to boil, the vapor does not, in general, have the same composition as the liquid. Instead, the substance with the lower boiling temperature (or higher vapor pressure) will have a vapor concentration higher than that with the higher boiling temperature, though both will be present in the vapor. A similar argument applies when a vapor mixture condenses.

The concentrations of the vapor and liquid when the *overall* concentration and one of the temperature or pressure are fixed can easily be read off of a phase diagram. In order to read and understand a phase diagram, it is necessary to understand the concepts of *bubble point* and *dew point* for a mixture.

Bubble Point and Dew Point

In order to be able to predict the phase behavior of a mixture, scientists and engineers examine the *limits* of phase changes, and then utilize the laws of thermodynamics to determine what happens in between those limits. The limits in the case of gas-liquid phase changes are called the bubble point and the dew point.

The names imply which one is which :

1. The bubble point is the point at which the first drop of a liquid mixture begins to vaporize.
2. The dew point is the point at which the first drop of a gaseous mixture begins to condense.

If you are able to plot both the bubble and the dew points on the same graph, you come up with what is called a Pxy or a Txy diagram, depending on whether it is graphed at constant temperature or constant pressure. The "xy" implies that the curve is able to provide information on both liquid *and* vapor compositions, as we will see when we examine the thermodynamics in more detail.

Txy and Pxy Diagrams

The easier of the two diagrams to calculate (but sometimes harder to grasp intuitively) is the Pxy diagram, which is shown below for an idealized Benzene-Toluene system :

In order to avoid getting confused about what you're looking at, think : what causes a liquid to vaporize? Two things should come to mind :

*   Increasing the temperature
*   *Decreasing* the pressure

Therefore, the region with the *higher* pressure is the liquid region, and that of *lower* pressure is vapor, as labeled. The region in between the curves is called the two-phase region.

Note : You may be tempted to try and memorize something like the dew point line is on the bottom in a Pxy diagram and on the top in a Txy diagram. This is, however, strongly discouraged, as you will very likely become confused if you depend on this type of memorizing. Instead think : which half of the graph will contain liquid and which half will be vapor? Then use the definitions of "dew" and "bubble" points to determine which line is which.

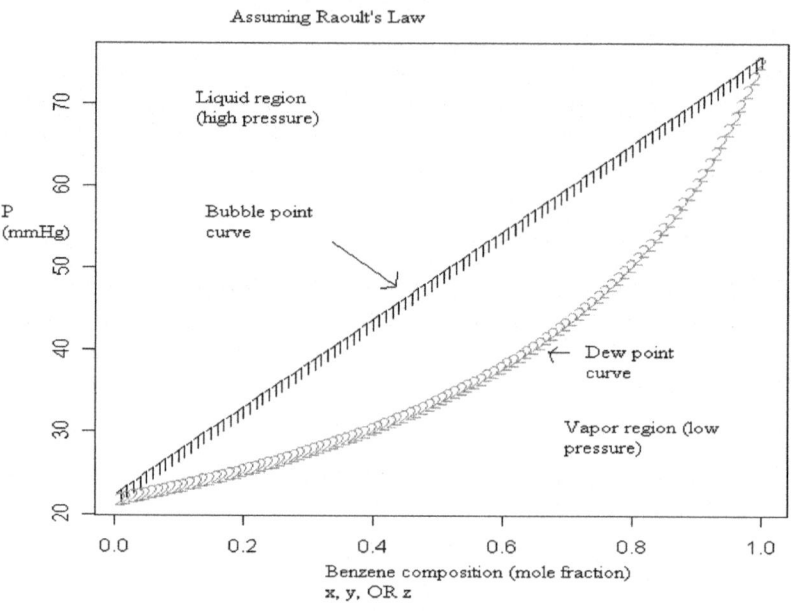

Benzene-Toluene Pxy Diagram at T=20oC
Assuming Raoult's Law

Now that we have this curve, what can we do with it? There are several critical pieces of information we can gather from this graph by simple techniques, which have complete analogies in the Txy diagram. First, note that the two lines intersect at $x_{Benzene} = 0$ and at $x_{Benzene} = 1$. These intersections are the *pure-component vapor pressures* at $T = 20°C$, since a pure component boils at its vapor pressure.

We can determine, given the mole fraction of one component and a pressure, whether the system is gas, liquid, or two-phase, which is critical information from a design standpoint. For example, if the Benzene composition in the Benzene-Toluene system is 40% and the pressure is 25 mmHg, the entire mixture will be vapor, whereas if the pressure is raised to 50 mmHg it will all condense. The design of a flash evaporator at 20°C would require a pressure between about 30 and 40 mmHg (the 2-phase region).

We can also determine the composition of each component in a 2-phase mixture, if we know the overall composition and the vapor pressure. First, start on the x-axis at the overall composition and go up to the pressure you want to know about. Then from this point, go left until you reach the bubble-point curve to find the *liquid* composition, and go to the right until you reach the dew-point curve to find the *vapor* composition. See the below diagram.

This method "works" because the pressure must be constant between each phase while the two phases are in equilibrium. The bubble and dew compositions are the only liquid and vapor compositions that are stable at a given pressure and temperature, so the system will tend toward those values.

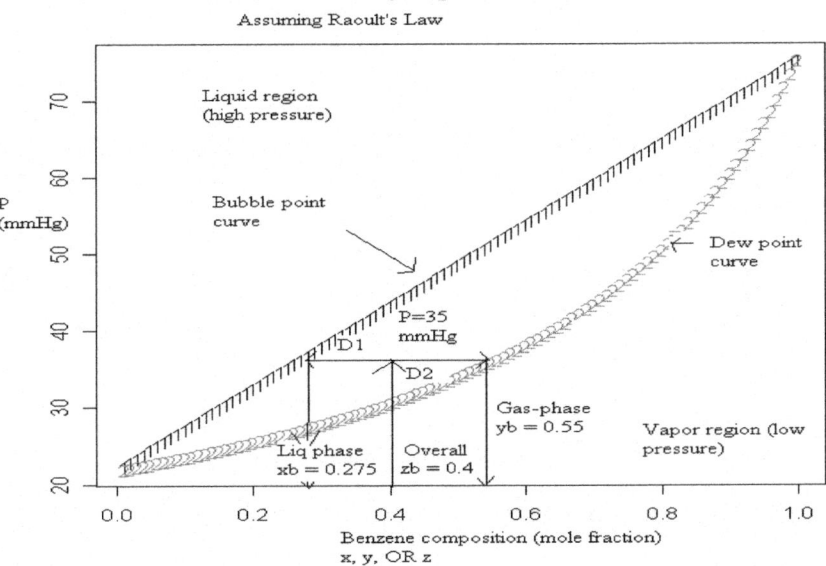

Benzene-Toluene Pxy Diagram at T=20oC
Assuming Raoult's Law

Another useful rule is the lever rule which can be used to calculate the *percentage of all the material* that is in a given phase (as opposed to the composition of the vapor). The lever rule equation is :

$$\%Liquid = \frac{D_2}{D_1 + D_2}$$

- Lever Rule

and therefore,

$$\%Vapor = \frac{D_1}{D_1 + D_2}.$$

The phase whose percent you're calculating is simply the one which you are going towards for the line segment in the numerator; for example, D_1 is going from the point of interest to the liquid phase, so if D_1 is in the numerator then you're calculating percent of liquid.

Txy diagrams have entirely analogous rules, but just be aware that the graph is "reversed" somewhat in shape. It's somewhat harder to calculate even in an ideal case, requiring an iterative solution, but is more useful for isobaric (constant-pressure) systems and is worth the effort. The extreme ends of the txy diagram are the boiling temperatures of pure toluene ($xb = 0$) and benzene ($xb = 1$) at 760 mmHg.

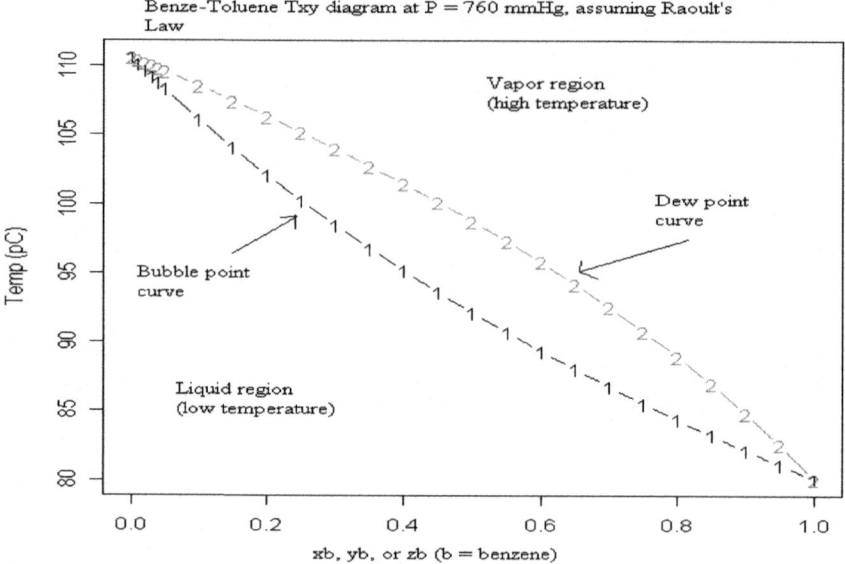

Benze-Toluene Txy diagram at P = 760 mmHg, assuming Raoult's Law

VLE Phase Diagram Summary

To summarize, here's the information you can directly garner from a phase diagram. Many of these can be used for *all* types of phase diagrams, not just VLE.

1. You can use it to tell you what phase(s) you are in at a given composition, temperature, and/or pressure.

2. You can use it to tell you what the *composition* of each phase will be, if you're in a multiphase region.

3. You can use it to tell you how *much* of the original solution is in each phase, if you're in a multiphase region.

4. You can use it to gather some properties of the *pure materials* from the endpoints (though these are usually the best-known of all the mixture properties).

This data is invaluable in systems design which is why you'll be drilled with it before you graduate.

RAOULT'S LAW : THE SIMPLEST CASE

The simplest case (by far) to analyze occurs when an ideal solution is in equilibrium with an ideal gas. This is potentially a good approximation when two very similar liquids (the archetypal example is benzene and toluene) are dissolved in each other. It is also a good approximation for solvent properties (NOT solute properties; there is another law for that) when a very small amount of a solute is dissolved.

In an ideal liquid, the pressure exerted by a certain component on the gas is proportional to the vapor pressure of the *pure liquid*. The only thing that may prevent the liquid from exerting this much pressure is the fact that another component is present. Therefore, the partial pressure of the *liquid component* on the *gas component* is :

$$P_A = x_A * P^*_A$$

where P^*_A is the vapor pressure of pure component A.

Note : For all VLE equations, x_A denote a liquid mole fraction of component A, and y_A is the vapor mole fraction </math> of component A.

Recall that the partial pressure of an ideal gas in a mixture is given by :

$$P_A = y_A * P$$

Therefore, since the partial pressures must be equal at equilibrium, we have the Raoult's Law equation for each component :

Raoult's Law for component A

$$y_A * P = x_A * P^*_A$$

Vapor Pressure Correlations

Unfortunately, life isn't that simple even when everything is ideal. Vapor pressure is not by any stretch of the imagination a constant. In fact, it has a very strong dependence on temperature. Therefore, people have spent a good deal of time and energy developing correlations with which to predict the vapor pressure of a given substance at any reasonable temperature.

One of the most successful correlations is called the Antoine Equation, which uses three coefficients, A, B, and C, which depend on the substance being analyzed. The Antoine Equation is as follows.

Antoine Equation

$$\log(P^*_A) = A - \frac{B}{T + C}$$

Note : In the external link provided in the appendix, the logarithm is to base 10, T is in degrees Celsius, and P* is in mmHg. Other sources use different forms (for example, natural log or P* in bars) so be wary.

Bubble Point and Dew Point with Raoult's Law

Key Concept

When calculating either a bubble point or a dew point, one quantity is key, and this is the *overall composition*, denoted with the letter z. This is to distinguish it from the *single-phase composition* in either the liquid or the gas phase. It is necessary to distinguish between them because the composition of the two phases will almost always be different at equilibrium.

It is important to remember that the dew and bubble points of a multi-component mixture are *limits*. The bubble point is the point at which a very small amount of the liquid has evaporated - so small, in fact, that in essence, *the liquid phase composition remains the same as the overall composition*. Making this assumption, it is possible to calculate the composition of that single bubble of vapor that has formed.

Similarly, the dew point is the point at which a very small amount of the vapor has condensed, so that *the gas phase composition remains the same as the overall composition,* and thus it is possible to calculate the composition of the single bubble of liquid.

Bubble Point

Recall that the dew point of a solution is the set of conditions (either a temperature at constant pressure or a pressure at constant temperature) at which the first drops of a vapor mixture begin to condense.

Let us first consider how to calculate the bubble point (at a constant temperature) of a mixture of 2 components A and B, assuming that the mixture follows Raoult's Law under all conditions. To begin, write Raoult's Law for each component in the mixture.

$$y_A * P = x_A * P^*_A$$
$$y_B * P = x_B * P^*_B$$

We can add these two equations together to yield :

$$P(y_A + y_B) = P^*_A x_A + P^*_B x_B$$

Now since y_A and y_B are mole fractions and A and B are the only components of the mixture, $y_A + y_B = 1$. In addition, recall that since we are considering the bubble point, the liquid composition is essentially equal to the overall composition. Therefore, $x_A = z_A$ and $x_B = z_B$.

Note : This is only true at the bubble point, not in general.

Hence we have the following equation valid at the bubble point for an ideal equilibrium :

Bubble Point Equation for Two Components under Raoult's Law

$$P = z_A * P^*_A + z_B * P^*_B$$

Therefore, if the temperature and overall composition are known, the bubble pressure can be determined directly.

If the pressure is held constant and the bubble point temperature is required, it is necessary to calculate the temperature by an iterative method. The temperature dependence is contained in the Antoine equation for vapor pressure of each component. One method to solve for the temperature is to :

1. Guess a temperature
2. Use the guess and the Antoine equation to calculate the vapor pressure of each component in the mixture.
3. Calculate an equilibrium pressure using the bubble point pressure equation.
4. If the calculated pressure does not equal the known pressure, it is necessary to change the temperature and try again.

This process is ideally suited to spreadsheet functions such as Excel's "goal-seek" routine. An example calculation will be shown in the next section.

If there is more than one component, a similar derivation yields the following :

Bubble Point Equation for Multiple Components Under Raoult's Law

$$P = \Sigma(z_i * P_i^*) \text{ (summed over all components i)}$$

Dew Point

The Dew Point calculation is similar, although the equation that results from the derivation is somewhat more complex. The starting point is the same : assume that Raoult's Law applies to each component.

$$y_A * P = x_A * P_A^*$$
$$y_B * P = x_B * P_B^*$$

Now we want to eliminate the liquid compositions in a similar manner to how we eliminated the vapor compositions in the previous derivation. To do this we need to divide by the vapor pressures :

$$\frac{y_A * P}{P_A^*} = x_A$$

$$\frac{y_B * P}{P_B^*} = x_B$$

Adding the equations and recalling that $x_A + x_B = 1$, we have :

$$\frac{y_A * P}{P_A^*} + \frac{y_B * P}{P_B^*} = 1.$$

Since this is the dew point, the gas-phase composition is essentially the overall composition, and therefore we have the following dew point equation :

Dew Point Equation for Two Components Under Raoult's La

$$\frac{1}{P} = \frac{z_A}{P_A^*} + \frac{z_B}{P_B^*}$$

Note : This is only valid at the dew point, just as the other equation was only valid at the bubble point.

For multiple components, the equation is similarly :

Dew Point Equation for Multiple Components Under Raoult's Law

$$\frac{1}{P} = \Sigma(\frac{z_A}{P_A^*})$$

Phase Diagrams Resulting from Raoult's Law

By holding one variable constant, varying a second, and calculating the other two, it is possible to calculate a phase diagram from Raoult's Law. Typical Pxy and Txy diagrams derived from Raoult's Law were shown in the previous section for the benzene-toluene system.

Diagrams for systems that follow Raoult's Law are relatively "nice"; it can be shown that they will never have azeotropes, which would be indicated by intersection of the bubble and dew point lines. In addition, since only one parameter in the equation depends on the temperature (the vapor pressure) and the pressure dependence is explicit, the dew and bubble point lines are relatively easy to calculate.

Non-ideal VLE

Deviations from Raoult's Law occur because not all solutions are ideal, nor are all gas mixtures. Therefore, methods have been developed in order to take these nonidealities into account.

Henry's Law

The third non-ideal method, Henry's Law, is especially useful for dilute solutions, and states that at very low concentrations, the partial pressure of the dilute component over a liquid mixture is proportional to the concentration :

Henry's Law

For a dilute component A, $P_A = H_A * x_A$ where H_A is a constant and x_A is the liquid-phase mole fraction of A.

This law is very similar to Raoult's Law, except that the proportionality constant is not the pure-component vapor pressure but is empirically determined from VLE data. Like the pure-component vapor pressure, the Henry's constant is dependent on temperature and the nature of component A. Unlike the pure-component vapor pressure, *it also depends on the solvent*, so when utilizing tables of Henry's constants, make sure that the solvents match.

Note : If Henry's Law applies to one component of a two-component mixture, the other component is often concentrated enough for Raoult's Law to apply to a reasonable approximation. Therefore, for a mixture of components A and B, where A is dilute and B is concentrated, a system similar to the following is common :

$$y_A * P = x_A * H_A$$

$$y_B * P = x_B * P_B^*$$

Henry's Law constants are generally very small, and are most useful when the concentration is less than 10% (depending on how accurate you want it, the concentration may need to be less than this).

Excess Gibbs Energy

The other two commonly-used correction parameters, the activity coefficient and the fugacity coefficient, are based on how non-ideal a given phase is. For a gas, the degree of non-ideality present is called the residual Gibbs energy, while for a liquid it is called excess Gibbs energy. The distinction is made because the Gibbs energy of an ideal gas and the Gibbs energy of an ideal solution are very different, as are the natures of how real solutions and real gasses deviate from ideality.

The residual Gibbs energy is based on the ideal gas and is defined as follows :

Residual Gibbs Energy definition (for a gas phase)

$$G^R = G_{real} - G_{ideal\ gas}$$

The excess Gibbs energy of a *liquid* phase is based on an ideal solution and is defined as :

Excess Gibbs Energy definition (for a liquid phase)

$$G^E = G_{real} - G_{ideal\ solution}$$

Activity Coefficients

The activity coefficient takes into account variation from Raoult's Law due to liquid excess Gibbs energy. It may be defined as :

$$\ln \gamma_i = \frac{G_i^E}{RT}$$

where γ is a composition-dependent value which is also different for each component. It therefore is a measure of the effect of each component in contributing to the nonideality of the mixture.

Raoult's Law can be extended using activity coefficients as follows :

EXTENDED RAOULT'S LAW

$$y_A * P = x_A * P_A^* * \gamma_i$$

The extended Raoult's law may be used to calculate activity coefficients : the vapor pressure and equilibrium composition are measured at a low pressure (to avoid gaseous nonideality). Then, since the activity coefficient is only weakly dependent on pressure (liquid properties often change very little with pressure), the same values of the activity coefficient may be used at higher pressures to aid in determining the change in equilibrium properties.

Once activity coefficients are determined at a wide variety of concentrations, it is often desired to condense the information into one equation. See this publication for an interesting read on this topic, though it will probably make more sense after you take thermodynamics, it offers a good description of what you will see.

Note : The definition of the activity coefficient implies that an ideal solution will have an activity coefficient equal to 1 (since its excess Gibbs energy is 0). Thus for an ideal solution the equilibrium equation reduces back to Raoult's law.

Fugacity Coefficients

The fugacity coefficient of a gas is defined in a similar way to the activity coefficient, but it is based on the *residual* Gibbs energy :

$$\ln \phi_i = \frac{G_i^R}{RT}$$

The fugacity coefficient of a gas depends on temperature, as can be seen clearly from the definition. It also depends heavily on the pressure. Indeed, if you have data available that relates the compressibility of a pure gas, Z, as a function of pressure at constant temperature, the fugacity can be computed using calculus or estimated (roughly) using the following equation if the change in pressure between each set of points is constant :

$$\ln \phi_i = \Delta P \Sigma (\frac{Z_k - 1}{P_k})$$

where "k" referring to a data point

Note : Since the compressibility (Z_k) of an ideal gas is 1 regardless of what the pressure is, the fugacity coefficient of an ideal gas is 1 as well. Therefore, like the activity coefficient, the fugacity coefficient provides us with a measure of how nonideal a given gas or mixture of gasses is.

To do this calculation, it is necessary to extrapolate so that the first data point is taken at P = 0..

Example :

Given the following data :

P (atm) Z

0.1 0.98

0.2 0.96

0.3 0.95

Calculate the fugacity coefficient at 0.2 atm and 0.3 atm.

Solution : It is necessary to first extrapolate to zero pressure :

$P = 0 \rightarrow Z\ 1.0$

Then insert the data into the formula :

$$\ln \phi_i = (0.2 - 0.1) * (\frac{0.98 - 1}{0.1} + \frac{0.96 - 1}{0.2})$$

Raoult's law can be modified to account for non-ideal gasses in a similar way to its modification for non-ideal liquids :

VLE Equation for non-ideal gasses and non-ideal liquids

$$\phi_i * y_i * P = \gamma_i * x_i * P_i^*$$

GENERALIZED CORRELATIONS

Critical Constants

At room temperature (about 298K), it is possible to add enough pressure to carbon dioxide to get it to liquify (some fire extinguishers work by keeping liquid carbon dioxide in them under very high pressure, which rapidly vaporizes when the pressure is relieved . However, if the temperature is raised to higher than 304.2 K, it will be *impossible* to keep carbon dioxide in a liquid form, because it has too much kinetic energy to remain in the liquid phase. No amount of pressure can turn carbon dioxide into a liquid if the temperature is too high.

This threshold temperature is called a critical temperature. Any pure stable substance (not just carbon dioxide) will have a single characteristic critical temperature. Pure stable substances will also have a single characteristic critical pressure, which is the pressure needed to achieve a phase transition at the critical temperature, and a critical specific volume which is the specific volume (volume per mass) of the fluid at this temperature and pressure.

Critical pressures are typically extremely large, ranging from 2.26 atm for helium to 218.3 atm for water, and about 40 atm on average. Critical temperatures typically range from 5.26 K (for helium) to the high 600s K for some aromatic compounds.

A substance which is at a temperature higher than the critical temperature and a pressure higher than its critical pressure is called a supercritical fluid. Supercritical fluids have some properties in common with gasses and some in common with liquid, as may be expected since it they are not observed to be liquid but would be expected to be liquefied at extreme pressures.

Law of Corresponding States

Recall from the last section that the *compressibility* of any substance (but most useful for gasses) is defined as :

$$Z = \frac{P * \hat{V}}{RT}$$

The compressibility of a gas is a measure of how non-ideal it is; an ideal gas has a compressibility of 1. At the critical point, in particular, the compressibility is :

$$Z_C = \frac{P_c * \hat{V}_c}{R * T_c}$$

Critical constants are important because it has been found experimentally that the following rule is true for many substances :

The Law of Corresponding States

Many substances behave in similar manners to each other depending on how far the system conditions are from the critical temperature and pressure of the substance. In particular, the compressibility of a substance is strongly correlated to its variance from the critical conditions.

It has been found experimentally that many substances have very similar compressibility at their critical point. . Most nonpolar substances in particular have a critical compressibility of about 0.27. The similarity of the critical compressibility between substances is what gives some weight to the law of corresponding states. However, the fact that the critical compressibility is not exactly the same for all substances leads to potential estimation errors if this method is used.

The critical constants are able to effectively predict the properties of a substance without gathering a large amount of data. However, it is necessary to define *how* the properties of the substance change as the system variables become closer to or farther from the critical point of the substance. These methods are discussed in the following sections.

Compressibility Charts

Recall that many substances have similar critical compressibility values near 0.27. Therefore, charts have been developed which relate compressibility at other conditions to those at the critical point. In order to use these charts, the system parameters are *normalized* by dividing by the critical constants to yield reduced temperature, pressure, and volume :

Reduced Parameters

$$T_r = \frac{T}{T_c}, P_r = \frac{P}{P_c}, \hat{V}_r = \frac{\hat{V}}{\hat{V}_c}$$

Chapter 6

ENERGY BALANCES ON NON-REACTING SYSTEMS

STEADY STATE ENERGY BALANCE

General Balance Equation Revisited

Recall the general balance equation that was derived for *any* system property :

In − Out + Generation − Consumption = Accumulation

When we derived the mass balance, we did so by citing the law of conservation of mass, which states that the total generation of mass is 0, and therefore *Accumulation = In − Out.*

There is one other major conservation law which provides an additional equation we can use : the law of conservation of energy. This states that if E denotes the entire amount of energy in the system,

Law of Conservation of Energy

$$E_{in} - E_{out} = E_{accumulated}$$

Types of Energy

In order to write an energy balance, we need to know what kinds of energy can enter or leave a system. Here are some examples (this is not an exhaustive list by any means) of the types of energy that can be gained or lost.

1. A system could gain or lose *kinetic energy*, if we're analyzing a moving system.
2. Again, if the system is moving, there could be *potential energy* changes.
3. *Heat* could enter the system via conduction, convection, or radiation.
4. *Work* (either **expansion work** or **shaft work**) could be done on, or by, the system.

The total amount of energy entering the system is the sum of all of the different types entering the system. Here are the expressions for the different types of energy :

1. From physics, recall that $KE = \frac{1}{2}mv^2$. If the *system* itself is not moving, this is zero.

2. The *gravitational* potential energy of a system is $GPE = mgh$ where g is the gravitational constant, m is mass in kg and h is the height of the center of mass of the system. If the system does not change height, there is no change in GPE.

3. The heat entering the system is denoted by Q, regardless of the mechanism by which it enters (the means of calculating this will be discussed in a course on transport phenomenon). According to this book's conventions, *heat entering a system is positive and heat leaving a system is negative*, because the system in effect gains energy when heat enters.

4. The work done by or on the system is denoted by W. *Work done BY a system is negative* because the system has to "give up" energy to do work on its surroundings. For example, if a system expands, it loses energy to account for that expansion. Conversely, *work done ON a system is positive*.

Energy Flows due to Mass Flows

Accumulation of *anything* is 0 at steady state, and energy is no exception. If, as we have the entire time, we assume that the system is at steady state, we obtain the energy balance equation :

$E_{in} = E_{out}$

This is the starting point for all of the energy balances below.

Consider a system in which a mass, such as water, enters a system, such as a cup, like so :

The mass flow into (or out of) the system carries a certain amount of energy, associated with how fast it is moving (kinetic energy), how high off the ground it is (potential energy), and its temperature (internal energy). It is possible for it to have other types of energy as well, but for now let's assume that these are the only three types of energy that are important. If this is true, then we can say that the total energy carried *in the flow itself* is :

$$\dot{E}_i = (\frac{1}{2}\dot{m}v^2 + \dot{m}gh + \dot{U})_i$$

However, there is one additional factor that must be taken into account. When a mass stream flows into a system it expands or contracts and therefore performs work on the system. An expression for work due to this expansion is :

$$W_{exp} = P * \dot{V}_i$$

Since this work is done *on* the system, it enters the energy balance as a positive quantity. Therefore the total energy flow into the system due to mass flow is as follows :

$$\dot{E}_i = (\frac{1}{2}\dot{m}v^2 + \dot{m}gh + \dot{U})_i + P * \dot{V}_i$$

Now, to simplify the math a little bit, we generally don't use internal energy and the PV term. Instead, we combine these terms and call the result the enthalpy of the stream. Enthalpy is just the combination of internal energy and expansion work due to the stream's flow, and is denoted by the letter H :

Definition of enthalpy

$H = U + PV$

Therefore, we obtain the following important equation for energy flow carried by mass :

In stream i, if only KE, GPE, internal energy, and expansion work are considered, the energy carried by mass flow is :

$$\dot{E}_i = (\frac{1}{2}\dot{m}v^2 + \dot{m}gh + \dot{H})_i$$

Note : Kinetic energy and potential energy are generally very small compared to the enthalpy, except in cases of very rapid flow or when there are no significant temperature changes occurring in the system. Therefore, they are often neglected when performing energy balances.

Other Energy Flows into and Out of the System

The other types of energy flows that could occur in and out of a system are *heat* and *work*. Heat is defined as energy flow due to a change in temperature, and always flows from higher temperature to lower temperature. Work is defined as an energy transferred by a force (see here for details).

- If there is no heat flow into or out of a system, it is referred to as **adiabatic**.
- If there are no mechanical parts connected to a system, and the system is not able to expand, then the work is essentially 0.

Some systems which have mechanical parts that perform work are turbines, mixers, engines, stirred tank reactors, agitators, and many others. The type of work performed by these parts is called shaft work to distinguish it from work due to expansion of the system itself (which is called *expansion work*).

An "insulated system" is generally interpreted as being essentially adiabatic, though how good this assumption is depends on the quality of the insulation. A system that cannot expand is sometimes described as "rigid".

The notation for these values are as follows :

- Heat flows : $\dot{Q}_j$, at the "j"th location.
- Shaft work : $\dot{W}_s$
- Expansion work : $P * \dfrac{\Delta V}{\Delta t}$

Note that the above implies that there is no expansion work at steady state because at steady state nothing about the system, including the volume, changes with time, *i.e.* $\dfrac{\Delta V}{\Delta t} = 0$ at steady state.

Overall Steady - State Energy Balance

If we combine all of these components together, remembering that heat flow into a system and work done *on* a system are positive, we obtain the following :

Steady State Energy Balance on an Open System

$$\Sigma(\frac{1}{2}\dot{m}v^2 + \dot{m}gh + \dot{H})_{i,in} - \Sigma(\frac{1}{2}\dot{m}v^2 + \dot{m}gh + \dot{H})_{i,out} + \Sigma\dot{Q}_j + \dot{W}_s = 0$$

Some important points :

1. If the system is **closed AND at steady state** that means the total heat flow must equal the total work done in magnitude, and be opposite in sign. However, according to another law of thermodynamics, the second law, it is impossible to change ALL of the heat flow into work, even in the most ideal case.

2. In an adiabatic system with no work done, the total amount of energy carried by mass flows is equal between those flowing in and those flowing out. However, that DOES NOT imply that the temperature remains

the same, as we will see in a later section. Some substances have a greater capacity to hold heat than others, hence the term **heat capacity**.

3.   If the conditions *inside the system* change over time, then we CANNOT use this form of the energy balance. The next section has information on what to do in the case that the energetics of the system change.

UNSTEADY STATE ENERGY AND MASS BALANCES

What is Accumulation?

Recall that so far in this text it has been assumed that all systems are at steady state, which means that there is no buildup of mass, energy, or other conserved quantities. However, there are many situations, such as whenever operating levels change, that a system will not be at steady state, and mass and energy will be accumulated over time.

The most important thing to remember about accumulation is that *it deals with the actual amount of stuff in the system, not any sort of flow rate*. If you remember if you're dealing with actual system properties rather than flow rates, it will help keep the terms straight in unsteady-state balances.

Unsteady-state Mass Balance

Lets begin the derivation of an unsteady-state mass balance with the general balance equation which you should know and love by now :

In − Out + Generation = Accumulation

Substituting the terms we usually used for in, out, and generation, we obtain :

$\Sigma \dot{m}_{i,in} - \Sigma \dot{m}_{i,out} + \dot{m}_{i,gen} = Accumulation$

Now we have to come up with a mathematical formulation for the accumulation. Unlike all of the other terms in this equation, which deal with energy *flows* into the system, the accumulation deals with the amount of energy that is already in the system at a certain point of time, and more specifically how it *changes* with time.

The rate of accumulation of energy *will not be constant* unless it is zero (otherwise every reactor in the world would either blow up from excessive mass and energy buildup or would cease operating because all of the reactants and products would be drained out). Recall that if the accumulation reaches zero, the system is at steady state. Most systems tend to move towards a steady state (it is possible to have more than one set of steady state conditions, but it won't be covered here) over long periods of time, as shown below :

Such a system is called self-regulating (or naturally stable). If a system is not self-regulating then special control techniques (see Control systems) must be utilized to force the system into a steady state. In order to take into account variation in the accumulation rate, we must consider the rate of change over a *very small* amount of time, so small in fact that it is practically zero, and the accumulation vs. time curve resembles a straight line. The slope of this line at time t is approximately :

Slope = Accumulation rate at time t = $\dfrac{M_{sys,t+\Delta t} - M_{sys,t}}{\Delta t}$

Therefore we could write the following :

Accumulation = $\dfrac{M_{sys,t+\Delta t} - M_{sys,t}}{\Delta t}$

We then write our mass balance by substituting this accumulation into the mass balance above :

$$\frac{M_{sys,t+\Delta t} - M_{sys,t}}{\Delta t} = \Sigma \dot{m}_{i,in} - \Sigma \dot{m}_{i,out}$$

For practical applications, this equation is generally multiplied by Δt. Then, rather than dealing with flow rates, a new quantity is defined :

$$\Delta m_i = \dot{m}_i * \Delta t$$

This quantity is the *total amount of mass that enters the system in a finite amount of time*. Substituting this definition into the mass balance yields the following :

Unsteady State Mass Balance

$$\Sigma \Delta m_{i,in} - \Sigma \Delta m_{i,out} + m_{i,gen} = M_{sys,t+\Delta t} - M_{sys,t}$$

Example :

A feed stream with $50\dfrac{kg}{h}$ of water and $1\dfrac{kg}{h}$ of ethanol enters a distillation column. A distillation column generally has two outlet streams called the bottoms and the condensate. At steady state, the condensate is 12% ethanol by mass and the total condensate flow rate is $9\dfrac{kg}{h}$.

One day, the boss calls and says that she needs more production, so you turn up the feed to $60\dfrac{kg}{h}$. Two hours later, the distillation column floods.

a. What was the cause of the flooding?

b. Assuming that the total outlet mass flow rates remained the same throughout the process, what was the total mass accumulation in the column?

c. Describe two methods by which the flow rates may be modified to reach a new steady state. Will the new steady state produce the same outlet concentrations as the old steady state? Explain. (hint : how is the separation effectiveness related to the ratio of the two outlet flowrates? You may need to do some research on this)

Unsteady-state Energy Balance

Lets start by examining what we have so far, but with the accumulation term (yet to be defined mathematically) added in the right side, since we're not at steady state any more :

$$\Sigma(\frac{1}{2}\dot{m}v^2 + \dot{m}gh + \dot{H})_{i,in} - \Sigma(\frac{1}{2}\dot{m}v^2 + \dot{m}gh + \dot{H})_{i,out} + \Sigma\dot{Q}_j + \dot{W}_s = Accumulation$$

Following the logic from the mass balance, we obtain for the accumulation :

$$\frac{M_{sys,t+\Delta t} - E_{sys,t}}{\Delta t}$$

Therefore, we have :

$$\Sigma(\frac{1}{2}\dot{m}v^2 + \dot{m}gh + \dot{H})_{i,in} - \Sigma(\frac{1}{2}\dot{m}v^2 + \dot{m}gh + \dot{H})_{i,out} + \Sigma\dot{Q}_j + \dot{W}_s = \frac{M_{sys,t+\Delta t} - E_{sys,t}}{\Delta t}$$

Like in the case of the mass balance, we can only consider the total energy change over a total amount of time using this equation. To do this, we multiply the entire equation above by the time change from some starting point to the point of interest.

Now we need some definitions :

1. $Q = \dot{Q} * \Delta t$ is the TOTAL heat flow over the time period.

2. $W_s = \dot{W}_s * \Delta t$ is the TOTAL shaft work over the time period.

3. $\dot{m}_i * \Delta t = \Delta m_i$ is the TOTAL mass flow into (or out of) the system due to stream i during the time period.

4. $\dot{H}_i * \Delta t = H$ is the TOTAL enthalpy carried into (or out of) the system due to stream i during the time period.

The major assumption here is that the enthalpies, heat flow rates, and shaft work on the left hand side of the equals sign must either be constant, or the average value over the whole time period must be used, in order for this equation to be valid. Whether this assumption is valid or not depends on the situation (for example, it depends on whether the process feeding mass to your process is itself at steady state or not).

With these in mind, we multiply by delta t in order to obtain the following, unsteady state energy balance.

Unsteady State Energy Balance

$$\Sigma(\frac{1}{2}\Delta mv^2 + \Delta mgh + H)_{i,in} - \Sigma(\frac{1}{2}\Delta mv^2 + \Delta mgh + H)_{i,out} + \Sigma Q_j + W_s = M_{sys,t+\Delta t} - E_{sys,t}$$

Chapter 7

DESIGN REACTORS

In chemical engineering, chemical reactors are vessels designed to contain chemical reactions. One example is a pressure reactor. The design of a chemical reactor deals with multiple aspects of chemical engineering. Chemical engineers design reactors to maximize net present value for the given reaction. Designers ensure that the reaction proceeds with the highest efficiency towards the desired output product, producing the highest yield of product while requiring the least amount of money to purchase and operate. Normal operating expenses include energy input, energy removal, raw material costs, labor, *etc.* Energy changes can come in the form of heating or cooling, pumping to increase pressure, frictional pressure loss (such as pressure drop across a 90° elbow or an orifice plate) or agitation.

Chemical reaction engineering is the branch of chemical engineering which deals with chemical reactors and their design, especially by application of chemical kinetics to industrial systems.

OVERVIEW

There are a couple of main basic vessel types :

- A tank
- A pipe or tubular reactor (laminar flow reactor(LFR))

Both types can be used as continuous reactors or batch reactors, and either may accommodate one or more solids (reagents, catalyst, or inert materials), but the reagents and products are typically fluids. Most commonly, reactors are run at steady-state, but can also be operated in a transient state. When a reactor is first brought into operation (after maintenance or in operation) it would be considered to be in a transient state, where key process variables change with time.

There are three main basic models used to estimate the most important process variables of different chemical reactors :

- *batch reactor* model (batch),
- *continuous stirred-tank reactor* model (CSTR), and
- *plug flow reactor* model (PFR).

Furthermore, catalytic reactors require separate treatment, whether they are batch, CST, or PF reactors, as the many assumptions of the simpler models are not valid.

Key process variables include

- Residence time (τ, lower case Greek tau)
- Volume (V)
- Temperature (T)
- Pressure (P)
- Concentrations of chemical species (C_1, C_2, C_3, ... C_n)
- Heat transfer coefficients (h, U)

A chemical reactor, typically tubular reactor, could be a packed bed. The packing inside the bed may have catalyst to catalyze the chemical reaction. A chemical reactor may also be a fluidized bed; see Fluidized bed reactor.

Chemical reactions occurring in a reactor may be exothermic, meaning giving off heat, or endothermic, meaning absorbing heat. A chemical reactor vessel may have a cooling or heating jacket or cooling or heating coils (tubes) wrapped around the outside of its vessel wall to cool down or heat up the contents.

TYPES

CSTR (Continuous Stirred-Tank Reactor)

Checking condition inside the case of a continuous stirred tank reactor (CSTR). Note the impeller (or agitator) blades on the shaft for mixing. Also note the baffle at the bottom of the image which also helps in mixing.

In a CSTR, one or more fluid reagents are introduced into a tank reactor (typically) equipped with an impeller while the reactor effluent is removed. The impeller stirs the reagents to ensure proper mixing. Simply dividing the volume of the tank by the average volumetric flow rate through the tank gives the *residence time*, or the average amount of time a discrete quantity of reagent spends inside the tank. Using chemical kinetics, the reaction's expected percent completion can be calculated. Some important aspects of the CSTR :

- At steady-state, the mass flow rate in must equal the mass flow rate out, otherwise the tank will overflow or go empty (transient state). While the reactor is in a transient state the model equation must be derived from the differential mass and energy balances.

- The reaction proceeds at the reaction rate associated with the final (output) concentration, since the concentration is assumed to be homogenous throughout the reactor.

- Often, it is economically beneficial to operate several CSTRs in series. This allows, for example, the first CSTR to operate at a higher reagent concentration and therefore a higher reaction rate. In these cases, the sizes of the reactors may be varied in order to minimize the total capital investment required to implement the process.

- It can be demonstrated that an infinite number of infinitely small CSTRs operating in series would be equivalent to a PFR.

The behavior of a CSTR is often approximated or modeled by that of a Continuous Ideally Stirred-Tank Reactor (CISTR). All calculations performed with CISTRs assume perfect mixing. If the residence time is 5-10 times the mixing time, this approximation is considered valid for engineering purposes. The CISTR model is often used to simplify engineering calculations and can be used to describe research reactors. In practice it can only be approached, particularly in industrial size reactors in which the mixing time may be very large.

PFR (Plug Flow Reactor)

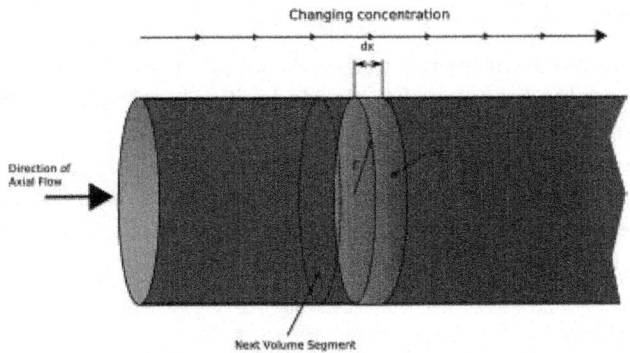

Simple diagram illustrating plug flow reactor model

In a PFR, one or more fluid reagents are pumped through a pipe or tube. The chemical reaction proceeds as the reagents travel through the PFR. In this type of reactor, the changing reaction rate creates a gradient with respect to distance traversed; at the inlet to the PFR the rate is very high, but as the concentrations of the reagents decrease and the concentration of the product(s) increases the reaction rate slows. Some important aspects of the PFR :

- All calculations performed with PFRs assume no upstream or downstream mixing, as implied by the term "plug flow".

- Reagents may be introduced into the PFR at locations in the reactor other than the inlet. In this way, a higher efficiency may be obtained, or the size and cost of the PFR may be reduced.

- A PFR typically has a higher efficiency than a CSTR of the same volume. That is, given the same space-time (or residence time), a reaction will typically proceed to a higher percentage completion in a PFR than in a CSTR. This is not always true for reversible reactions.

For most chemical reactions of industrial interest, it is impossible for the reaction to proceed to 100% completion. The rate of reaction decreases as the reactants are consumed until the point where the system reaches dynamic equilibrium (no net reaction, or change in chemical species occurs). The equilibrium point for most systems is less than 100% complete. For this reason a separation process, such as distillation, often follows a chemical reactor in order to separate any remaining reagents or byproducts from the desired product. These reagents may sometimes be reused at the beginning of the process, such as in the Haber process. In some cases, very large reactors would be necessary to approach equilibrium, and chemical engineers may choose to separate the partially reacted mixture and recycle the leftover reactants.

Continuous oscillatory baffled reactor (COBR) is a tubular plug flow reactor. The mixing in COBR is achieved by the combination of fluid oscillation and orifice baffles, allowing plug flow to be achieved under laminar flow conditions with the net flow Reynolds numberjust about 100.

Semi-batch Reactor

A semi-batch reactor is operated with both continuous and batch inputs and outputs. A fermenter, for example, is loaded with a batch of medium and microbes which constantly produce carbon dioxide that must removed continuously. Analogously, driving a reaction of gas with a liquid is usually difficult, since the gas bubbles off. Therefore, a continuous feed of gas is injected into the batch of a liquid. One chemical reactant is charged to the vessel and a second chemical is added slowly (for instance, to prevent side reactions).

Catalytic Reactor

Although catalytic reactors are often implemented as plug flow reactors, their analysis requires more complicated treatment. The rate of a catalytic reaction is proportional to the amount of catalyst the reagents contact, as well as the concentration of the reactants. With a solid phase catalyst and fluid phase reagents, this is proportional to the exposed area, efficiency of diffusion of reagents in and products out, and efficacy of mixing. Perfect mixing usually cannot be assumed. Furthermore, a catalytic reaction pathway often occurs in multiple steps with intermediates that are chemically bound to the catalyst; and as the chemical binding to the catalyst is also a chemical reaction, it may affect the kinetics. Catalytic reactions often display the so-called *falsified kinetics, i.e.* the apparent kinetics differ from elementary chemical kinetics due to physical transport effects.

The behavior of the catalyst is also a consideration. Particularly in high-temperature petrochemical processes, catalysts are deactivated by sintering, coking, and similar processes.

A common example of a catalytic reactor is the catalytic converter following an engine. However, most petrochemical reactors are catalytic, and are responsible for most of industrial chemical production in the world, with extremely high-volume examples such as sulfuric acid, ammonia, reformate/BTEX (benzene, toluene, ethylbenzene and xylene) and alkylate gasoline blending stock.

MEMBRANE REACTOR

A membrane reactor is a piece of chemical equipment that combines a catalyst-filled reaction chamber with a membrane to add reactants or remove products of the reaction.

Chemical reactors making use of membranes are usually referred to as membrane reactors. The membrane can be used for different tasks :

- Separation
- Selective extraction of reactants
- Retention of the catalyst
- Distribution/dosing of a reactant
- Catalyst support (often combined with distribution of reactants)

Membrane reactors are an example for the combination of two unit operations in one step *e.g.* membrane filtration with the chemical reaction.

Examples

Biological Systems

In biological systems membranes fulfil a number of essential functions. The compartmentalisation of biological cells is achieved by membranes. The semi-permeability allows to separate reactions and reaction environments. A number of enzymes are membrane bound and often mass transport through the membrane is active rather than passive as in artificial membranes allowing the cell to keep up gradients for example by using active transport of protons or water.

The use of a natural membrane is the first example of the utilisation for a chemical reaction. By using the selective permeability of apig's bladder water could be removed from a condensation reaction to shift the equilibrium position of the reaction towards the condensation products according to the principle of Le Châtelier.

SIZE EXCLUSION : ENZYME MEMBRANE REACTOR

As enzymes are macromolecules and often differ greatly in size from reactants they can be separated by size exclusion membrane filtration with ultra- or nanofiltration [artificial membranes]. This is used on industrial scale for the production of enantiopure amino acids by kinetic racemic resolution of chemically derived racemic amino acids. The most prominent example is the production of L-methionine on a scale of 400t/a. The advantage of this method over other forms of immobilisation of the catalyst is that the enzymes are not altered in activity or selectivity as it remains solubilised.

The principle can be applied to all macro-molecular catalysts which can be separated from the other reactants by means of filtration. So far, only enzymes have been used to a significant extent.

Reaction Combined with Pervaporation

In P. dense membranes are used for separation. For dense membranes the separation is governed by the difference of the chemical potential of the components in the membrane. The selectivity of the transport through the membrane is dependent by the difference insolubility of the materials in the membrane and their diffusivity through the membrane. For example for the selective removal of water by using lipophilic membranes. This can be used to overcome thermodynamic limitations of condensation *e.g.* esterification reactions by removing water.

Dosing : Partial oxidation of methane to methanol

In the STAR process for the catalytic conversion of from methane from natural gas and air oxygen to methanol by the partial oxidation $2CH_4 + O_2 \rightarrow 2CH_3OH$.

The partial pressure of oxygen has to be low to prevent the formation of explosive mixtures and to suppress the successive reaction tocarbon monoxide, carbon dioxide and water. This is achieved by using a tubular reactor with an oxygen-selective membrane. The membrane allows the uniform distribution of oxygen as the driving force for the permeation of oxygen through the membrane is the difference in partial pressures on the air side and the methane side.

Selective Removal : Hydrogen

A number of metal membranes are highly hydrogen selective at higher temperatures. Especially palladium and platinum can therefore be used for the production of highly purified hydrogen from steam reforming of gases. The equilibrium limited reaction gives :

$$CH_4 + H_2O \leftrightarrow 3H_2 + CO$$

$$CO + H_2O \leftrightarrow H_2 + CO_2 \text{ or}$$

$$CH_3OH + H_2O \leftrightarrow CO_2 + 3 H_2$$

Ultra pure hydrogen, generated from these reactions, is extracted by use of thin dense metallic membranes that are 100% selective to hydrogen. The mechanism of the transport is the separation of hydrogen into protons and electrons at the surface and recombination on the filtrate or raffinate side. Other high temperature membranes are being considered for hydrogen generation where the purity requirements are not as great; for example for clean coal power generation. Hydrogen, produced from coal gas in the membrane reactor would be used for power generation, while the carbon dioxide would remain at high pressure for carbon capture and storage.

An alternative application of membrane reactors, developed at University Laval was to convert methane into benzene by the following reaction :

$$6 CH_4 \rightarrow C_6H_6 + 9 H_2$$

As with the other reactions, hydrogen extraction drives the conversion forward, but for this reaction, the desired product is the benzene, and not the hydrogen.

Benefits

Making a gaseous product in a membrane reactor generally affects the way that pressure affects the extent of reaction at thermodynamic pseudo-equilibrium. In an ordinary flow reactor, the composition of the exhaust gas is determined by the composition of the feed gas and the extent of reaction. As a result, at pseudo-equilibrium, the extent of reaction is entirely determined by the feed composition and the exhaust equilibrium constant, the latter being determined by the temperature and pressure of the exhaust. In a membrane reactor, the partial pressure of the components at psueudo-equilibrium are not uniquely determined by the total pressure, exit temperature, and feed composition. There is also a significant (and beneficial) effect that derives from the controlled removal of a product or addition of reactant.

CHEMICAL DESIGN REACTOR

Chemical reactor design requires the integration of heat and mass transfer with thermodynamics and kinetics.

At bench-top scales (for example 1 g of reactants) reactions tend to be homogenous and isothermal. Simple thermodynamics and kinetics are normally sufficient to describe the performance of these chemical reactions.

As the scale is increased, temperature, concentration, pressure and phase can vary within different regions of the reactor and with time.

Bulk effects such as heat and mass transfer become important. Many chemical engineering processes favourable at the small scale are not viable at a larger scale because of heat and mass transfer issues.

The bulk effects of heat and mass transfer become more important as the scale of chemical reactions increases.

We have expertise in heat and mass transfer, and combine this with the thermodynamics and kinetics knowledge developed by other groups to design, construct and operate chemical reactors. We have particular experience in continuous gas-solid reactors such as fluidised beds.

Our group also has a strong understanding of how complex reactions may lead to operational challenges such as particle sintering, accretion and attrition due to materials behaviour under extreme operating conditions.

Our expertise in chemical reactor design is currently being used in projects such as :

- development of a continuous fluidised bed process for the production of titanium to replace the existing commercial batch process – see Cutting the cost of making titanium metal
- development of a novel reactor involving supersonic quenching to produce magnesium - see Direct carbothermal reduction of magnesium
- development of a circulating fluidised bed process for using a metal oxide to provide the oxygen necessary for fossil fuel combustion.

Reactor Design

Because they represent the heart of a chemical plant — in which high-value products are produced through chemical transformation — reactors are a crucial component, and their high performance must be ensured. Reaction engineers are concerned with each reactor's specific yield, selectivity, safety, environment, quality and purity — as well as the degree to which reactors support overall plant economic viability and optimal operational conditions.

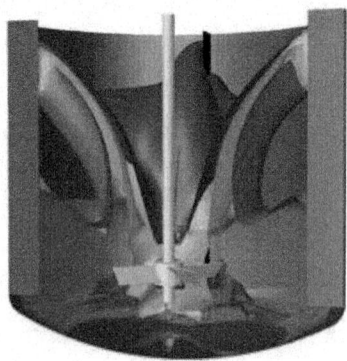

Contours of species mixing in a stirred tank reactor.

Engineering simulation software from ANSYS is a powerful, reliable tool that complements conventional reactor engineering practices. By bringing comprehensive multiphysics capabilities to reaction engineering, ANSYS has created a simulation platform that encompasses fluid mechanics, structural mechanics, impact and safety analyses, customized mixing tools, pressure vessel analysis, electromagnetic and control systems, and heat and mass transfer analyses. ANSYS solutions can model a diverse range of reactions, including gaseous and liquid, single-phase and multiphase, and homogenous and heterogeneous.

ANSYS simulation tools enable reaction engineers to answer what-if questions as they design and enhance chemical reactors. Users can optimize reactor performance by better understanding the effects and impacts of feed locations, vessel geometries and internals, vibrations, failures, dead spots, shear rates, resident time distributions, hot spots, and particle size distributions.

ANSYS software helps a wide range of customers to simulate and improve a wide variety of reactors and reaction types.

- Stiff chemistry
- Competing and parallel reactions
- Catalytic reactions
- Heterogeneous and homogeneous reactions
- Surface and volumetric reactions
- Laminar and turbulent flows
- Single-phase and multiphase reactions
- Fluidized bed reactions
- Multi-tube reactors
- Membrane reactors
- Microreactors
- Stirred tank reactors
- Fixed reactors
- Autoclave reactors
- Emulsion
- Hydrogenation
- Chlorination
- Polymerization
- Hydrocracking
- Crystallization and precipitation

The reactors, in which chemicals are made in industry, vary in size from a few cm^3 to the vast structures that are often depicted in photographs of industrial plants. For example, kilns that produce lime from limestone may be over 25 metres high and hold, at any one time, well over 400 tonnes of materials.

The design of the reactor is determined by many factors but of particular importance are the thermodynamics and kinetics of the chemical reactions being carried out.

The two main types of reactor are termed *batch* and *continuous*.

BATCH REACTORS

Batch reactors are used for most of the reactions carried out in a laboratory. The reactants are placed in a test-tube, flask or beaker. They are mixed together, often heated for the reaction to take place and are then cooled. The products are poured out and, if necessary, purified. This procedure is also carried out in industry, the key difference being one of size of reactor and the quantities of reactants.

- Following reaction, the reactor is cleaned ready for another batch of re-actants to be added.
- Batch reactors are usually used when a company wants to produce a range of products involving different reactants and reactor conditions. They can then use the same equipment for these reactions.

CONTINUOUS REACTORS

An alternative to a batch process is to feed the reactants continuously into the reactor at one point, allow the reaction to take place and withdraw the products at another point. There must be an equal flow rate of reactants and products. While *continuous reactors* are rarely used in the laboratory, a water-softener can beregarded as an example of a continuous process. Hard water from the mains is passed through a tube containing an ion-exchange resin. Reaction occurs down the tube and soft water pours out at the exit.

Continuous reactors are normally installed when large quantities of a chemical are being produced. It is important that the reactor can operate for several months without a shutdown.

The residence time in the reactor is controlled by the feed rate of reactants to the reactor. For example, if a reactor has a volume of 20 m^3 and the feed rate of reactants is 40 m^3 h^{-1} the residence time is 20 m^3 / 40 m^3 h^{-1} = 0.5 h. It is simple to control accurately the flow rate of reactants. The volume is fixed and therefore the residence time in the reactor is also well controlled.

The product tends to be of a more consistent quality from a continuous reactor because the reaction parameters (*e.g.* residence time, temperature and pressure) are better controlled than in batch operations.

They also produce less waste and require much lower storage of both raw materials and products resulting in a more efficient operation. Capital costs per tonne of product produced are consequently lower. The main disadvantage is their lack of flexibility as once the reactor has been built it is only in rare cases that it can be used to perform a different chemical reaction.

Types of Continuous Reactors

Industry uses several types of continuous reactors.

(a) Tubular Reactors

In a tubular reactor, fluids (gases and/or liquids) flow through it at high velocities. As the reactants flow, for example along a heated pipe, they are converted to products (Figure 4). At these high velocities, the products are unable to diffuse back and there is little or no back mixing. The conditions are referred to as plug flow. This reduces the occurrence of side reactions and increases the yield of the desired product.

With a constant flow rate, the conditions at any one point remain constant with time and changes in time of the reaction are measured in terms of the position along the length of the tube.

The reaction rate is faster at the pipe inlet because the concentration of reactants is at its highest and the reaction rate reduces as the reactants flow through the pipe due to the decrease in concentration of the reactant.

A further example of a fixed bed reactor is in catalytic reforming of naphtha to produce branched chain alkanes, cycloalkanes and aromatic hydrocarbons using usually platinum or a platinum-rhenium alloy on an alumina support.

(C) Fluid Bed Reactors

A fluid bed reactor is sometimes used whereby the catalyst particles, which are very fine, sit on a distributor plate. When the gaseous reactants pass through the distributor plate, the particles are carried with the gases forming a fluid (Figure 6). This ensures very good mixing of the reactants with the catalyst, with very high contact between the gaseous molecules and the catalyst and a good heat transfer. This results in a rapid reaction and a uniform mixture, reducing the variability of the process conditions.

One example of the use of fluid bed reactors is in the oxychlorination of ethene to chloroethene (vinyl chloride), the feedstock for the polymer poly(chloroethene) (PVC). The catalyst is copper(II) chloride and potassium chloride deposited on the surface of alumina. This support is so fine, it acts as a fluid when gases pass through it.

Loop reactors are used, for example, in the manufacture of poly(ethene) and the manufacture of poly(propene). Ethene (or propene) and the catalyst are mixed, under pressure, with a diluent, usually a hydrocarbon. A slurry is produced which is heated and circulated around the loops. Particles of the polymer gather at the bottom of one of the loop legs and, with some hydrocarbon diluent, are continuously released from the system. The diluent evaporates, leaving the solid polymer, and is then cooled to reform a liquid and passed back into the loop system, thus recirculating the hydrocarbon.

HEAT EXCHANGERS

Most chemical reactions are faster at higher temperatures and heat exchangers are frequently used to provide the heat necessary to increase the temperature of the reaction.

A common heat exchanger is the shell and tube type (Figures 12 and 13) where one part of the process flows through a tube and the other part around the shell.

A good example where heat exchange is important is in the manufacture of sulfur trioxide from sulfur dioxide in the Contact Process where the excess heat is used to warm incoming gases.

The heat from the reaction is transferred to incoming gases across the tube wall (Figure 12) and the rate of heat transfer is proportional to :

i.   The temperature difference between the hot gases and the incoming gases and

ii. The total surface area of the tubes

Chapter 8

BIOREACTORS AND NONLINEAR SYSTEM

INTRODUCTION

A bioreactor may refer to any manufactured or engineered device or system that supports a biologically active environment. In one case, a bioreactor is a vessel in which a chemical process is carried out which involves organisms or biochemically active substances derived from such organisms. This process can either be aerobic or anaerobic. These bioreactors are commonly cylindrical, ranging in size from litres to cubic metres, and are often made of stainless steel.

A bioreactor may also refer to a device or system meant to grow cells or tissues in the context of cell culture. These devices are being developed for use in tissue engineering or biochemical engineering.

On the basis of mode of operation, a bioreactor may be classified as batch, fed batch or continuous (*e.g.* a continuous stirred-tank reactor model). An example of a continuous bioreactor is the chemostat.

Organisms growing in bioreactors may be suspended or immobilized. Immobilization is a general term describing a wide variety of the cell or the particle attachment or entrapment. It can be applied to basically all types of biocatalysts including enzymes, cellular organelles, animal and plant cells.

Large scale immobilized cell bioreactors are :

- Moving media, also known as Moving Bed Biofilm Reactor (MBBR)
- Packed bed
- Fibrous bed
- Membrane.

BIOREACTOR DESIGN

Bioreactor design is a relatively complex engineering task, which is studied in the discipline of biochemical engineering. Under optimum conditions, the microorganisms or cells are able to perform their desired function with a 100 percent rate of success. The bioreactor's environmental conditions like gas (*i.e.*, air, oxygen,nitrogen, carbon dioxide) flow rates, temperature, pH and dissolved oxygen levels, and agitation speed/circulation rate need to be closely monitored and controlled.

Most industrial bioreactor manufacturers use vessels, sensors and a control systemnetworked together.

Fouling can harm the overall sterility and efficiency of the bioreactor, especially the heat exchangers. To avoid it, the bioreactor must be easily cleaned and as smooth as possible (hence the round shape)..

A heat exchanger is needed to maintain the bioprocess at a constant temperature. Biological fermentation is a major source of heat, therefore in most cases bioreactors need refrigeration. They can be refrigerated with an external jacket or, for very large vessels, with internal coils.

In an aerobic process, optimal oxygen transfer is perhaps the most difficult task to accomplish. Oxygen is poorly soluble in water — even less in fermentation broths — and is relatively scarce in air (20.95%). Oxygen transfer is usually helped by agitation, which is also needed to mix nutrients and to keep the fermentation homogeneous. There are, however, limits to the speed of agitation, due both to high power consumption (which is proportional to the cube of the speed of the electric motor) and to the damage to organisms caused by excessive tip speed. In practice, bioreactors are often pressurized; this increases the solubility of oxygen in water.

Photobioreactor

Moss photobioreactor with *Physcomitrella patens*

A photobioreactor (PBR) is a bioreactor which incorporates some type of light source. Virtually anytranslucent container could be called a PBR, however the term is more commonly used to define a closed system, as opposed to an open tank or pond. Photobioreactors are used to grow small phototrophic organisms such as cyanobacteria, algae, or moss plants. These organisms use light through photosynthesis as their energy source and do not require sugars or lipids as energy source. Consequently, risk of contamination with other organisms like bacteria or fungi is lower in photobioreactors when compared to bioreactors for heterotroph organisms.

Sewage Treatment

Bioreactors are also designed to treat sewage and wastewater. In the most efficient of these systems, there is a supply of a free-flowing, chemically inert medium which acts as a receptacle for the bacteria that break down the raw sewage. Examples of these bioreactors often have separate, sequential tanks and a mechanical separator or cyclone to speed the separation of water and biosolids. Aerators supply oxygen to the sewage and medium, further accelerating breakdown. Submersible mixers provide agitation in anoxic bioreactors to keep the solids in suspension and thereby ensure that the bacteria and the organic materials "meet". In the process, the liquid's Biochemical Oxygen Demand (BOD) is reduced sufficiently to render the contaminated water fit for reuse. The biosolids can be collected for further processing, or dried and used as fertilizer. An extremely simple version of a sewage bioreactor is a septic tank whereby the sewage is left in situ, with or without additional media to house bacteria. In this instance, the biosludge itself is the primary host (activated sludge) for the bacteria. Septic systems are best suited where there is sufficient landmass, and the system is not subject to flooding or overly saturated ground, and where time and efficiency are not prioritized.

Because they are the engine that drives biological wastewater treatment, it is critical to closely monitor the quantity and quality of microorganisms in bioreactors. One method for this is via 2nd Generation ATP tests.

NASA Tissue Cloning Bioreactor

In bioreactors in which the goal is to grow cells or tissues for experimental or therapeutic purposes, the design is significantly different from industrial bioreactors. Many cells and tissues, especially mammalian ones, must have a surface or other structural support in order to grow, and agitated environments are often destructive to these cell types and tissues. Higher organisms, being auxotrophic, also require highly specialized growth media.

NASA has developed a new type of bioreactor that artificially grows tissue in cell cultures. NASA's tissue bioreactor can grow heart tissue, skeletal tissue, ligaments, cancer tissue for study, and other types of tissue.

For more information on artificial tissue culture, see tissue engineering.

DIFFERENT TYPES OF BIOREACTOR

Bioreactor : Bioreactor is a vessel, which is used to carry out one or more biochemical reactions to convert raw materials to products through the action of biocatalyst, enzyme microorganisms, cells of animal or plants. The raw material could be an organic compound *e.g* sugar or an inorganic chemical *e.g.,* CO_2 or even complex material such as meat, animal manure or waste stream. The product of conversion may be biomass (*e.g,* Bakers yeast growth associated primary metabolites (*e.g.* Ethanol, Citric acid) or non growth associated metabolites (Antibiotics, bioactive compounds for plants) *etc.* These products may be extra cellular or intracellular as well. A large number of bioreactor designs are therefore needed to accommodate great diversity of substrate product and biocatalyst. Cultivation of the cells or biocatalyst is done in a perfectly mixed (submerged) mode or non mixed mode (surface) or via solid state cultivation.

Stirred Tank Reactors: These are the vessels with four baffles, round sparger for aeration (if required) and contain flat blade turbine impeller for agitation. These reactors are operated with the modified designs of impellers for the cultivation of shear sensitive plant or animal cells. Such typical impeller designs include setric impeller helical impeller & marine impeller *etc.*

Bubble Column Reactor : This type of reactor consists of long cylindrical column with or without horizontal perforated plates. Aeration and agitation is effected by sparging gas from the base via perforated pipe/plates or sintered glass. These reactors are particularly useful for hairy root culture of the plant cells.

Air Lift Reactor: In an air lift reactors the fluid volume is divided by providing a draft tube in the reactor. Sparging is done either inside or outside the draft tube. The sparger zone (having less density) is known as riser and the zone receiving no gas (having high relative density) is known as down comer. This density difference between the two zones leads to circulation of broth and keep the heterogeneous cell mass under homogenous conditions. No power is consumed for agitation in this type of the reactor and so they are regarded as highly energy efficient system. These reactors are suitable for bacteria yeast fungi and particularly for animal and / or plant cells however, these can be not be applied to viscous systems.

Fluidized Bed Reactor : In this an up flowing liquid stream is used to suspend or fluidize dense solid particles. It is more or less equivalent to bubble column except that cross section area is expanded near the top to reduce super facial velocity of fluidizing liquid to a value below that needed to keep the solids or gas to produce gas – liquid – solid fluidized bed Typical application of these reactor is in waste treatment.

Packed Bed Bioreactors : A bed of solid particles usually with compressing walls constitute packed bed. Biocatalyst is supported in porous or non porous bed. The particle may be randomly shaped wood chips or rocks or sphere cubes Fluid comprising of dissolved nutrient and substrate flows through the solid bed. The Flow rate and in term the residence time of substrate is manipulated to increase or decrease substrate contact with the bed (microorganisms).

Photo Bioreactors : Photobioreactors are used for photosynthetic culture of cyano bacteria, micro algae and plants. Due to high dependence on light, these types of reactors should have high surface to volume ratio and this greatly affects the design of the bioreactor. These reactors are generally open to atmosphere *e.g.* ponds, lagoon *etc.* In ponds and lagoons plastic lined channels are used. The culture is mixed and circulated by a paddle wheel. Closed photo bioreactors are also in use now which are made up of glass / plastic tubing which are arranged in a ladder configuration. Instead of tubes a conventional vessel with low surface to volume ratio may be illuminated by optical fibers to convey light inside from an external source.

Rotary Drum Reactors : In this, the drum is filled with 40% of its volume and rotated by means of rollers. It is particularly suitable for the cultivation of the plant cell cultures.

Mist Bioreactor : This is suitable for hairy root cultivation of plant cells. Static root mass is contained in a chamber that is mostly empty. Nutrients are supplied as mist of fine droplets suspended in circulating air currents that penetrates the spaces between the roots.

Solid State Reactor : The substrates of solid state fermentation are solid having little or no water. Small particles with large surface to volume ratio are preferred. Small heaps may be prepared to do this effectively. Deep bed requires forced aeration with moistened air. Aeration may vary widely. Occasional turning and mixing improves oxygen transfer. Solid state process generally use mixed culture. The product formed is extracted from the bed by solvent extraction.

Bioreactors are used for carrying out biochemical processes which employ microbes, fungus, plant cells or mammalian cell systems for production of biological products. The bioreactors provide a controlled environment for the production of metabolites which can help to achieve the optimal growth of microbes. The term fermentor is used as synonym to bioreactors.

CLASSIFICATION OF BIOREACTORS

There are numerous types of bioreactors - batch, sequence, continuously stirred tanks, anaerobic contact processes, anaerobic filters, *etc.*

1. They can be conveniently classified into three major types based on the presence or absence of oxygen and requirement of stirring.
   - Non stirred non aerated bioreactors are used for production of traditional products such as wine, beer, cheese *etc.*
   - Non stirred aerated reactors are used much rarely.
   - Stirred and aerated reactors are most often used for production of metabolites which require growth of microbes which require oxygen. Most of the newer methods are based on this type of bioreactors.
2. Based on mode of operation, the bioreactors can be classified into three types.
   - Batch reactors
   - Fed batch
   - Continuous *e.g.* : chemo stat
3. Based on the method of growing of microbes, bioreactors can be either
   - Suspended or
   - Immobilized

The Petri dish is the simplest immobilized bioreactor. The large scale immobilized bioreactors are used for commercial manufacturing of metabolites. They include

- Moving bed

- Fibrous bed

- Packed bed

- Membrane

4.   On the basis of the microbial agent used, the bioreactors can be classified into
   • Those based on living cells
   • Which employ enzymes

5.   Based on the process requirements, bioreactors can be classified into
   a. Aerobic
   b. Anaerobic
   c. Solid state
   d. Immobilized

I. Aerobic Fermentation

These reactors should have adequate provisions for supply of sterile air and also need a mechanism of stirring up and mixing the medium and cells. These can be either :

a. Stirred tank or

b. Air lift type

Generally, they are either closed type or batch reactors. Some special cases use continuous flow reactors also.

1. Stirred Tank Bioreactor

This is the conventional mixing reactor which is made of either glass or stainless steel. The stirrer can be either at the top or bottom of the reactor. The dimensions of the reactor depend on the amount of heat to be removed from the vessel. Baffles in the centre of the tank prevent formation of vortex and effective mixing of the ingredients.

Advantages

• Low investment needs

• Low operating costs

Disadvantages

• Foaming is often a problem. But this can be overcome using proper antifoaming agents. However, this has to be exercised with caution since some antifoaming agents inhibit the growth of microbes.

2. Air Lift Bioreactors

The stirred tank bioreactors lack well defined flow of air. In these, air is pumped from below. This creates the bubbles in the medium which rises up through the draught tube by buoyancy and drags the surrounding fluid up. The air that is used to lift up is sufficient to stir up the contents.

Advantages

- Low friction
- Less energy requirements
- The mechanical parts are easy to construct. There is no need of special aseptic seals.
- Scaling up is easier
- Metabolic performance does not drastically reduce on scale up.

Disadvantages

- Capital needed is more
- Difficulty of sterilization
- Efficiency of mixing is low

II. Anaerobic Fermentation

These reactors do not require aeration except in a few where initial preparation of inoculums requires aeration. Once the fermentation starts off, the gas released from the media is sufficient to provide mixing.

In case of enzyme production, the recovery has to be strictly under anaerobic conditions since for most of the enzymatic activity is sensitive to the presence of oxygen.

III. Immobilized Cell Bioreactors

These are based on immobilized cells.

Advantages

- Useful fro manufacture of intracellular enzymes.
- When the extracted enzymes are unstable
- For preparing low weight products which are released into the medium.
- Reduction of pollution
- Allow continuous operation of bioreactors
- Suitable for production of amino acids, organic acids *etc.*

Commonly fluidized bed reactors and hollow fiber membrane bioreactors are used.

1. Fluidized Bed Reactors

These reactors can utilize high density of particles and reduce bulk fluid density.

Advantages

- Heat and mass transfer are efficient
- The mixing of the media between the liquid, solid and gaseous phases are effective.
- The reactor requires less energy.
- Low shear rates and hence suitable for cells which are more sensitive to friction like the plant cells and mammalian cells.

2. Hollow Fiber Membrane Bioreactors

These reactors have hollow fibers are made from cellulose acetate, acrylic polymers, polysulphone *etc.*

Advantages

- Extracellular products can be separated from cells at the same time.
- The productivity is high.
- Scale up is easy since several parallel fiber units can be added.

Disadvantages

- Sometimes, the pores get plugged.
- Cell growth around the lumen can sometimes distort and rupture the fibers.
- Nutrients and products can diffuse through the membrane and limit the growth of microbes.
- If the toxic products happen to accumulate in the fiber it may inhibit the growth of microbes.

OTHER TYPES

Bioreactor Types

Numerous reactor designs for biological sulfate reduction have been reported, such as batch reactors, sequencing batch reactors, continuously stirred tank reactors, anaerobic contact processes, anaerobic baffled reactors, anaerobic filters, fluidized-bed reactors, gas lift reactors, upflow anaerobic sludge blanket reactors and anaerobic hybrid reactors.

The reactor configuration has implications for the ratio of sludge retention time/hydraulic retention time (SRT/HRT) in continuous flow reactors. The loading rates of a process are largely dictated by the biomass retention in the reactor. Maximal sludge retention or biomass retention is desirable for process stability and minimal sludge production. Minimal HRT minimizes the reactor volume and thus reduces capital costs. Continuously stirred tank reactors (CSTR) are subjected to washout of active biomass (Figure 1). Biomass retention has been enhanced by employing internal sedimentation systems and cationic flocculants. Anaerobic contact process (ACP) relies on biomass separation and recycling to increase the SRT/HRT. Several methods have been suggested for recovering biomass from the reactor effluent, including sedimentation, flocculation, centrifugation and magnetic separation of sulfate-reducing bacteria.

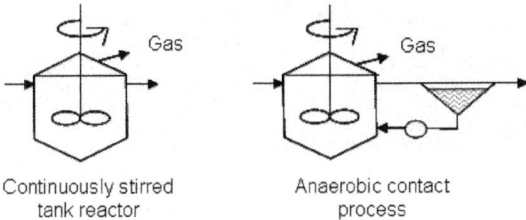

Continuously stirred
tank reactor

Anaerobic contact
process

Figure 1. Continuously stirred tank reactor (CSTR) and anaerobic contact process (ACP).

Due to the slow growth rate and low biomass yield of SRB, various immobilized biomass reactors have gained increasing attention. In anaerobic filter reactors (AFR) (or packed bed reactors, PBR) biomass is retained as a biofilm on packing material as well as unattached in the packing interstices. AFRs have been operated in horizontal, upflow or downflow modes (Figure 2). The downflow AFR allows the utilization of gravity and, thus, passive operation. Packing materials used in AFRs include cobbles, polypropylene pall rings, glass beads and alkaline minerals. Biological sulfate reduction has been enhanced with solid organic materials as well as liquid substrates. Solid substrates have a limited lifetime and have to be replaced or supplemented with liquid substrates once the original substrate is depleted. The main shortcomings of AFRs are the channeling of the flow and clogging of the bed by precipitates.

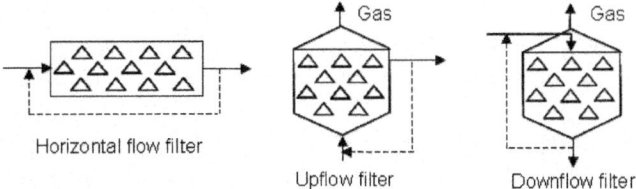

Horizontal flow filter

Upflow filter

Downflow filter

Figure 2. Anaerobic filter reactors (AFR) used in horizontal, upflow and downfow modes.

In the fluidized-bed reactor (FBR), channeling and clogging are avoided by fluidizing the inert biomass carrier (Figure 3). Fluidization can be carried out with recycle water or using a gas stream in which case the reactor is called a gas lift reactor. Carrier materials used include iron chips, synthetic polymeric granules covered with iron dust, pumice particles, porous glass beads, and carbon dust. The fluidized carrier enables efficient mass transfer and provides a large surface area for biofilm formation.

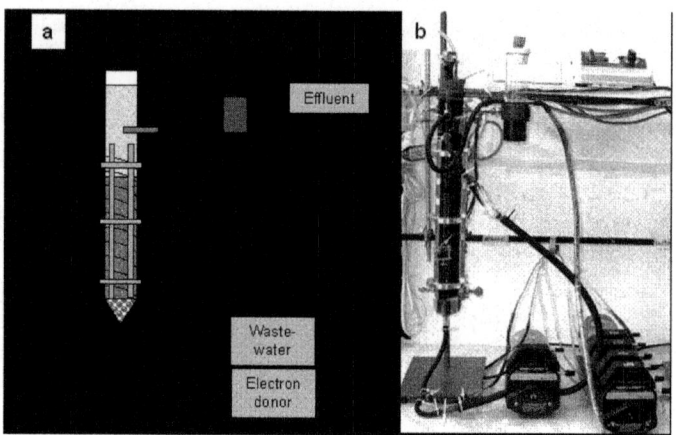

Figure 3. A schematic diagram and a photo of a laboratory scale
fluidized-bed reactor (FBR).

In upflow anaerobic sludge blanket (UASB) reactors, biomass retention is
based on good settling characteristics of granular sludge (Figure 4). The presence
of methanogens in the biomass can enhance granulation. The produced biogas
is trapped by a hood located below the water surface and can be periodically
burned in a flare. Due to the biomass granulation, no packing or carrier material
is needed which reduces the start-up costs of the UASB compared to AFR and
FBR. However, extensive biogas production may require extra instrumentation
which increases capital costs. Moreover, methanogens compete for substrates
(acetate and $H_2 + CO_2$) with sulfate reducers, resulting in decrease in the yields of
H_2S and alkalinity per amount of substrate added. Other problems encountered
with UASB reactors are poor or slow granulation and the rapid disintegration of
the granular sludge under certain conditions.

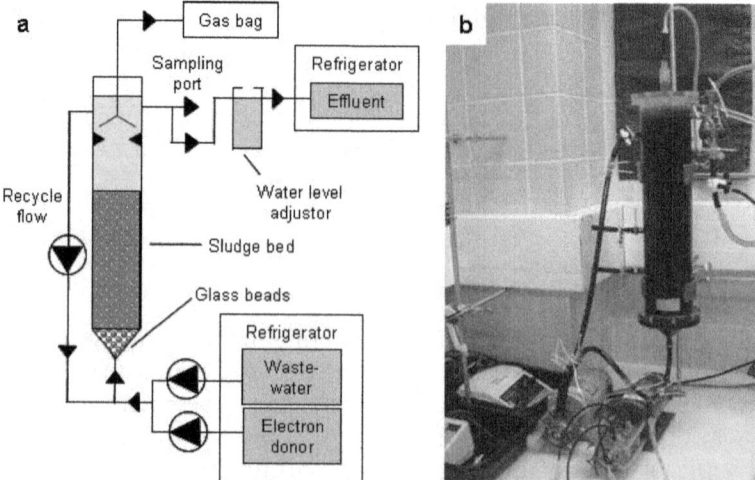

Figure 4. A schematic diagram and a photo of a laboratory scale upflow anaerobic sludge
blanket reactor (UASB).

The anaerobic hybrid reactor (AHR) is a combination of UASB and AFR, where the granular sludge bed is in the lower section of the reactor and packing material in the upper section (Figure 5). The packing material improved the separation of solids from the reactor effluent. Another modification of the UASB reactor is an anaerobic baffled reactor (ABR) which is a staged reactor where biomass retention is enhanced by forcing the water through several compartments (Figure 5).

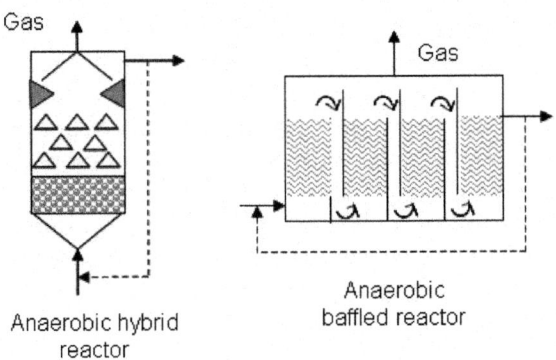

Figure 5. Anaerobic hybrid reactor (AHR) and anaerobic baffled reactor (ABR)

Depending on the reactor type and process configuration, the metal sulfide sludge can be recovered from the bottom of the bioreactor, such as AHR, by back washing the AFR at regular intervals or with a clarifier downstream of the precipitation unit.

NON LINEAR SYSTEM

For nonlinear systems that are uniformly observable for any U(.) (*i.e.* the states of the system can be determined from the output of the system and its derivatives, independently of the input)(Gauthier and Bornard, 1981), a high-gain observer has been suggested in (Tornarnbk, 1992).

One of the advantages of this observer are its excellent robustness properties (Tornambk, 1992). By choosing the observer gain k large enough (therefore the name "high-gain") the observer error can be made arbitrarily small. The difficulty in practical applications is, however, the determination of an appropriate value for the observer gain. For values too low, the desired bounds on the observer error cannot be achieved. For values unnecessarily high, the sensitivity to noise increases, thus limiting the practical use.

High-Gain Observer

The theory of non-adaptive high-gain observers as in (Tornarnbk, 1992) assumes that the system is given in observability normal form (Zeitz, 1989), also called the generalized controller canonical.

Input multiplicities mean more than one value of input variable produces the same value of output variable in a single input and single output (SISO) process. For a process with Input multiplicities, the value of steady state gain of the process changes as the input variable changes and after certain value of the input, the sign of the gain also changes. Proportional Integral (PI) controller for such nonlinear process with Input multiplicities may give unstable, less economical or oscillatory responses. Model based nonlinear controller controllers are reported to give superior performance for process with Input multiplicities. In the present work, the nonlinear internal model controller based on pseudo dynamic model is designed to a continuous bioreactor which exhibits input multiplicities in dilution rate on productivity *i.e.*, two values of dilution rate will give the same value of productivity. As the NIMC provides always two values of dilution rate for control action and by selecting the value nearer to the operating point, it is found to give stable and better performance than the linear PI controller. Whereas, linear PI controller results in washout condition when linear PI designed for lower dilution rate is used for higher dilution rate. Thus, NIMC is found to overcome the control problems of PI controller due to the input multiplicities.

What is a System of Equations?

A system of equations is any number of equations with more than one total unknown, such that the same unknown must have the same value in every equation. You have probably dealt a great deal, in the past, with *linear systems of equations*, for which many solution methods exist. A linear system is a system of the form :

Linear Systems

$$C_1 = a_1 x_1 + a_2 x_2 + \cdots$$
$$C_2 = b_1 x_1 + b_2 x_2 + \cdots$$

And so on, where the a's and b's are constant.

Any system that is not linear is nonlinear. Nonlinear equations are, generally, far more difficult to solve than linear equations but there are techniques by which some special cases can be solved for an exact answer. For other cases, there may not be any solutions (which is even true about linear systems!), or those solutions may only be obtainable using a *numerical method* similar to those for single-variable equations. As you might imagine, these will be considerably more complicated on a multiple-variable system than on a single equation, so it is recommended that you use a computer program if the equations get too nasty.

Solvability

A system is solvable if and only if there are only a finite number of solutions. This is, of course, what you usually want, since you want the results to be somewhat predictable of whatever you're designing.

Here is how you can tell if it will *definitely* be impossible to solve a set of equations, or if it merely *may* be impossible.

Solvability of systems :

1. If a set of n **independent** equations has n unknowns, then the system has a finite (possibly 0) number of solutions.

2. If a set of n **independent** equations has *less than* n unknowns then the system has an infinite number of solutions.

3. If a set of n **independent or dependent** equations has *more than* n unknowns then the system has no solutions.

4. Any dependent equations in a system do not count towards n.

Note that even if a system is solvable it doesn't mean it has solutions, it just means that there's not an infinite number.

Methods to Solve Systems

As you may recall there are many ways to solve systems of *linear* equations. These include :

- **Linear Combination** : Add multiples of one equation to the others in order to get rid of one variable. This is the basis for Gaussian elimination which is one of the faster techniques to use with a computer.

- **Cramer's rule** which involves determinants of coefficient matrices.

- **Substitution** : Solve one equation for one variable and then substitute the resulting expression into all other equations, thus eliminating the variable you solved for.

The last one, substitution, is most useful when you have to solve a set of nonlinear equations. Linear combination can only be employed if the same type of term appears in all equations (which is unlikely except for a linear system), and no general analogue for Cramer's rule exists for nonlinear systems. However, substitution is still equally valid. Let's look at a simple example.

Example of the Substitution Method for Nonlinear Systems

Example :

Solve the following system of equations for X and Y

$X + Y^2 = 4$

$X^2 - Y^2 = 22$

Solution : We want to employ substitution, so we should ask : which variable is easier to solve for?. In this case, X (in the top equation) is easiest to solve for so we do that to obtain :

$X = 4 - Y^2$

Substituting into the bottom equation gives :

$4 - 8Y^2 + Y^4 - Y^2 = 22$

$Y^4 - - 18 = 0$

This can be solved by the method of substitution :

Let $U = Y^2$. Plugging this in :

$$U^2 - 9U - 18 = 0^2$$

Note : All Ys must be eliminated for this method to be valid. Do something like this when the same variable (or set of variables) appears *in the same for every time*. If it's not in the same form every time, *i.e.* if the equation was something like

$Y^4 - 9Y^2 - 18 + e^Y$, then the method would not simplify your calculations enough to make it worth doing.

Solving by factoring :

$$(U - 6)(U - 3) = 0$$

$$U = 3, 6$$

Thus since $U = Y^2$ we obtain **four** solutions for Y!

$$Y = \pm\sqrt{3}, Y = \pm\sqrt{6}$$

Notice, however, that depending on where this system *came* from, the negative solutions may not make sense, so think before you continue!

Let's take into account all of them for now. Since we have Y we can now solve for X :

$$X = 4 - Y^2 = 4 - U$$

$$X = 1, -2$$

Note : Again, it may be true that only positive values of X make sense If only positive values of X and Y make sense then the ONLY solution that is of any worth to us is the solution (X, Y) since $Y = \sqrt{6}$ results in a negative value for X.

Notice that even a small system like this has a large number of solutions and, indeed, some systems will have an infinite number, such as :

1. $y = sin(x)$
2. $y = cos(x)$

NUMERICAL METHODS TO SOLVE SYSTEMS

There are numerical equivalents in multiple variables to *some* of the methods demonstrated in the previous section. Many of them in their purest forms involve the use of calculus (in fact, the Taylor method does as well), but as before, they can be reduced to approximate algebraic forms at the expense of some accuracy.

Shots in the Dark

If you can solve all of the equations explicitly for the same variable (say, y) you can guess all but one and then compare how different the resulting values of y are in each equation. This method is entirely brute-force, because if there are more than two equations, it is necessary to guess all of the variables but one using this method, and there is no way to tell what the next guess should be. Trying to guess multiple variables at once from thin air gets to be a hastle even with a computer.

Since there are so many problems with this method, it will not be discussed further.

Fixed-point Iteration

Again, the multivariate form of fixed-point iteration is so unstable that it generally can be assumed that it will not work. Weighted iteration is also significantly more difficult.

Looping Method

This is one method that *does* work, and that is somewhat different from any single-variable method. In the looping method technique, it is necessary to be able to solve *each equation for a unique variable*, and then you'll go around in a loop essentially, starting with an initial guess on (ideally) a *single* variable, say y, and then evaluating all equations until you return to your original variable with a new value y'. If the result is not the same as the guess(es) you started with, you need to make a new guess based on the *trends in the results*.

Note : What kind of trends am I talking about? If you have a well-behaved system, an increase in y will consistently lead to either an increase or a decrease in y', so you can take advantage of this to see which way you need to adjust your original guess. DO NOT attempt to use the value for y' as a new guess!

More specifically, here is an algorithm you can use :

1. Solve all equations for a **different variable.**
2. Make a guess on one variable (or as many as necessary to evaluate a second one, if it's more than one it gets harder though, so it is recommended to use another method)
3. Go through all of the equations until you end up recalculating the variable (or all of the variables) which you had originally guessed. Note whether the result is higher or lower than your guess.
4. Make another guess on the variable(s). Go through the loop again.
5. After these two guesses, we know whether increasing or guess will increase or decrease the recalculated value. Therefore, we can deduce whether we need to increase or decrease our guess to get a recalculated value equal to the guess.
6. Keep guessing appropriately until the recalculated value equals the guess.

This technique is often necessary in engineering calculations because they are based on data, not on explicit equations for quantities. As we'll see, however, it can be difficult to get it to converge, and this method isn't that convenient to do by hand (though it is the most reliable one to do realistically). It is great, however, for inputting guesses into a computer or spreadsheet until it works.

Example :

Solve this system :

$y = e^{-x}$

$y = ln\,(x)$

First we need to solve one of them for x, let's choose the first one :

$y = -ln\,(y)$

- To start off, we make a guess : $y = 0.1$ Then from the first equation, $x = 2.303$
- Plug this back into the second equation and you'll come out with $y' = 0.834$. The recalculated value is **too high**.
- Now make a new guess on y : say, $y = 0.5$. This results in $x = 0.6931$
- Plugging back into the second equation gives $y' = -0.3665$. The recalculated value is **too low**.

Note : Now we know that increasing the guess decreases the recalculated value y' and vice versa. Since the second value of y' is too low this means that we need the guess to be smaller than 0.5; likewise, since the first y' was too high we need it to be greater than 0.1.

- Lets now try $y = 0.25$.
- This results in $x = 1.386$ from the first equation and $y' = 0.326$ from the second. **Too high** so we need to increase our guess.
- Let's guess $y = 0.3$
- This yields $x = 1.204$ and thus $y' = 0.185$, which is **too low** indicating the guessed value was too high.
- Guess $y = 0.28$, hence $x = 1.273$ and $y' = 0.241$. The guess is therefore still too high.
- Guess $y = 0.27$, hence $x = 1.309$ and $y' = 0.269$. Therefore we have now converged :

$x = 1.309, y = 0.27$

Looping Method with Spreadsheets

We can do the guessing procedure more easily by programming it into a spreadsheet. First set up three rows like so :

A B C

1 y guess x y′

2 =-ln(A2) =ln(B2)

In B2 we put the first function solved for x, and in C2 we have the second function solved for y. Now all we need to do is type in guesses in A2 until the value in C2 is the same as our guess (the spreadsheet will automatically calculate B2 and C2 for you). To make things even easier, put the line = A2 − C2 into cell D2. Since we want y' to equal y, just keep guessing until the value in D2 is as close to zero as you like.

As a more in-depth example (which would be significantly more difficult to do by hand), consider the system :

Example :

Solve :

$$T = \frac{2P^2 X^2 - 3e^{-X/T}}{T - 2}$$

$$X^2 = T^3 - P$$

$$10P = T$$

In order for this to work, we only need to solve each equation for a unique variable, the expression need not be explicit! The following will work (assuming that X is a positive quantity), and this will be evident shortly :

- $T = \dfrac{2 * P^2 * X^2 - 3 * e^{-X/T}}{T - 2}$

- $X = \sqrt{T^3 - P}$

- $P = 0.1T$

Now we need to ask : which variable would be the best to guess to start the iteration procedure? In this case the best answer is T because from this guess, we can calculate P from equation 3, then X from equation 2, and finally a new guess on T from equation 1, and use this new value as a gauge of our old guess.

Note : Generally you want to start the loop with a variable that allows you to calculate a second value with only that one guess. Try to algebraically manipulate your equations so that this is the case before solving, because we want to avoid guessing on multiple variables if at all possible.

Lets program this into the spreadsheet :

A B C D E

1 T guess P X T' T' - T guess

2 =0.1*A2 =sqrt(A2^3 - B2) =(2*B2^2*C2^2 - 3*exp(-C2/ A2))/ (A2 - 2) =D2 - A2

Once all this is programmed in, you can just input guesses as before, with the eventual result that :

$P = 0.2453,\ X = 3.8098,\ T = 2.453$

Multivariable Newton Method

Note : You may want to skip this section if you don't know how to invert matrices, add them, or multiply them.

There is a multivariate extension to Newton's method which is highly useful. It converges quickly, like the single-variable version, with the downside that, at least by hand, it is tedious. However, a computer can be programmed to do this with little difficulty, and the method is not limited only to systems which can be explicitly solved like the looping method is. In addition, unlike the looping method, the Newton method will actually give you the next set of values to use as a guess.

The method works as follows :

1. Solve all of the equations for 0, *i.e.* let $0 = F(x_1, x_2, ...)$ for all functions F in the system.

2. Guess a value for all variables, and put them into a matrix (X). Calculate the value of all functions F at this guess, and put them into a matrix (F).

3. We need to find estimates for all the partial derivatives of the function at the guessed values, which is described later.

4. Construct a matrix (to become the Jacobian) as follows : make an empty matrix with n rows and n columns, where n is the number of equations or the number of variables (remember, a solvable system generally has the same number of equations as variables. Then label the columns with the names of variables and the rows with the names of your functions. It should look something like this :

$$\begin{bmatrix} -- & x_1 & y_2 & \cdots \\ F_1 & & & \\ F_2 & & & \\ \cdots & & & \end{bmatrix}.$$

5. Put the appropriate partial derivative in the labeled spot. For example, put the partial derivative with respect to x1 from function 1 in the first spot.

6. Once the Jacobian matrix is completely constructed, find the inverse of the matrix. There are multiple computer programs that can do this including this one (WARNING :Not tested software, use at your own risk!). Or you can do it by hand if you know how.

7. Matrix-multiply the inverse Jacobian with the transpose function matrix F (to make it a column matrix), then subtract this from the transposition of X (again, make it a column matrix) :

$$X_{n+1}^T = Xn^T - J^{-1} * F_n^T$$

MULTI-VARIABLE NEWTON METHOD FORMULA

8. The result is your next guess. Repeat until convergence.

Estimating Partial Derivatives

A Partial derivative is, in its most basic sense, the slope of the tangent line of a function with more than one variable when all variables except one are held constant. The way to calculate it is :

Now we need to stay organized, so let's introduce some notation :

$\dfrac{\delta F_i}{\delta x_j}$ is the partial derivative of function i with respect to variable j.

To calculate it :

1. Calculate one function F at your guess.
2. Increase *one* variable, x, by a very small amount δ. **Leave all other variables constant.**
3. Recalculate F at the modified guess to give you F'.

The partial derivative of the function F with respect to x is then

$$\frac{\delta F_1}{\delta x} = \frac{F_1(x+\delta,y)-F_1(x,y)}{\delta}.$$

Example of Use of Newton Method

Let's go back to our archetypal example :

- $y = e^{-x}$
- $y = ln(x)$

Step 1 : We need to solve each for zero :

- $F_1 = 0 = e^{-x} - y$
- $F_2 = 0 \; ln(x) - y$

Step 2 : Lets guess that $x = 2.303$ and $y = 0.1$ (it's a good idea to choose guesses that satisfy one of the equations). Then :

$X = [2.303, 0.1]$

The values of F at this guess are $F_1 = 0$, $F_2 = 0.734213$, and hence by definition :

$F = [0, 0.724213]$

Step 3-5 : Calculate the partial derivatives Lets choose δ = 0.01. Then :

- $\dfrac{\delta F_1}{\delta x} = \dfrac{F_1(x+\delta,y)-F_1(x,y)}{\delta}$

- $= \dfrac{(e^{-2.303+0.01}-0.1)-(e^{-2.303}-0.1)}{0.01} = -0.1036$

- $\dfrac{\delta F_1}{\delta y} = \dfrac{F_1(x+\delta,y)-F_1(x,y)}{\delta}$

- $$= \frac{(e^{-2.303} - (0.1 + 0.01)) - (e^{-2.303} - 0.1)}{0.01} = -1$$

The partial derivatives of F_2 can be similarly calculated to be

$\delta F_2 \delta x = 0.433$ and $\delta F_2 \delta y = -1$

Therefore, the Jacobian of the system is :

$$\begin{bmatrix} -- & x & y \\ F_1 & -0.1036 & -1 \\ F_2 & +0.433 & -1 \end{bmatrix}.$$

Step 6 : Using any method you know how to do, you can come up with the inverse of the matrix :

$$J^{-1} = \begin{bmatrix} -- & x & y \\ F_1 & -1.3636 & 1.8636 \\ F_2 & -0.8069 & -0.1931 \end{bmatrix}.$$

Step 7 : The transposition of F is simply :

$$F^T = \begin{bmatrix} F \\ 0 \\ 0.734213 \end{bmatrix}.$$

Therefore by doing matrix multiplication you can come up with the following *modifying matrix* :

$$J^{-1} * F^T = \begin{bmatrix} \Delta \\ 1.3682 \\ 0.1418 \end{bmatrix}.$$

Therefore, we should subtract 1.3682 from x and 0.1418 from y to get the next guess :

$x = 0.9373, y = 0.2418$

Notice how much closer this is to the true answer than what we started with. However, this method is generally better suited to a computer due to all of the tedious matrix algebra.

Linearization

Many of an engineer's tools for analyzing dynamic systems apply only to linear systems. The Laplace transform, for instance, only works if the equations to be transformed are linear.

What makes an equation "linear"?

- all variables present only to the first power
- no product terms where variables are multiplied (constants are ok)
- no square roots, exponentials, products, *etc.* involving variables

These can be understood by looking at some examples.

$a \dfrac{dx}{dt} = m(t)$ is linear as long as a is a constant and $m(t)$ is linear.

$a_1 x_1 + a_2 \sqrt{x_2} = m(t)$ is nonlinear because of the square root.

$a_1 \dfrac{dx_1}{dt} + a_2 x_1 x_2 + a_3 \dfrac{dx_2}{dt} = m(t)$ is nonlinear because of the cross-product

$x_1 x_2$ term, while $a_1 \dfrac{dx_1}{dt} + a_2 \dfrac{dx_2}{dt} + x_3 = m(t)$ is linear when m(t) is linear.

Linearity is useful, because of the mathematical theorems that state :

$f(x)$ is a linear differential equation :

1. If x_1 is a solution to the equation and c_1 a constant, then $c_1 x_1$ is also a solution

2. If x_1 and x_2 are solutions to the equation, then $x_1 + x_2$ is also a solution.

The latter means that for a linear process, the result of two input changes is the sum of the results of the individual changes.

Making a Model Linear

Many chemical engineering systems are highly nonlinear and general methods for working with nonlinear models are few, so it is important to know how to approximate nonlinear equations with linear ones.

The approach is really pretty straightforward :

- Expand all nonlinear terms in a Taylor series, usually around the steady state value
- Truncate the expansion after the 1st order terms

This gives a general result for linearizing equations :

$$f(x_1, x_2) \cong f(x_{1ss}, x_{2ss}) + \left(\frac{\partial f}{\partial x_1} \right)_{ss} (x_1 - x_{1ss}) + \left(\frac{\partial f}{\partial x_2} \right)_{ss} (x_2 - x_{2ss})$$

Notice that when you linearize, you do so around a specific point. Choice of this point is important. If the linear version of your model is to work, you must be operating close to the chosen point, so that you remain within the region where the linear approximation is valid. The steady state value is the usual choice since control systems are most often used to reject disturbances moving the plant away from steady state.

Example : Linearize $k(T) = Ae^{\frac{-E}{RT}}$

$$k(T) \cong k(T_{ss}) + \left(\frac{\partial}{\partial T}k(T)\right)_{ss}(T - T_{ss})$$

$$= k(T_{ss}) + Ae^{\frac{-E}{RT_{ss}}}\frac{E}{RT_{ss}^2}(T - T_{ss})$$

$$= k(T_{ss}) + \frac{E(T_{ss})}{RT_{ss}^2}(T - T_{ss})$$

$$= k(T_{ss})\left(1 + \frac{E}{RT_{ss}^2}(T - T_{ss})\right)$$

Example : Linearize $y = xz$

$$y \cong y(x_{ss}, z_{ss}) + \left(\frac{\partial y}{\partial x}\right)_{ss}(x - x_{ss}) + \left(\frac{\partial y}{\partial z}\right)_{ss}(z - z_{ss})$$

$$= x_{ss}z_{ss} + z_{ss}(x - x_{ss}) + x_{ss}(z - z_{ss})$$

Linearization in combination with perturbation variables has particular advantages. Recall that a deviation variable is defined as

$$x_p(t) = x(t) - x_{ss}$$

so if we are at the steady state, the value of the deviation variable will be zero. All of the "pure constant" terms will then vanish from the linearized deviation variables. Moreover, when we switch to perturbation variables and then linearize around the steady-state, the initial conditions are zero. This means we can drop the $x(0)$ terms as we take Laplace transforms.

Example : Take the two variable case above into perturbation variables.

$$y \cong y(x_{ss}, z_{ss}) + \left(\frac{\partial y}{\partial x}\right)_{ss}(x - x_{ss}) + \left(\frac{\partial y}{\partial z}\right)_{ss}(z - z_{ss})$$

$$= x_{ss}z_{ss} + z_{ss}(x - x_{ss}) + x_{ss}(z - z_{ss})$$

Notice how the terms containing only steady-state information vanish from the expression. If there were any constants (not functions of x, y, or z) they would also have vanished.

The last example showed that by combining linearizatio with perturbation variables, you effectively change the linearization equation to

$$f(x_1, x_2) \cong \left(\frac{\partial f}{\partial x_1}\right)_{ss}x_{p1} + \left(\frac{\partial f}{\partial x_2}\right)_{ss}x_{p2}$$

Chapter 9

TWO PHASE SYSTEM AND INTERFACIAL MASS TRANSFER

INTRODUCTION

Interfacial resistance to solute mass transfer between two unstirred immiscible fluids is theoretically calculated. Solute molecules are modeled as Brownian particles, bathed by homogeneous fluid continua when wholly immersed in either fluid, or else by heterogeneous fluid continua when instantaneously straddling the interface. These diffusing particles are assumed to be subjected to either repulsive or attractive conservative forces exerted on them by the interface. Additionally, their mobility is supposed affected by proximity to the interface. Circumstances are found to exist under which the interface may offer significant resistance to interphase transport. Surprisingly, conditions also exist in which the interface may actually offer a negative resistance to such solute transfer. In such cases, the presence of the interface enhances the overall interphase mass transfer rate.

Mass transfer occurs in mixtures containing local concentration variation. For example, when dye is dropped into a cup of water, mass-transfer processes are responsible for the movement of dye molecules through the water until equilibrium is established and the concentration is uniform. Mass is transferred from one place to another under the influence of a concentration difference or concentration gradient in the system.

Gas-liquid mass transfer is extremely important in bioprocessing because many processes are aerobic, oxygen must first be transferred from gas bulk through a series of steps onto the surfaces of cells before it can be utilized. The solubility of oxygen within broth is very poor. Therefore, the enhancement of gas-liquid mass transfer during aerobic cultures and fermentations is always put into priority.

Molecular Diffusion

Molecular diffusion is the movement of component molecules in a mixture under the influence of a concentration difference in the system. Diffusion of molecules occurs in the direction required to destroy the concentration gradient. If the gradient is maintained by constantly supplying material to the region of high concentration and removing it from the region of low concentration, diffusion will be continuous. This situation is often exploited in mass-transfer operations and bioreaction system.

Role of Diffusion in Bioprocessing

Mixing

As discussed before, turbulence in fluids produces bulk mixing on a scale equal to the smallest eddy size. Within the smallest eddies, flow is largely stream-line so that further mixing must occur by diffusion of fluid components. Mixing on a molecular scale therefore completely relies on diffusion as the final step in the mixing process. Solid-phase reaction

In biological systems, reactions are sometimes mediated by catalysts in solid form, *e.g.* clumps, flocs and films of cells and immobilized-enzyme and -cell particles. When cells or enzyme molecules are clumped together into a solid particle, substrates must be transported into the solid before reaction can take place. Mass transfer within solid particles is usually unassisted by bulk fluid convection; the only mechanism for intraparticle mass transfer is molecular diffusion. As the reaction proceeds, diffusion is also responsible for removing of product molecules away from the site of reaction, this will be discussed more fully in heterogeneous bioreaction kinetics. When reaction is coupled with diffusion, the overall reaction rate can be significantly reduced if diffusion is low.

Mass transfer across a phase boundary Mass transfer between phases occurs often in bioprocesses. Oxygen transfer from gas bubbles to fermentation broth, penicillin recovery from aqueous to organic liquid, and glucose transfer from liquid medium into mould pellets are typical examples. When different phases come into contact, fluid velocity near the phase interface is significantly decreased and diffusion becomes crucial for mass transfer across the phase interface.

Mass transfer coupled with fluid flow is a more complicated process than diffusive mass transfer. The value of the mass-transfer coefficient reflects the contribution to mass transfer from all the processes in the system that affect the boundary layer depends on the combined effects of flow velocity, geometry of equipment,and fluid properties such as viscosity and diffusivity. Because the hydrodynamics of most practical systems are not easily characterized.

INTRODUCTION, DROPLET-BASED MICROFLUIDICS

Droplet-based microfluidics has emerged as an invaluable tool that facilitates unique and varied research. For example, droplet-based microfluidics can be used to create droplet flows with precise control over drop size, frequency and flow conditions . This control over droplet characteristics and flow conditions has facilitated diverse experimentation. For instance, droplet-based microfluidic devices have been utilized to create double emulsion drops , nanoliter-sized chemical reaction vessels (drops), microcapsules for drug delivery A complete review of droplet-based microfluidics is beyond the scope of this work, and the interested reader is referred to a recent review on the subject .

This technology presents an ideal arena to study the dynamics of emulsions because of the precise control of droplet characteristics and flow conditions. Emulsions are omnipresent in and essential to everyday life, *e.g.* foods, pharmaceuticals, cosmetics, paints, oil recovery, *etc.* The morphology and stability of such systems depend on dynamic interfacial processes and properties. Measurement of these properties and the coefficients associated with these processes is therefore fundamental and important for application. Typical methods used to measure such properties often rely on a balance of forces that dictates drop sizes of the order of millimeters. Furthermore, these typical measurement methods often employ flows much simpler than those encountered in typical processing applications. For example, the pendant drop method relies on a balance of surface tension and gravity forces and is quiescent. This force balance requires that drops be approximately millimeters in size, whereas droplets in typical processing applications (*e.g.* detergents) are of the order of micrometers in size. Most importantly, the droplet size governs surfactant mass transfer mechanisms . It is therefore critical that dynamic interfacial studies be conducted on droplets with sizes comparable to those found in typical applications and in appropriate flow conditions. Droplet-based microfluidics are able to address these needs. Introduction and fine-tuning of flow complexity is relatively straightforward in droplet-based microfluidic devices. Advances in the ability to create monodisperse, stable droplet flows) and tailored, complex flow fields in microfluidic devices have facilitated the study of interfacial phenomena at reduced length scales and complex flows. For example, in past work a microfluidic device capable of measuring dynamic, multi-component interfacial tension was realized.

Here, we use this microfluidic interfacial tensiometer as a means to probe interfacial dynamics and surfactant mass transfer processes at reduced length scales that are relevant in drop microfluidic and emulsion processing applications, *i.e.* tens of micrometers. The interfacial tension in a two-phase, surfactant-containing system is used as a direct measure of the surfactant concentration adjacent to the interface. The tension is measured dynamically and mass transfer kinetics are determined through modeling the surfactant diffusion and interfacial kinetics. Simultaneous internal circulation data using particle tracers measures Marangoni effects, *i.e.* interfacial immobilization. These data are combined to evaluate the rate limiting mass transfer mechanisms and to measure associated parameters. Although much experimental research has been focused on studying dynamic interfacial tension and surfactant mass transfer in two-phase systems a great deal of this work uses relatively large drops (millimeter-sized) in the presence of simple or no flow. Our microfluidic approach allows us to create and measure the dynamics of many drops with sizes typical of those in industrial applications. Since the drop size governs the mass transfer mechanisms, a shift from diffusion-limited to kinetic-limited (interfacial adsorption/desorption) mass transfer is expected Furthermore, droplets can be subjected to easily tunable complex flows, component concentrations can be easily changed through manipulation of inlet flows, and the total volume of experimental material needed is very small. Most importantly, our approach gives us the unique ability to perform high-throughput measurements of interfacial tension, surfactant mass transfer kinetics and Marangoni effects in a single experiment.

2. BACKGROUND

2.1. Interfacial Tension, Surfactant Sorption Dynamics and Interfacial Mobility

In two-phase systems with surface-active solutes, many techniques have been employed to measure the interfacial tension as a function of time. For example, bubble pressure tensiometry utilizes measurements of the capillary pressure of bubbles forced out of a capillary to determine the interfacial tension. This technique can be employed to study dynamic interfacial tensions from the sub-millisecond range to times of the order of tens of seconds common methods of measuring dynamic interfacial tension include sessile and pendant drop methods, whose dynamic range is from a few seconds to a very long time Although our microfluidic technique has limited time range (here 0.5–20 s), it has a number of other features advantageous for emulsion characterization, which will be discussed in this section.

Because of the balance of forces involved, the common methods (of bubble pressure and pendant drop tensiometry) must employ droplets with sizes of the order of millimeters or larger. As mentioned previously, many typical applications have drop sizes less than 100 μm. Recent studies have shown that when considering surfactant-containing two-phase systems with micrometer-sized drops, the size of the droplet is important in determining the governing mass transfer mechanisms (*e.g.* diffusion limited versus interfacially limited kinetics) More specifically, Jin *et. al.* assuming Langmuir kinetics and a planar interface, define an intrinsic length scale given by

$$R_{D-K} = \frac{D}{\beta \Gamma_{max}},$$ (1)

where D is the surfactant diffusion coefficient, β is the interfacial adsorption kinetic constant, and Γ_{max} is the interfacial concentration corresponding to maximum packing of surfactant. Jin *et. al.* state that mass transfer is diffusion-controlled when the droplet radius $a \gg R_{D-K}$ and kinetically controlled when $a \ll R_{D-K}$. Nevertheless diffusion often remains important when $a \ll R_{D-K}$, since diffusion and sorption processes are in series. When this is the case, the limiting mechanism of mass transport is mixed (kinetic and diffusion). The value of R_{D-K} depends on the properties of the surfactant/two-phase system, and typical values were found to be of the order of tens of microns Alvarez *et. al.* have recently recalculated R_{D-K} for a spherical geometry (when curvature of the interface is important), and found R_{D-K} to be somewhat larger than given in equation (1) For local equilibration near a planar interface, molecules are displaced from a depth $\delta = \Gamma/c_0$ near the interface, where c_0 is the initial bulk surfactant concentration. In our experiments, as noted below, the surfactant concentration is sufficiently high that δ is very small, approximately 10 to 80 nm, and the planar interface approximation is appropriate. R_{D-K} (equation (1)) is thus predicted to be approximately [80, 200] μm for our system, depending on surfactant concentration. Our microfluidic approach allows generation of droplets with diameters of the order of tens of micrometers, facilitating experimentation at droplet length scales smaller than R_{D-K}. At these length scales, the transition from diffusion-controlled to mixed kinetically controlled surfactant mass transfer can be verified.

Beyond the ability to measure dynamic interfacial tension on droplets with diameters of the order of tens of micrometers, our microfluidic approach has the advantage of facilitating the study of drops in complex flow conditions like that which emulsions encounter in many industrial processes. This is accomplished by introducing flow complexity through variation of the microchannel geometry (see the appendix). Interfacial dynamics have typically been studied in simple, symmetric flows and it is unknown if the breaking of symmetry influences dynamics. Typical processing applications involve complex flows and different symmetries and it is unknown if interfacial dynamics in complex flows are perhaps just a superposition of simpler effects.

When studying surfactant-containing drops under flow, one must consider that the flow can convect surfactant adsorbed on the interface and interfacial surfactant concentration gradients can arise. These concentration gradients can in turn induce Marangoni effects, which retard the interface and lead to interfacial immobilization . Generally, interfacial immobilization through Marangoni effects is maximized when the quantity $\partial\sigma/\partial\Gamma$, where σ is the interfacial tension and Γ is the interfacial surfactant concentration, is large and when surface diffusion and surfactant exchange to and from the interface are slow. (Physically, the quantity $\partial\sigma/\partial\Gamma$ is a measure of the ability of a surfactant to change the interfacial tension through a change in interfacial concentration.) This interfacial immobilization arrests internal droplet fluid motion. Our microfluidic approach allows us to image drops under flow and, by placing particle tracers inside the drops, directly measure interfacial immobilization caused by Marangoni effects.

2.2. MASS TRANSFER

The increased surface-to-volume ratio that arises in microfluidic applications can have significant effects on mass transfer. In general, flows encountered in microfluidic devices are typically laminar, and as a result, mass transfer between phases typically occurs primarily through diffusion (convection can become important in droplet/slug systems). However, this limitation is somewhat mitigated by the fact that mass transfer occurs over small length scales inherent in microfluidic devices.

Mass transfer between phases in microfluidics have been studied for a number of systems including exchange to and from droplets and slugs , and counter-flow and co-flow schemes . In co-flow devices, two streams, miscible or immiscible, are brought into contact and flow side by side down a channel, with mass transfer between the streams occurring by diffusion ; whereas in counter-flow devices, two streams flow toward each other in a channel, meet and exit through channels perpendicular to the initial flow direction. In the case of droplets and slugs, mass transfer occurs between the continuous phase and droplets or slugs as they flow through the channel Most of the aforementioned work utilizing co-flow and counter-flow schemes involves single-species diffusion from one stream or phase into a second stream or phase, with a chemical reaction taking place in the initially solute-free stream or phase. Typically, some optical parameter of the fluid changes as the reaction proceeds, allowing for measurement of solute diffusion and reaction rate constants For immiscible phases, interfacial mobility and kinetics and their effect on microfluidic mass transfer processes needs further study.

In the experiments reported here, aqueous droplet flows in mineral oil are created, and a surfactant begins only in the drop and escapes into the surrounding medium. Since the concentrations of surfactant adjacent to the interface are quasi-equilibrated by kinetic interfacial sorption processes (equations (5) and (13)), the interfacial tension measured at a given droplet or interface age is a direct measure of these local surfactant concentrations. Thus by measuring interfacial tension, the local surfactant concentration can be inferred.

At larger length scales, surfactant transport occurs through diffusion and convection, with the relative importance of each depending on experimental conditions. For diffusion, the appropriate timescale is Dt/a^2, where t is time, while for convection it is ut/a, with the relevant velocity, u, differing for convection inside and outside the drop whose radius is a. When considering convection outside the drop, the appropriate velocity is the velocity of the drop u_d itself moving relative to the fixed channel. For convection inside the drop, the appropriate velocity is the velocity of the drop relative to that of the adjacent streamlines, which when the drop is on the centerline is approximately $u_d(2a/h)^2$, where h is the channel height. For a drop off of the centerline, the relative velocity of the adjacent streamlines can be calculated similarly. We note that even when convection is rapid, diffusion remains essential if the flow is laminar and non-chaotic. Chaotic flow inside drops may occur even in simple circumstances when translational and simple shear flows are superposed e.g. when a droplet rises or falls while moving in a microchannel. However, it has been shown that symmetric flows, e.g. Poiseuille, possess invariants (functions constant along streamlines) which are a barrier to convective transport inside droplets so that transport across these streamlines remains diffusion limited. We will consider cases when the flow is non-chaotic both inside and outside of the drop. Let us now consider the internal circulation in more detail.

The flow fields inside and outside of nearly spherical drops have been determined for simple shear and extensional flow by Taylor. The flow fields for nonlinear flow were calculated by Hetsroni and Haber and expressed in simplest terms by Nadim and Stone . In a drop on the channel centerline in Poiseuille flow, the scaling of the internal circulation velocity at the midplane of the drop relative to its frame of reference is given by

$$\hat{u} = \frac{u_c / u_d}{(2a/h)^2},$$ (2)

where u_c is the circulation velocity inside the droplet at a point of interest. In this paper, we consider $\hat{u}$ only at the center of the drop. In wide rectangular channels, $\hat{u}$ there ranges from 0 to 0.5 depending on drop viscosity and interfacial retardation . (When the aspect ratio of the channel cross section w/h decreases, $\hat{u}$ increases, e.g. $\hat{u} \cong 0.85$ when the aspect ratio is unity). In our experiments, the relative viscosity of the drop is nearly zero (giving maximum $\hat{u}$), but the interfacial retardation is variable depending on fluid formulation. For example, Marangoni effects caused by surfactants reduce $\hat{u}$ through interfacial immobilization.

As mentioned above, surfactant transport occurs through diffusion and convection. Let us consider four limits : firstly, when convection is negligible; secondly, where convection inside the drop is slow, but fast outside the drop; thirdly, when convection inside and out are fast; and fourthly, when the kinetics at the interface become a limiting factor.

Firstly, when convection is negligible, the diffusion problem has been described by Liggieri et. al. an extension of Ward and Tordai's analysis . When diffusion limited and with no partitioning across the interface, the concentration of surfactant is given as a function of radius r and time t by the equation

$$\frac{c}{c_0} = \frac{1}{2}\left(\operatorname{erf}\left(\frac{a-r}{2\sqrt{Dt}} \right) + \operatorname{erf}\left(\frac{a+r}{2\sqrt{Dt}} \right) \right) - \frac{2\sqrt{Dt/\pi}}{r}\left(\exp\left(\frac{-(a-r)^2}{4Dt} \right) - \exp\left(\frac{-(a-r)^2}{4Dt} \right) \right).$$ (3)

When surfactant transport from a sphere is rate limited by diffusion, almost all (92%) of the surfactant is lost by the characteristic time a^2/D. The time to lose 50% of the surfactant is approximately one-fifth of this, which in the experiments here is somewhat less than 1 s.

Next, let us consider slow convection inside the drop, and fast outside the drop. The external Peclet number is given by $Pe_{ext} = au_d/D$, which in our experiments is $O(100)$. Since the flow is laminar ($Re = O(0.1)$), it convects components only along, and not transverse to, the channel. Thus, if the drops are very far apart, then the external convection, i.e. shear dispersion, would be significant. In such a case, the dilution effect is complicated. However, in our experiments the drops follow relatively closely in one another's wakes, and thus dilution is essentially diffusion controlled. Typically, the ratio of the distance between drops to the drop radius x_s/a is between 6 and 10. The ratio of the time it takes for a surfactant molecule to diffuse a distance equal to the drop radius a to the time it takes for a drop in the train to translate a distance of x_s is therefore $u_d a^2 / x_s D$. At the experimental conditions studied here, this ratio $u_d a^2 / x_s D$ = [20, 50]. Therefore, except for very short times, external convection and external shear dispersion have a weak effect on dilution, and diffusive mass transfer transverse to the streamlines is dominant, and thus this case is essentially like the first.

When the convection inside and outside the drop are fast, the fountain flow inside the drop constantly brings solute from the droplet center to near the interface. (In linear shear, the streamlines in the drop are more or less parallel to the interface, and thus not effective for transport from the center.) The internal Peclet number of the fountain flow is given by

$$Pe_{int} = 4a^3 \hat{u} u_d / h^2 D = Pe_{ext} \hat{u} (2a/h)^2. \tag{4}$$

When Pe_{int} is more than unity, the mass transport at short times is significantly accelerated in comparison to the first two cases). Since this is analogous to increased diffusivity, the size R_{D-K} (equation (1)) effectively increases.

As mentioned above in section 2.1, when diffusion is fast compared to the surfactant adsorption/desorption kinetics at the interface, its mass transfer becomes limited by the kinetics at the interface. Because the drop size in our system a_0 = [30, 50] μm is less than but still somewhat comparable to R_{D-K}, a mixed diffusion-kinetic model is appropriate According to the Langmuir model, the interfacial sorption kinetics are governed by the following equation :

$$\frac{d\Gamma}{dt} = \beta_w c_w (\Gamma_{max} - \Gamma) - \alpha_w \Gamma + \beta_{oil} c_{oil} (\Gamma_{max} - \Gamma) - \alpha_{oil} \Gamma, \tag{5}$$

where α_i are the surfactant interfacial desorption kinetic constants and c_i are the concentrations adjacent to the interface. $c_i^* = \alpha_i / \beta_i$ are characteristic concentrations (noted below) that measure interfacial activity. The characteristic time for desorption from the interface is proportional to $1/\alpha_i$, as appropriate, independent of drop size.

In summary, mass transport may be mixed mode or limited by either interfacial kinetics or diffusion, depending on the magnitude of the droplet radius a with respect to R_{D-K} (equation (1)). In addition, convection inside the drop may accelerate bulk transport there if Pe_{int} is substantial, increasing the effective size of R_{D-K}, and causing interfacial kinetics to play a greater role.

3. EXPERIMENTAL

3.1. Fluids and Particles

Fluids used in this study are mineral oil (white, heavy, Aldrich)[Note2] , n-butanol (Mallinckrodt) and distilled water. Aqueous solutions of n-butanol of (0.20, 0.50, 1.0, 2.0 and 5.0)% mass fraction were prepared. A small amount (less than 0.2% mass fraction) of polystyrene (PS) spheres (diameter 2.092 ±0.095 μm, Polybead, Polysciences, Inc.) were placed into the aqueous drops to serve as particle tracers. The mineral oil, n-butanol and PS spheres were used as received. The viscosity of the mineral oil is 0.17 Pa s at 22 °C.

3.2. Pendant Drop Tensiometry

Interfacial tension was measured by pendant drop tensiometry (ITConcepts, France), using a 20 gauge needle. The pixel dimensions in each direction were calibrated to give an air water surface tension of 72.0 mN m^{-1} at 25 °C, independent of drop size (surface area ranging from 7 to 30 mm^2). Thus calibrated, the interfacial tension between the heavy mineral oil and water was also independent of drop size and found to be 52 ±1.5 mN m^{-1} at 22 °C. The interfacial tension was measured as a function of time from several seconds to several days, as discussed in section 4.4.

3.3. Device Fabrication and Design

Fabrication of the microfluidic device from soft lithography and polydimethylsiloxane (PDMS, Dow Sylgard) replication has been described in detail previously . Briefly, a master channel pattern was fabricated on a Si wafer using SU-8, 2075 photoresist (MicroChem) and traditional photolithographic techniques. A PDMS device is created from the pattern, bonded to a glass slide by briefly exposing both to O$_2$ plasma, then fitted with tubing. Aqueous drops are formed at a T-junction and travel into the main channel which contains multiple constrictions/expansions. The droplet formation zone and a detailed geometry of constrictions/expansions can be seen in figures 1 and 3. Three inlets are provided for the continuous fluid (two leading directly into the main channel, numbers one and two, and one for drop formation, number three in figure 1) so that the drop formation rate, spacing, size and overall flow rate can be easily adjusted. Multiple inlets are also provided for the aqueous phase (numbers four and five) so that the surfactant and tracer particle concentrations can be controlled. (The circular zones in channels 3, 4 and 5 are strictly for stability of the SU-8 master.) Total flow rates, q_{tot}, were dependant on the desired size, frequency and velocity of the drops; typical operating values were between 2 and 3.5 mL h^{-1}. Multiple constrictions/expansions at various positions along the channel allow for probing of dynamics at varying drop ages. There are slight variations in channel height h throughout the channel, with the average height being 270 ±19 µm. All other device dimensions can be inferred from figure 1.

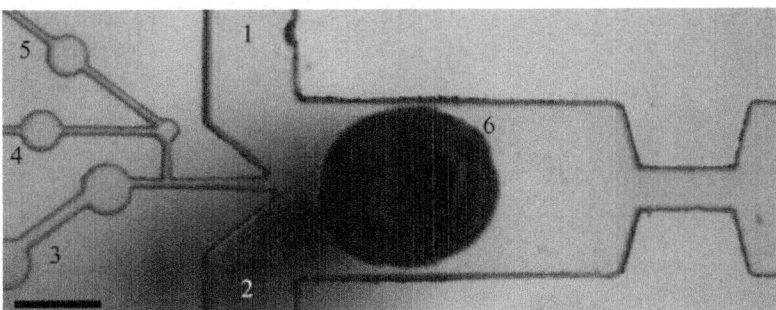

Figure 1. Optical micrograph showing droplet mixing zone, formation zone, inlet for controlling droplet height (dark circle) and a constriction (at right). Scale bar represents 500 µm.

3.4. Adjusting Drop Height

As mentioned previously, an advantage of our device is that the flow kinematics can be easily adjusted by means of channel geometry, drop size and drop height. Unfortunately, because of small variations in geometry during fabrication, each device forms droplets at different vertical heights. Therefore, it is advantageous to be able to exert precise control over the droplet height. The droplet height, y', is given by

$$y' = \frac{y}{h} - 0.5, \tag{6}$$

where y is the actual height of the drop in the channel. y' is measured by first focusing the microscope on the channel floor and ceiling, then the drop midplane, and noting the distances travelled by the objective. Precise control of the droplet height y' is achieved easily (see figure 2) through the injection or withdrawal of the continuous fluid through a sixth inlet (flow rate q_6) positioned above the main channel directly downstream of the droplet formation zone (see dark circle in figure 1). Continuous phase fluid is pumped either into or out of the channel to raise or lower the drop to the desired height, which in this study is the midplane of the channel. At this height, the flow in the wide channel (between constrictions) is mainly quadratic. Since this quadratic flow is relatively weak, it is overwhelmed and the flow is primarily planar extension upon entry to and exit from the constriction (other flow kinematics may also be obtained as discussed in the appendix). The inlet q_6 allows for precise control of droplet height over a wide range of y', which facilitates precise control of flow complexity (see the appendix). The uncertainties shown in figure 2, and all figures, are standard uncertainties.

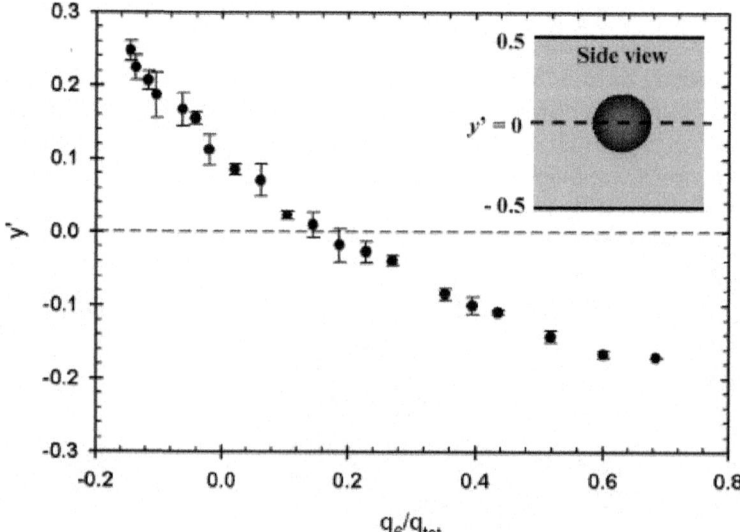

Figure 2. The dimensionless drop height versus relative flow rate into or out of the sixth inlet (see text and figure 1). Schematic of non-dimensional drop height is shown in insert.

3.5. Microscope, Pump Control and Data Acquisition

Microfluidic devices are mounted on an automated translating XY stage (Prior H107) fitted on an Olympus IX71 inverted microscope. Inlets are connected to micro-stepping syringe pumps (New Era) capable of delivering fluid volumes with accuracy better than 0.1% (with known syringe diameters). The pumps and data acquisition scheme are computer-controlled though a LabVIEW routine. Eight-bit grayscale images (1504 × 400 pixels) are acquired with a Redlake HG-100K high-speed camera at 1000 Hz. To avoid blurring, exposure times are chosen (typically 20–50 µs) so that the motion of the droplets is less than one pixel per exposure. Image analysis allows recording of instantaneous drop center of mass, deformation, orientation and detailed shape with time.

3.6. Analysis of Drop Deformation and Motion

As described previously the interfacial tension can be determined by plotting the transient response of an isolated drop to deformation in an unbounded extensional flow field (Taylor analysis) :

$$\kappa\eta_c \left(5\frac{\dot{\varepsilon}_1 - \dot{\varepsilon}_2}{4\hat{\eta} + 6} - \frac{dD}{dt} \right) = \sigma\frac{D}{a_0} \tag{7}$$

where the viscosity ratio $\hat{\eta} = \eta_d / \eta_c$, and η_d and η_c denote the droplet and continuous phase viscosities, respectively. $\dot{\varepsilon}_i$ denotes a principal extension rate, a_0 is the undeformed drop radius, and D is the deformation, given by

$$D = \frac{L - B}{L + B}, \tag{8}$$

where L and B are the major and minor axes of the droplet, respectively. The term $\kappa\eta_c$ is an 'effective viscosity', where κ is a function of viscosity ratio given by the equation

$$\kappa = \frac{(2\hat{\eta} + 3)(19\hat{\eta} + 16)}{40(\hat{\eta} + 1)}. \tag{9}$$

It is important to note that for rapidly accelerating flows (entrance or exit to a constriction), the material rate of change of the extension rate becomes non-negligible and the instantaneous droplet deformation lags the steady state deformation. Equation (7) takes into account this effect.

During data acquisition, the Taylor analysis is applied in the following manner. First, a threshold is applied to the raw image resulting in a binary image, from which the droplets are extracted. The transit time of the drops $t(x)$ (x is the droplet center of mass) and the deformation D are computed from the binary images. The deformation D is calculated from the moments of inertia of the drops rather than directly measuring the major and minor drop axes. Calculating D in this manner is a more accurate description of the droplet deformation in our system and has virtually no effect on the computation time. The transit time and deformation are then fit to polynomials (generally, the fit is insensitive to polynomial order above 5), and the extension rate along the x-direction is calculated by

$$\dot{\varepsilon} = \frac{du_d}{dx} = -\left(\frac{dt}{dx}\right)^{-2} \frac{d^2 t}{dx^2}, \tag{10}$$

where t is the time since the droplet entered the frame, and u_d is the droplet velocity dx/dt (the droplet is a marker of the flow). $\partial D / \partial x$ is computed from the data, and the material time derivative of D is calculated by the equation

$$\frac{dD}{dt} = u_d \frac{\partial D}{\partial x}, \tag{11}$$

which is true for unidirectional, time-invariant flow. Knowing the component viscosities, a plot of the left-hand side of equation (7) versus D/a_0 is made and the slope of this plot yields the interfacial tension σ.

Because we are working with surfactant-containing systems, we must keep in mind that deviations from ideal behavior (non-spherical and non-ellipsoidal shapes) can occur. Depending on various properties of the surfactant, its concentration, the droplet history, and the external and internal flow fields, gradients in surfactant concentration along the interface can develop. These gradients can lead to deviations from ideal behavior, *e.g.* deviations from ideal shapes and partial or full immobilization of the interface An ellipsoidal droplet shape is assumed in the determination of the interfacial tension; therefore, non-ellipsoidal shapes can introduce errors in σ. However, these non-ideal effects are expected to be minimized at small deformations where the Taylor analysis is applicable. A comprehensive discussion on other sources of error (*e.g.* focus, threshold and confinement) in the determination of σ through droplet deformation has been given previously

The ability to directly measure the aforementioned non-ideal surfactant effects is one of the primary goals of this paper (*e.g.* measurement of interfacial immobilization through internal circulation velocity). Therefore, it would also be advantageous to be able to precisely measure non-ideal droplet shapes by determining the position of the droplet interface to a high precision (sub-pixel resolution). A detailed shape analysis routine capable of such measurements was written and incorporated into the LabVIEW primary control and analysis program. This shape analysis was not incorporated into the current study, but will be utilized in future studies.

3.7. Examination of Flow Inside Drops

Flow fields inside of the droplets are visualized by the motion of the PS particle tracers. A series of images (focused on the droplet mid-plane) are saved and each drop is extracted from each image, yielding a single series of (smaller) images in the reference frame of the drop for each drop passing through the channel. To quantify the circulation in the droplet, we have chosen to use the velocity in the center of the droplet, given by equation (2) and illustrated schematically in figure 5. The circulation velocity u_c is calculated by tracking the position of a particle x_p near the drop center as it moves a distance comparable to the droplet radius, and fitting x_p versus t. At a frame rate of 1000 Hz and for a droplet in the channel, a particle on the centerline typically travels a distance comparable to the droplet radius over a period of about 200 images.

3.8. Mass Transport Kinetic Calculations

The mixed diffusion-kinetic model was modeled with a one-dimensional finite element model (COMSOL 3.2, COMSOL AB). For convenience, we followed the one-dimensional mapping of Liggieri *et. al.* (1997). Parameters in the model include the drop size (determined experimentally), butanol diffusivity in water and oil (assumed to be 9.5×10^{-10} m^2 s^{-1} in each phase (Li and Ong 1990)), and coefficients of equation (5) (Γ_{max} and c_i^* were determined by pendant drop tensiometry). Diffusion limited predictions were simulated artificially by either increasing the rate constants sufficiently, or reducing D and rescaling t accordingly.

4. RESULTS AND DISCUSSION

4.1. Analysis of Drops

The fundamental basis of this investigation is particle and drop tracking. Quantities derived from these primary measurements of particle position, and drop position and shape (sample data are shown in figure 3) are the drop velocity, the rate of strain in the continuous fluid, the interface age, the internal circulation rate and the interfacial tension, as described in the previous section. As also noted in section 3.4, the drop height is adjusted to the center line, and the drop velocity matches well the predicted centerline velocity u_{max} of a rectangular channel (figure 3)

$$u_{max} = \frac{3}{2} \frac{q_{tot}}{wh} \left(1 - 0.630 \frac{h}{w}\right)^{-1}. \tag{12}$$

This velocity is plotted in figure 3 for the sections of the channel that are of constant cross section (w, h) or (w_c, h). The slip velocity of the drop (difference from u_{max}) is expected to be small ($O(1\%)$) since the drop is relatively small.

The interface age is the drop passage time to travel a distance x, approximately $t_{passage} = \int_0^x dx / u_d$, which for constant flow velocity u_0 is simply x/u_0. Integrating the curve in figure 3, we find for this device that $_{passage} \cong \dfrac{x - 1.38n}{u_0}$,

where x and u_0 are expressed in mm and mm s^{-1}, respectively, and n is the number of constrictions passed. Droplet residence times ranging from approximately 0.5–20 s can be realized using our current experimental setup.

As mentioned previously, the circulation of fluid inside the drop is measured by tracking the tracer particles in the drop. This circulation is noteworthy because it directly probes the mobility of the interface, a property of at least as much interest as the tension itself. In the wide portion of the channel, a relatively slow and

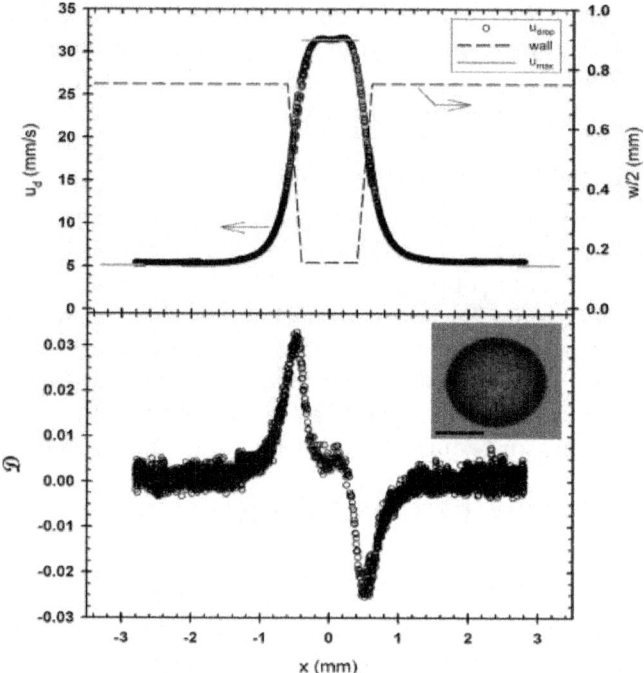

Figure 3. Droplet velocity and deformation through a constriction. Total flow qtot = 4.81 mL h–1; a0 = 29 μm; confinement 2a/h = 0.19;w = 1500 μm; wc = 300 μm; h = 278 ±18 μm. Inset image shows a droplet with particle tracers, scale bar represents 50 μm. Droplet velocity matches the predicted centerline velocity umax for a rectangular channel. The standard uncertainty of ud and D are ±0.091 mm s–1 and ±0.0047, respectively.

simple fountain flow occurs inside the drop. The flow inside the drop is faster and more complex when it passes the constriction, and some internal mixing takes place which further facilitates mass transfer. As the drop approaches or exits the constriction (*i.e.* when the extension rate is relatively high), four vortices in the drop are observed, and as expected the extension of the fluid inside the drop is transverse to the stretching direction outside. In later sections, we report the circulation velocity measured as the forward velocity at the center of the drop and only in the wide portion of the channel far from constrictions.

From the aforementioned measurements, *e.g.* time-dependant interfacial tension, we can evaluate various parameters noted in section 2 and estimate the dominant mechanism for surfactant mass transfer. The internal Peclet number ranges in our experiments from approximately 1 to 20 in the channel and 20 to 100 in the constrictions. As a consequence, convective transport inside the drop is significant, and we anticipate that the drop is effectively well mixed. Diffusion outside the drop across streamlines occurs in approximately $a^2/D = [0.5, 3]$ s. As mentioned previously, R_{D-K} (equation (1)) is predicted to be approximately [80, 200] μm, depending on butanol concentration. Therefore, mass transport of butanol is expected to be governed by kinetic interfacial transport and mixed kinetic-diffusion transport. The question of the limiting mass transfer process in these microchannel flows will be explored more quantitatively in later sections. In addition, the influence of interfacial concentration on interfacial mobility will also be addressed.

4.2. Interfacial Tension versus Drop Age

Following the procedure outlined in section 3.6, the interfacial tension is measured at various constrictions (interface ages) along the channel. Because of the nature of our microfluidic system, we have the advantage of performing a large number of measurements in a relatively short time. The interfacial tension values listed in figure 4 are averages of at least five measurements (error bars are one standard deviation), with the typical uncertainty in a single measurement (fit to Taylor analysis, equation (7)) being approximately 0.5–3 mN m^{-1}. The interfacial tension depends weakly on interface age (figure 4), and the tension increases modestly with increasing interface age, at a rate of nearly 1 mN m^{-1} s^{-1} for higher concentrations of butanol (2 and 5% mass fraction). The concentration continuously decreases as the butanol transfers to the oil and is dispersed. The drop size a_0 was varied by a factor of about 1.7 with surprisingly little effect on the interfacial tension. This size-independance is a signature of kinetically limited transfer (equation (5)). In the kinetic limit, the rate of decrease of the interfacial concentration is independent of a. A weak effect of a remains, since the maximum value of Γ during the mass transfer process depends on a and c_0. In contrast, the diffusion time depends strongly on a, $i.e.a^2/D$. The dotted and solid lines in figure 4 are the predictions from finite-element modeling, and will be described in more detail in a later section.

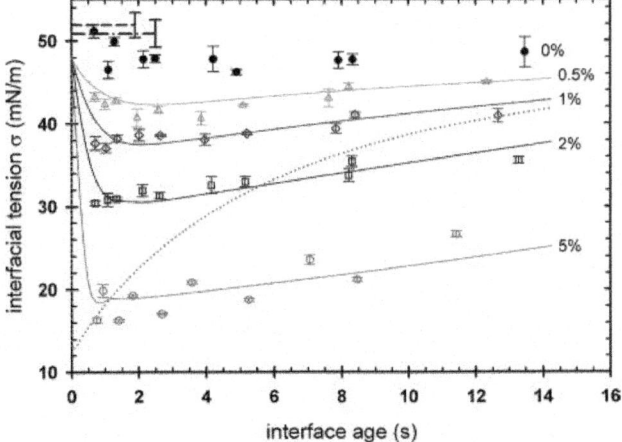

Figure 4. Interfacial tension as a function of interface age for various initial butanol concentrations for water in mineral oil. Butanol-free systems are shown as dash-dotted line from current microfluidic setup with no tracer particles, and dashed line from pendant drop. Solid lines represent model predictions from a mixed diffusion-kinetic model. The dotted line represents a diffusion-controlled model for 5% mass fraction butanol with the appropriate surfactant equilibrium distribution applied to the interface.

We note that the dispersion of butanol here is unlike a Taylor–Aris dispersion since the experimental time is not sufficiently long for the solute to fully sample the channel cross section. Moreover, when the solute reaches the wall it is not reflected or contained in the channel. Instead it may escape as it diffuses into and through the PDMS.

After complete extraction of butanol from the water to the surrounding oil (and elastomer PDMS device body) at times much longer than studied here, the interfacial tension of pure mineral oil and water is expected. If the channels were instead impenetrable to butanol, then the long time limit would be set by the dilution limit. Since the drop is relatively small (a approximately $0.1\,h$), this limit itself represents an exceedingly small concentration of butanol. As will be shown in section 4.4, the equilibrium distribution coefficient between the phases is not unity. Considering this factor, the dilution limit is approximately two orders of magnitude less than the starting butanol concentration.

4.3. Drop Circulation

For most of the interfacial tension measurements in figure 4, *i.e.* when particle tracers were present, the internal drop circulation was measured at the drop center and plotted in figure 5 (the circulation velocity $\hat{u}$ is plotted versus the interfacial tension so that the plot would be model-independent). A value of $\hat{u} = 0.5$ is expected for mobile interfaces Significant interfacial retardation was seen, however, with a rough trend of stronger retardation occurring at low surfactant

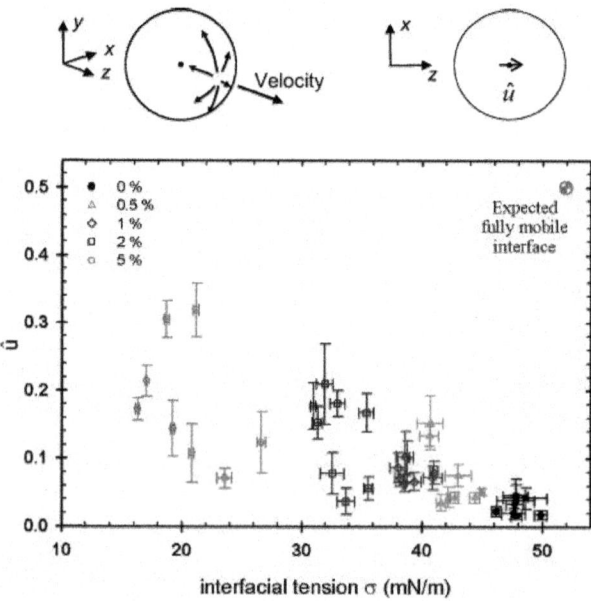

Figure 5. Internal circulation velocity $\hat{u}$ versus interfacial tension. Note at high tension the interface is immobilized. At low interfacial tension, where surfactant exchange is more rapid, the interface is partially remobilized. Schematics of the interfacial flow (left) and internal circulation $\hat{u}$ (right) are shown.

concentration (higher tension). At higher surfactant concentration, the surfactant exchange is more rapid, therefore the interfacial motion is much greater and approaches the mobility, *i.e.* $\hat{u} = 0.5$. This effect is termed remobilization.

When no butanol is added, we expect data on this plot at $\sigma = 52$ mN m^{-1} and $\hat{u} = 0.5$. Surprisingly, however, the retardation is substantial, and σ measured without butanol and with particle tracers is approximately 48±2 mN m^{-1}, measurably less than without tracers. These two results, interfacial tension reduction and interfacial retardation, indicate that the interface is contaminated with either the particles or an accompanying surface-active impurity. However, we have observed that the particles do not go to the interface, thus implicating an impurity, perhaps a residue from the particle synthesis. Pendant drop tensiometry also indicates the presence of this impurity. For the particle tracer solution, a substantial reduction in σ was observed by the pendant drop method, decreasing from 52 to approximately 40 mN m^{-1}, during a period of 30 min. In contrast, with pure water drops, σ is much more stable, decreasing only slightly to approximately 51.1 mN m^{-1} in the same period.

These results indicate that some surface-active impurity is present in the particle suspension, and it is this surface-active component that immobilizes the interface. However, when a large concentration of butanol is present, interfacial tension gradients are dominated by the distribution of butanol and not the surface active impurity, since Marangoni effects arise from gradients in interfacial tension not in interfacial concentration directly.

In spite of the severe retardation, the interface velocity remains significant: the internal Peclet number is still substantial (Pe_{int} is approximately 1–20). Therefore, convection in the drop, and the type of convection, *i.e.* a fountain flow, remains effective in bringing solute from the center of the drop to near the interface, even in the most severely retarded cases.

4.4. Pendant Drop Tensiometry and Equilibrium Interfacial Tension

To analyze the microfluidic data in terms of a kinetic model, it is helpful to measure the ratio of desorption to adsorption coefficients in both phases independently. Therefore, pendant drop tensiometry was used to measure the interfacial tension as a function of surfactant concentration in either the aqueous or oil phases (figure 6). We thereby demonstrate the suitability of the Langmuir model and measure the equilibrium distribution coefficient of butanol between the two phases. In this study, water drops in oil media were investigated, and butanol was added to either the water or oil phases. Because of the unequal partitioning of the butanol across the interface, care must be taken when analyzing pendant drop data from our system; therefore, we will describe the procedure in detail.

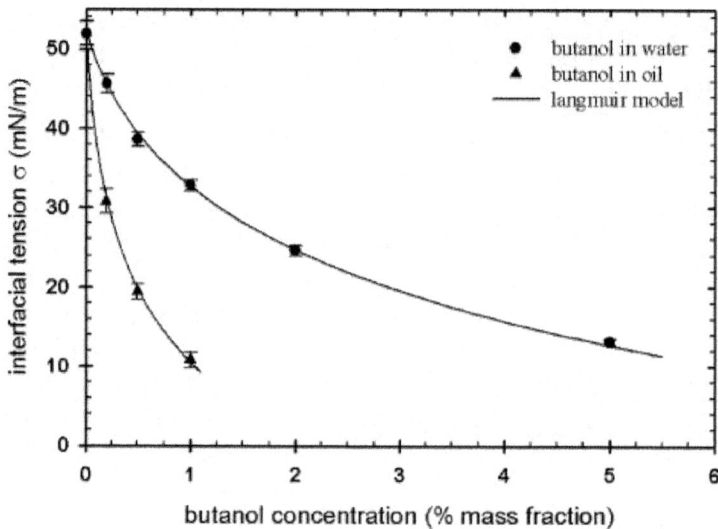

Figure 6. Equilibrium interfacial tension between water and heavy mineral oil as a function of butanol concentration in the water. Symbols represent data from the pendant drop method and the curves are fits according to the Langmuir model.

The data obtained is similar to that of Ferrari *et. al.* and Liggieri *et. al.* . Specifically, we find that when the butanol is in the water drop initially, the tension rapidly (< 30 s) decreases to a minimum and then slowly (several hours to days) increases. At the minimum, butanol has transferred across the interface and infused the oil adjacent to the interface at a concentration in equilibrium with the interface. Eventually the butanol diffuses in the oil over much longer distances, causing its overall concentration to decline and the interfacial tension to slowly rise. The value of this minimum tension is taken to be the equilibrium tension at the given concentration, since it occurs long before dilution. A major reason that the dilution is so slow is that the local equilibrium concentration of butanol in oil adjacent to the interface is substantially less than that in water, as determined below, and thus reducing concentration gradients markedly When the butanol is added to the oil phase, the tension decreases slowly (within 1 min to a couple of hours, depending on concentration) to an asymptote, which is taken to be the equilibrium tension since the concentration remains at nearly the initial value due to the fact that the relative volume of the oil is a few hundred times that of the water. At these quasi-equilibrium conditions, the concentrations of butanol adjacent to the interface are steady. The longer equilibration time observed when butanol is added to the oil phase is again a result of the unequal partitioning of the butanol across the interface. This partitioning is evident from the isotherms generated (figure 6), which demonstrates that at local equilibrium (equal tension), the concentration of butanol adjacent to the interface is substantially (approximately six times) lower in the oil phase, in comparison to that in the water phase.

According to the Langmuir model, the interfacial tension is given by the following expression :

$$\sigma = \sigma_0 - kT\,\Gamma_{\max}\ln(1 + c_w/c_w^*)$$
$$= \sigma_0 - kT\,\Gamma_{\max}\ln(1 + c_{\mathrm{oil}}/c_{\mathrm{oil}}^*)$$
$$= \sigma_0 + kT\,\Gamma_{\max}\ln(1 - \Gamma/\Gamma_{\max}),\tag{13}$$

where k is the Boltzmann constant and T is the temperature. This model approximates the data of figure 6 (indicating suitability of the Langmuir model) when $\Gamma_{\max} = 6\times10^{-6}$ mol m^{-2}, $c_w^* = 50$ mol m^{-3} (0.37% mass fraction), and $c_{\mathrm{oil}}^* = 8.6$ mol m^{-3} (0.063% mass fraction). These same values ($\Gamma_{\max}$ and c_w^*) were already found to apply to the effect of butanol on the tension of the air water interface). The interfacial structures at the two different interfaces (oil–water and air–water) are therefore similar, consistent with their comparable interfacial pressures. As noted above, the ratio $K = c_w^*/c_{\mathrm{oil}}^*$ is the equilibrium distribution coefficient, equal to approximately 5.8. Although $c_i^* = \alpha_i/\beta_i$ may be determined by pendant drop tensiometry, the kinetic coefficients α_i and β_i themselves are often, and in this case, inaccessible by pendant drop tensiometry. Therefore, in the subsequent section, the data of figure 4 will be analyzed in reference to that of figure 6 to evaluate the kinetic coefficients α_i for desorption.

4.5. Applying the Mixed Kinetic Model and Comparing to Experimental Data

Since both the interfacial tension and the drop circulation have been determined, we can now discuss the relevance of the various surfactant mass transfer parameters on the interfacial dynamics. First, a comparison between models will be made, and then comparisons to experimental data will be performed.

Concentration-time profiles for the mixed kinetic-diffusion model are calculated by finite-element modeling, and shown in figure 7 along with a diffusion-limited case (equation (3)) and a case where convection (internal mixing) is considered. The following parameters were used for these calculations. As noted above, the diffusivity in each phase was assumed to be $D = 9.5\times10^{-10}$ m^2 s^{-1}, and the drop radius was chosen from a typical experimental value, viz. $a = 43$ μm. $\Gamma_{\max} = 6\times10^{-6}$ mol m^{-2} was determined from pendant drop measurements. The distribution coefficient K was taken to be either 1 (no partitioning) or 5.8, to illustrate the effect of surfactant partitioning across the interface. Concerning the magnitude of the kinetic coefficients, $\alpha_{\mathrm{oil}} = \alpha_w$ were either 1×10^4 s^{-1} or 70 s^{-1} to illustrate, respectively, the diffusion limit or mixed kinetic transfer. Solid lines in figure 7 represent butanol concentrations in the droplet center, and dashed lines represent the concentration adjacent to the interface.

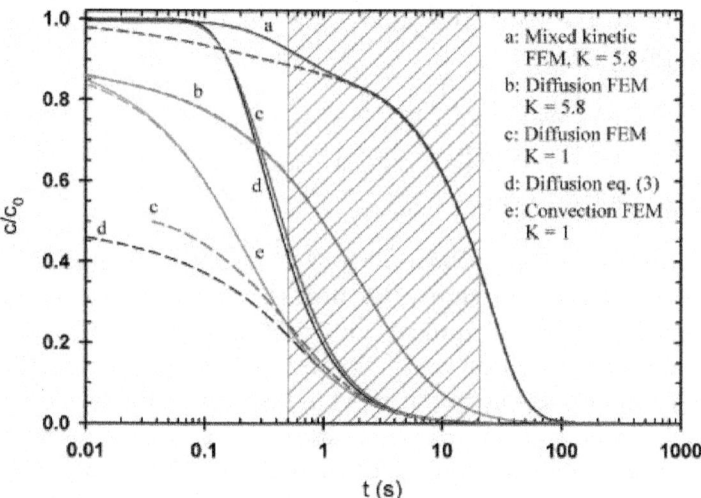

Figure 7. Comparison of the mixed kinetic model, and diffusion and convection limits without surfactant partitioning for a = 43 µm,c0/c* = 10 and D = 9.5×10-10 m2 s-1. Solid and dashed lines indicate butanol concentration at the droplet center and adjacent to the interface, respectively. The experimental window with the current setup is highlighted.

The various cases illustrated in figure 7 differ both qualitatively and quantitatively. The mixed kinetic model is more than an order of magnitude slower than the diffusion limited model (curve a versus d, respectively). The diffusion model of equation (3) (curve d) assumes no partitioning across the interface (*i.e.* $K = 1$); however, such partitioning has an influence on the diffusion limited transport (b) For example, if the surfactant is partitioned to the source, the transfer is slower than if it is partitioned to the destination. Two diffusion limited cases ($\alpha_{oil} = \alpha_w = 1 \times 10^4$ s⁻¹) with and without partitioning are plotted in figure 7 (curves b and c, respectively), and the non-partitioned case ($K = 1$) agrees well with equation (3) (curve d), as expected. The large differences caused by partitioning in the diffusion-limited cases are less evident in the mixed kinetic-diffusion case, likely due to the limiting kinetics at the interface (not shown). To simulate the effect of convective mixing in the drop, the diffusivity there can be increased artificially. This case is plotted as curve e in figure 7 ($K = 1$; $\alpha_{oil} = \alpha_w = 1 \times 10^4$ s⁻¹; and diffusivity in the drop increased by a factor of 100). In such a case, the concentration in the drop is essentially uniform and the solid and dashed lines merge together and follow an intermediate curve. When the transport across the interface is slowed by adsorption/desorption kinetics, a similar effect occurs. The bump in the mixed kinetic model (a) occurring between 0.1 and 1 s is due to a relatively rapid equalization of surfactant concentration in the drop. Internal convection simply causes this equalization at earlier times.

For comparison with experimental data, interfacial tension values predicted by the mixed kinetic-diffusion model with $K = 5.8$ are shown as solid lines in figure 4. Average values of the experimental drop size a_0 were used in the model for each concentration. We note that because α_{oil} and α_w are both fitting parameters in the mixed kinetic-diffusion model, a wide parameter space exists in which to conduct fitting to the data in figure 4. α_{oil} and α_w were both individually and systematically varied, with the best fit to all datasets occurring at $\alpha_{oil} = \alpha_w = 70$ s^{-1}. Both the minimum and slope of the curves in figure 4 can be changed by varying the desorption coefficients : they have similar effect on the slope (large α_i increase the slope) and compensating effect on the minimum (larger α_w lowers the minimum, while larger α_{oil} raises it). We estimate the uncertainty on α_{oil} and α_w to be approximately ± 50 s^{-1} : multiplying or dividing values of α_{oil} and α_w by $\sqrt{2}$ does not significantly change the quality of the fit in figure 4; however, multiplying or dividing $\alpha_{oil} = \alpha_w$ by a factor of 2 produces poor quality fits.

The mixed kinetic-diffusion model is in good agreement with the experimental data at all but very short times, verifying that our system indeed follows the mixed kinetic-diffusion model. The disagreement at very short times likely comes from neglecting convective effects that the drop experiences in the interval between the T-junction and the first measurement point. For comparison, the interfacial tension predicted by the partitioned diffusion-controlled model ($K = 5.8$; curve b in figure 7) is shown as a dotted line in figure 4 for 5% mass fraction of butanol. Interfacial kinetics are greatly accelerated in this model and thus mass transfer occurs more quickly and the interfacial tension is overpredicted at all but very short times, where it is underpredicted. Clearly, interfacial kinetics must be considered.

The diffusion and mixed kinetic models can be compared further by using them with the interfacial tension data of figure 4 to generate adsorption isotherms (figure 8). To construct the isotherms, concentration-time profiles were calculated by the finite-element models, and butanol concentrations were calculated corresponding to the times at which interfacial tension measurements were performed in the microfluidic device. The mixed kinetic-diffusion model produces a superior collapse of the data, and the fitting parameters are the desorption coefficients α_w and α_{oil}, determined from figure 4. Consistency with the microfluidic data depends significantly on the desorption coefficients. These coefficients, moreover, have an even more fundamental effect on the shape of the plot, so that even in the absence of reference data from pendant drop (which was used to determine the surfactant equilibrium distribution coefficients), the desorption coefficients could be determined, leading to an independent method to determine the adsorption isotherm. The deviations at low concentration in figure 8(b) can be explained by the presence of the surface-active impurity perceived in section 4.3, whereas deviations at high concentration may be due to the loss of butanol through diffusion into the PDMS in the input channel or dilution from shear dispersion (section 2.2), which were not accounted for.

To summarize, the kinetic constants for adsorption and desorption of butanol in a water/mineral oil system were determined through the use of a microfluidic approach combined with simple modeling. Langmuir kinetics were assumed at the interface, and the importance of surfactant partitioning was demonstrated.

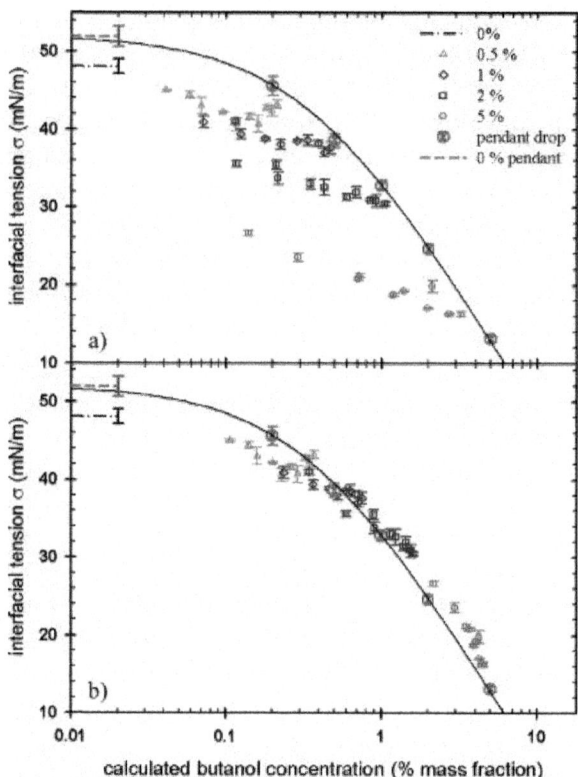

Figure 8. Interfacial tension as a function of butanol concentration for water in mineral oil : (a) for the diffusion limited case with surfactant partitioning ($\alpha w = \alpha oil = 1 \times 104$ s-1, K = 5. 8), and (b) taking into account adsorption/desorption kinetics ($\alpha w = \alpha oil = 70$ s-1, K = 5.8). Hourglass symbols represent steady-state pendant drop data, and the solid line represents a fit to the Langmuir model (equation (5) and figure 6).

5. CONCLUSIONS AND FUTURE DIRECTIONS

These experiments demonstrate that interfacial tension, surfactant mass transfer and interfacial retardation can be measured in a single experiment, so that interfacial properties and mobility can be correlated directly. At low concentration of butanol the interfacial tension is not reduced much, but the interfacial mobility is severely retarded. However, the interface is remobilized at higher surfactant concentration. The interfacial tension is described well by Langmuir kinetics and the parameters for interfaces with mineral oil (studied here) are similar to those previously found at air interfaces. The mass transfer of butanol from water drops into the surrounding flowing oil was shown to be well described by a mixed kinetic-diffusion limited model, and the desorption rate coefficient (from the interface to the oil) is measured to be approximately 70 s^{-1}.

The microfluidic approach employed here facilitates measurements of the interrelated quantities of interfacial tension, surfactant mass transfer kinetics and

Marangoni effects in a single experiment utilizing two-phase flows under conditions, particularly droplet size, that are relevant to industrial and microfluidic applications. At this reduced droplet length scale, the shift from diffusion-controlled to mixed kinetic-diffusion mass transfer was verified, stressing the importance of measurement at reduced length scales relating to processing applications. These measurements along with measurements of Marangoni effects (interfacial immobilization) are essential in understanding the dynamics of two-phase systems, *i.e.* emulsions, and moving toward the ultimate goal of accurately predicting emulsion performance and stability.

DIFFUSION

Diffusion is a result of random molecular motion in a fluid, a consequence of which is to tend to reduce differences in concentration. This is because there will be more molecules moving into an area of low concentration than moving out.

Diffusion may be illustrated by reference to the following diagrams (Fig 1.1)

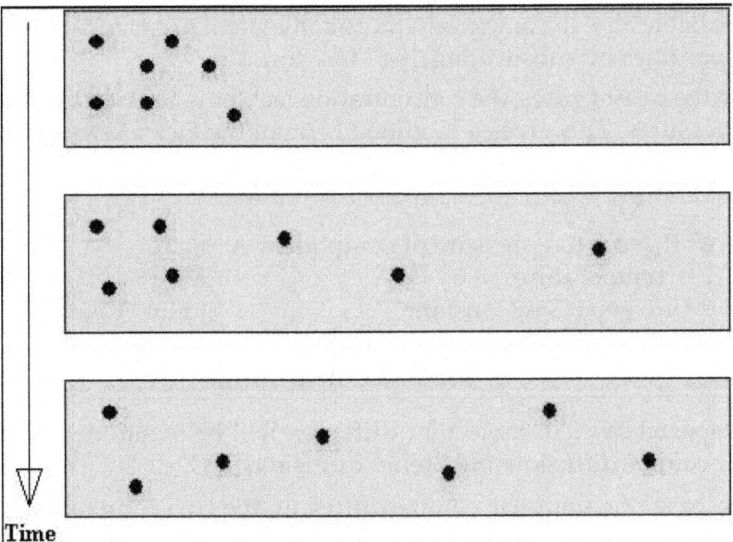

Fig 1.1

2. Fick's Law

The rate of diffusion is governed by Fick's law. This states that the rate of diffusion is proportional to the concentration gradient and to ease with which the molecules will move. The latter property is called the *Diffusivity*. Fick's law states that

$$J_A = -D_{AB}\frac{dC_A}{dz} \qquad\qquad 1.1$$

Where : J_A = net rate of diffusion or *molar flux* of component A kmol m^{-2} s^{-1}

D_{AB} = diffusivity of A through B m^2 s^{-1}

$\dfrac{dC_A}{dz}$ = concentration gradient along the axis of diffusion (kmol m^{-3}) m^{-1}

This equation is similar to the Fourier equation for heat transfer and is form of the general transfer equation which states that

$$\text{Rate of transfer} = \frac{\text{Driving force}}{\text{Resistance}}$$

In this case, the resistance is $1/D_{AB}$

Notes

1. The rate of diffusion is expressed in terms of flow/per unit area. This is known as the *Molar flux* and is obtained by dividing the rate of diffusion by the cross sectional area.

2. The units above are all expressed in molar quantities. However it is also possible and, in some cases, more convenient to express the units in mass quantities by substituting "kg" for "kmol".

3. In the case of gases, the concentration may be expressed in terms of partial pressures. This gives a modified form of the Fick's law equation

$$J_A = \frac{-D_{AB}}{RT} \frac{dP_A}{dz} \qquad\qquad 1.2$$

Where : P_A = Partial pressure of component A   Pa
        T = Temperature   K
        R = Universal Gas Constant   J mol^{-1} K^{-1}

3. TWO SPECIAL CASES

Two special cases of molecular diffusion will be mentioned here. They are equimolar counterdiffusion and Stefan diffusion.

In the case of equimolar counterdiffusion, the two components, A & B are diffusing in opposite directions at equal rates.

$$J_A = -J_B = -D_{AB} \frac{dC_A}{dz} \qquad\qquad 1.3$$

Stefan diffusion is, in effect, the opposite of equimolar counterdiffusion. One component diffuses, while the other remains stagnant

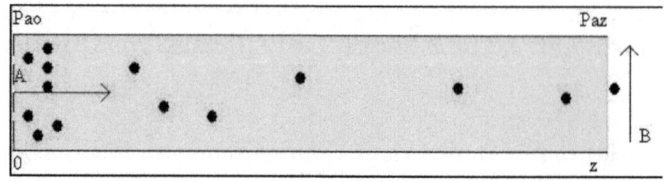

Fig 1.2

In Fig 1.2, point 0 is saturated with component A while a stream of pure B flows past the end of the tube removing any A that has reached point z. In this situation, the concentration gradient along the tube is exponential and the rate of diffusion is given by (in partial pressure terms)

$$J_A = \frac{PD_{AB}}{RTz}\ln\left(\frac{P - P_{AZ}}{P - P_{AO}}\right)$$ 1.4

4. DIFFUSION WITH BULK FLOW

In most practical situations, diffusion does not occur only as a result of molecular motion but is aided by the bulk motion of the fluid. For example animals, when hunting, try to remain downwind of their prey so that the odours from the prey are carried towards them and their own odours are carried away from their prey. This is an example of how bulk movement of the air assists the process of molecular diffusion in a particular direction.

In order to account for the effect of bulk flow of the fluid, it is necessary to extend the basic Fick's law equation.

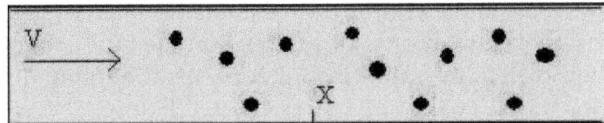

Fig 1.3

Consider a fluid flowing with Volume flow V along a pipe past a point X in the pipe. The fluid comprises two components A & B and the system is at steady state.

The molar flux of component A past X will be due to molecular diffusion **plus** the bulk flow of Component A. This molar flux is referred to as the mass transfer rate. The rate of transfer is now a combination of bulk flow and diffusivity. This combined term is called the mass transfer coefficient. Thus the basic Fick's law equation must be modified to give

$$NA = kA\Delta C_A$$ 1.5

N_A is the rate of mass transfer and is equal to $J_A A$, where A is the cross sectional area.

The quantity k is called the *mass transfer coefficient* and is a function of the diffusivity, the fluid properties and the physical geometry of the system.

5. DIFFUSION BETWEEN PHASES - PHASE EQUILIBRIA

When two immiscible phases are in contact, eg. gas-liquid, two immiscible liquids, it is possible for diffusing molecules to pass from one phase to the other across the interphase boundary.

When diffusion between two phases occurs, there are two factors to take into account;

1. equilibrium relationships between the two phases and
2. the rate at which the diffusion takes place

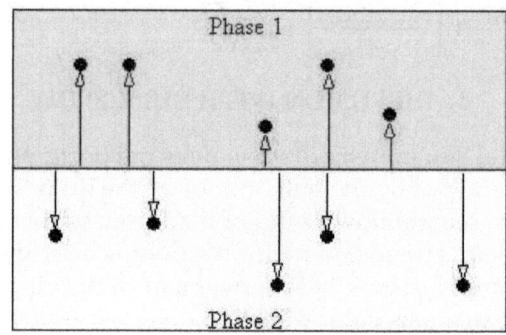

Fig 1.4

Referring to Fig 1.4 if a component, A is passing between phases 1 & 2, there will ultimately be a dynamic equilibrium established where the rate of diffusion of A is the same in both directions. Depending on the type of system (liquid – vapour, liquid – liquid *etc.*), the equilibrium concentration may often be expressed by simple relationships. Two such are mentioned below.

5.1. Distribution Coefficients

These most are often found in liquid-liquid systems and in vapour-liquid systems. The simplest form of law is a linear one of the form

$$YA = KX_A \qquad\qquad 1.6$$

Where : X_A = concentration of the component A in Phase 1

Y_A = concentration of the component A in Phase 2

K = constant

In more complex systems K is not a constant but varies with concentration.

5.2. Henry's Law

Henry's law is a special case of the distribution coefficient for gas-liquid systems. In the case of systems where one of the phases is a gas or vapour it is common to express the concentration in the gas phase terms of partial pressure. Thus Henry's Law states that

$$P_A = H\,C_A \qquad\qquad 1.7$$

Where: P_A = Partial pressure of A in the gas phase

C_A = Concentration of A in the liquid phase

H = constant - called Henry's law constant

6. RATE OF DIFFUSION BETWEEN PHASES - MASS TRANSFER

The theory discussed so far enables us to calculate the rates of diffusion within a single phase and to calculate the concentrations of a component in two phases when the system is in a state of equilibrium. However, many practical problems concern the rate of diffusion between two phases when the two phases are not in equilibrium.

The rate of mass transfer between two phases is dependant on a number of factors including

1. The diffusivity of the diffusing component in the two phases
2. How far the system is from equilibrium
3. The resistance to transfer across the interface between the two phases

There are a number of theories of mass transfer. Three widely used theories are the Whitman two film theory, the Penetration theory of Higbie and the surface renewal theory of Dankwaerts. Each of the theories has its strengths and weakness but ultimately all come up with an equation of the general type

$$N_A = kA\Delta C_A \qquad\qquad 1.5$$

Where : N_A = mass transfer rate of A across the phase boundary

K = mass transfer coefficient

C = concentration driving force

7. CONCENTRATION DRIVING FORCE

The concentration driving force is a measure of how far the system is from equilibrium and may be developed with reference to the diagram below. (Fig 1.5)

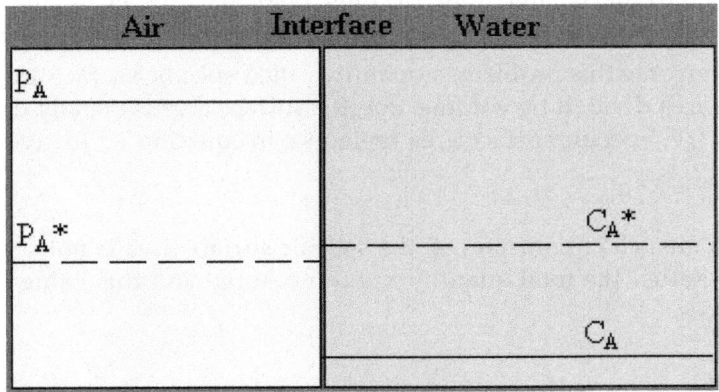

Fig 1.5

Consider a component, such as oxygen, diffusing from air to water. Assume the system is at steady state but *not at equilibrium*.

* P_A is the partial pressure of oxygen in the air and C_A is the concentration of oxygen in the water.

- C_A^* is the concentration of oxygen in water that will be in equilibrium with P_A
- P_{A^*} is the partial pressure of oxygen in air that will be in equilibrium with C_A

The rate of oxygen mass transfer between air and water may be expressed in two ways. One based on partial pressure of oxygen in air and one based on the concentration of oxygen in water.

$$N_A = K_G A(P_A - P_A^*) \qquad \qquad 1.6$$

$$N_A = K_L A(C_A^* - C_A) \qquad \qquad 1.7$$

Where : $P_A - P_A^*$ = partial pressure driving force

$C_A - C_A^*$ = concentration driving force

K_L = mass transfer coefficient based on the liquid phase.

K_G = mass transfer coefficient based on the gas phase.

The approriate equilibrium value P_A^* or C_A^* may be calculated using Henry's law

$$P_A^* = H\,C_A \qquad \qquad 1.8$$

or $$P_A = H\,C_A^* \qquad \qquad 1.9$$

It is usual in fermentation to use concentration driving force and liquid phase mass transfer coefficients (equation 1.7)

In practical mass transfer systems, there is a usually a problem of determining the interfacial area. Most practical mass transfer involves either passing the two phases over some form of packing or bubbling one phase through another. Either way the aim is to maximise the interfacial area. In such circumstances, it is not practically possible to determine the actual interfacial area.

To overcome this problem, a quantity called specific surface area is defined as surface area divided by volume. Specific surface area is usually denoted by a, hence $a = A/V$. Specific surface area replaces a in equation 1.7 to give.

$$N_A = K_L a(C_A^* - C_A) \qquad \qquad 1.10$$

Even in such circumstances, the specific surface area is not separately determined. Rather the total quantity $K_L a$ is measured and this value is quoted in most sources.

Problems

1. A tube 300 mm long and 6 mm diameter contains a mixture of ammonia and air at 20°C. The partial pressure of ammonia at one end of the tube is maintained at 10 000 Pa and at the other end the ammonia is removed at by a current of air so that its partial pressure is negligible. How long will it take to transfer 0.1mg of ammonia along the tube?

Data

Diffusivity,	PD_{AB}	$= 2.33$ Pa m^2 s^{-1}
Total Pressure,	P	$= 100$ kPa
Universal Gas constant,	R	$= 8.314$ J mol^{-1} K^{-1}
RMM ammonia,	NH_3	$= 17$

2. A tube 60 mm high having a cross sectional area of 3 mm^2 contains 3 mg water at 30°C. The partial pressure of water vapour above the water surface is 7370 Pa and at the top of the tube is 2950 Pa. Assuming only molecular diffusion, How long will it take to evaporate the water in the tube?

Data

Total Pressure,	P	$= 10^5$ Pa
Diffusivity	PD_{AB}	$= 2.76$ Pa m^2 s^{-1}
Universal Gas constant	R	$= 8.314$ J mol^{-1} K^{-1}

3. Oxygen is diffusing from air to water under conditions such that oxygen concentration in water is constant at 2.0 mg l^{-1} and the oxygen concentration in air is constant at 21 % vol. The temperature is 20°C and the pressure atmospheric ($= 10^5$ Pa). Calculate;

    a. The rate of diffusion of air into water given an gas phase mass transfer coefficient K_G of 1.05×10^{-7} mg m s^{-1} Pa^{-1}

    b. The liquid phase mass transfer coefficient K_L.

Values of Henry's law constant for oxygen-water are given in Fig 1.8

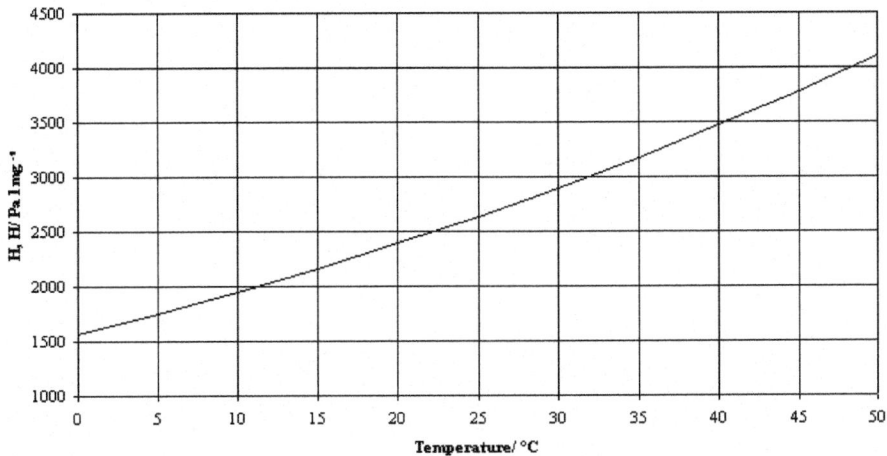

4. Oxygen gas is diffusing from air to water. Conditions are such that the oxygen concentration in the water is maintained at 10% of saturation. Under these conditions Henry's constant for oxygen-water is 2350 Pa l mg^{-1} and the liquid phase mass transfer coefficient, K_L is 1.7×10^{-3} m s^{-1}. The concentration of oxygen in air is 21% vol and the atmospheric pressure is 10^5 Pa. Calculate;

a. the rate of diffusion of the oxygen from air to water,

b. the gas phase mass transfer coefficient Kg,

c. The rate of diffussion if the oxygen concentration in the air is enriched to 30% v/v. The mass transfer coefficient remains constant and the oxygen content of the water is 10% of the saturation value.

Chapter 10

ENERGY BALANCE AND HEAT EXCHANGE

STEADY STATE ENERGY BALANCE

General Balance Equation Revisited

Recall the general balance equation that was derived for *any* system property :

In − Out + Generation − Consumption = Accumulation

When we derived the mass balance, we did so by citing the law of conservation of mass, which states that the total generation of mass is 0, and therefore *Accumulation = In − Out.*

There is one other major conservation law which provides an additional equation we can use : the law of conservation of energy. This states that if E denotes the entire amount of energy in the system,

Law of Conservation of Energy

$$E_{in} - E_{out} = E_{accumulated}$$

Types of Energy

In order to write an energy balance, we need to know what kinds of energy can enter or leave a system. Here are some examples (this is not an exhaustive list by any means) of the types of energy that can be gained or lost.

1. A system could gain or lose *kinetic energy*, if we're analyzing a moving system.

2. Again, if the system is moving, there could be *potential energy* changes.

3. *Heat* could enter the system via conduction, convection, or radiation.

4. *Work* (either **expansion work** or **shaft work**) could be done on, or by, the system.

The total amount of energy entering the system is the sum of all of the different types entering the system. Here are the expressions for the different types of energy :

1. From physics, recall that $KE = \frac{1}{2}mv^2$. If the *system* itself is not moving, this is zero.

2. The *gravitational* potential energy of a system is $GPE = mgh$ where g is the gravitational constant, m is mass in kg and h is the height of the center of mass of the system. If the system does not change height, there is no change in GPE.

3. The heat entering the system is denoted by Q, regardless of the mechanism by which it enters (the means of calculating this will be discussed in a course on transport phenomenon). According to this book's conventions, *heat entering a system is positive and heat leaving a system is negative*, because the system in effect gains energy when heat enters.

4. The work done by or on the system is denoted by W. *Work done BY a system is negative* because the system has to "give up" energy to do work on its surroundings. For example, if a system expands, it loses energy to account for that expansion. Conversely, *work done ON a system is positve*.

Energy Flows due to Mass Flows

Accumulation of *anything* is 0 at steady state, and energy is no exception. If, as we have the entire time, we assume that the system is at steady state, we obtain the energy balance equation :

$E_{in} = E_{out}$

This is the starting point for all of the energy balances below.

Consider a system in which a mass, such as water, enters a system, such as a cup, like so :

The mass flow into (or out of) the system carries a certain amount of energy, associated with how fast it is moving (kinetic energy), how high off the ground it is (potential energy), and its temperature (internal energy). It is possible for it to have other types of energy as well, but for now let's assume that these are the only three types of energy that are important. If this is true, then we can say that the total energy carried *in the flow itself* is :

$$\dot{E}_i = (\frac{1}{2}\dot{m}v^2 + \dot{m}gh + \dot{U})_i$$

However, there is one additional factor that must be taken into account. When a mass stream flows into a system it expands or contracts and therefore performs work on the system. An expression for work due to this expansion is :

$$W_{exp} = P * \dot{V}_i$$

Since this work is done *on* the system, it enters the energy balance as a positive quantity. Therefore the total energy flow into the system due to mass flow is as follows :

$$\dot{E}_i = (\frac{1}{2}\dot{m}v^2 + \dot{m}gh + \dot{U})_i + P * \dot{V}_i$$

Now, to simplify the math a little bit, we generally don't use internal energy and the PV term. Instead, we combine these terms and call the result the enthalpy of the stream. Enthalpy is just the combination of internal energy and expansion work due to the stream's flow, and is denoted by the letter H :

Definition of Enthalpy

$H = U + PV$

Therefore, we obtain the following important equation for energy flow carried by mass :

In stream i, if only KE, GPE, internal energy, and expansion work are considered, the energy carried by mass flow is :

$$\dot{E}_i = (\frac{1}{2}\dot{m}v^2 + \dot{m}gh + \dot{H})_i$$

Note : Kinetic energy and potential energy are generally very small compared to the enthalpy, except in cases of very rapid flow or when there are no significant temperature changes occurring in the system. Therefore, they are often neglected when performing energy balances.

Other Energy Flows into and Out of the System

The other types of energy flows that could occur in and out of a system are *heat* and *work*. Heat is defined as energy flow due to a change in temperature, and always flows from higher temperature to lower temperature. Work is defined as an energy transferred by a force (see here for details).

- If there is no heat flow into or out of a system, it is referred to as **adiabatic**.
- If there are no mechanical parts connected to a system, and the system is not able to expand, then the work is essentially 0.

Some systems which have mechanical parts that perform work are turbines, mixers, engines, stirred tank reactors, agitators, and many others. The type of work performed by these parts is called shaft work to distinguish it from work due to expansion of the system itself (which is called *expansion work*).

An "insulated system" is generally interpreted as being essentially adiabatic, though how good this assumption is depends on the quality of the insulation. A system that cannot expand is sometimes described as "rigid".

The notation for these values are as follows :

- Heat flows : $\dot{Q}_j$, at the "j"th location.

- Shaft work : $\dot{W}_s$

- Expansion work : $P * \dfrac{\Delta V}{\Delta t}$

Note that the above implies that there is no expansion work at steady state because at steady state nothing about the system, including the volume, changes with time, *i.e.* $\dfrac{\Delta V}{\Delta t} = 0$ at steady state.

Overall Steady - State Energy Balance

If we combine all of these components together, remembering that heat flow into a system and work done *on* a system are positive, we obtain the following :

Steady State Energy Balance on an Open System

$$\Sigma(\frac{1}{2}\dot{m}v^2 + \dot{m}gh + \dot{H})_{i,in} - \Sigma(\frac{1}{2}\dot{m}v^2 + \dot{m}gh + \dot{H})_{i,out} + \Sigma\dot{Q}_j + \dot{W}_s = 0$$

Some important points :

1. If the system is **closed AND at steady state** that means the total heat flow must equal the total work done in magnitude, and be opposite in sign. However, according to another law of thermodynamics, the second law, it is impossible to change ALL of the heat flow into work, even in the most ideal case.

2. In an adiabatic system with no work done, the total amount of energy carried by mass flows is equal between those flowing in and those flowing out. However, that DOES NOT imply that the temperature remains

the same, as we will see in a later section. Some substances have a greater capacity to hold heat than others, hence the term **heat capacity**.

3. If the conditions *inside the system* change over time, then we CANNOT use this form of the energy balance. The next section has information on what to do in the case that the energetics of the system change.

UNSTEADY STATE ENERGY AND MASS BALANCES

What is Accumulation?

Recall that so far in this text it has been assumed that all systems are at steady state, which means that there is no buildup of mass, energy, or other conserved quantities. However, there are many situations, such as whenever operating levels change, that a system will not be at steady state, and mass and energy will be accumulated over time.

The most important thing to remember about accumulation is that *it deals with the actual amount of stuff in the system, not any sort of flow rate*. If you remember if you're dealing with actual system properties rather than flow rates, it will help keep the terms straight in unsteady-state balances.

Unsteady-state Mass Balance

Lets begin the derivation of an unsteady-state mass balance with the general balance equation which you should know and love by now :

In − Out + Generation = Accumulation

Substituting the terms we usually used for in, out, and generation, we obtain :

$\Sigma \dot{m}_{i,in} - \Sigma \dot{m}_{i,out} + \dot{m}_{i,gen} = Accumulation$

Now we have to come up with a mathematical formulation for the accumulation. Unlike all of the other terms in this equation, which deal with energy *flows* into the system, the accumulation deals with the amount of energy that is already in the system at a certain point of time, and more specifically how it *changes* with time.

The rate of accumulation of energy *will not be constant* unless it is zero (otherwise every reactor in the world would either blow up from excessive mass and energy buildup or would cease operating because all of the reactants and products would be drained out). Recall that if the accumulation reaches zero, the system is at steady state. Most systems tend to move towards a steady state (it is possible to have more than one set of steady state conditions, but it won't be covered here) over long periods of time, as shown below :

Such a system is called self-regulating (or naturally stable). If a system is not self-regulating then special control techniques (see Control systems) must be utilized to force the system into a steady state. In order to take into account variation in the accumulation rate, we must consider the rate of change over a *very small* amount of time, so small in fact that it is practically zero, and the accumulation *vs.* time curve resembles a straight line. The slope of this line at time t is approximately :

Slope = Accumulation rate at time $t = \dfrac{M_{sys,t+\Delta t} - M_{sys,t}}{\Delta t}$

Therefore we could write the following :

Accumulation $= \dfrac{M_{sys,t+\Delta t} - M_{sys,t}}{\Delta t}$

We then write our mass balance by substituting this accumulation into the mass balance above :

$$\frac{M_{sys,t+\Delta t} - M_{sys,t}}{\Delta t} = \Sigma \dot{m}_{i,in} - \Sigma \dot{m}_{i,out}$$

For practical applications, this equation is generally multiplied by Δt. Then, rather than dealing with flow rates, a new quantity is defined :

$$\Delta m_i = \dot{m}_i * \Delta t$$

This quantity is the *total amount of mass that enters the system in a finite amount of time*. Substituting this definition into the mass balance yields the following :

Unsteady State Mass Balance

$$\Sigma \Delta m_{i,in} - \Sigma \Delta m_{i,out} + m_{i,gen} = M_{sys,t+\Delta t} - M_{sys,t}$$

Example :

A feed stream with $50\dfrac{kg}{h}$ of water and $1\dfrac{kg}{h}$ of ethanol enters a distillation column. A distillation column generally has two outlet streams called the bottoms and the condensate. At steady state, the condensate is 12% ethanol by mass and the total condensate flowrate is $9\dfrac{kg}{h}$.

One day, the boss calls and says that she needs more production, so you turn up the feed to $60\dfrac{kg}{h}$. Two hours later, the distillation column floods.

a.   What was the cause of the flooding?

b.   Assuming that the total outlet mass flow rates remained the same throughout the process, what was the total mass accumulation in the column?

c.   Describe two methods by which the flow rates may be modified to reach a new steady state. Will the new steady state produce the same outlet concentrations as the old steady state? Explain. (hint : how is the separation effectiveness related to the ratio of the two outlet flowrates? You may need to do some research on this)

Unsteady-state Energy Balance

Lets start by examining what we have so far, but with the accumulation term (yet to be defined mathematically) added in the right side, since we're not at steady state any more :

$$\Sigma(\frac{1}{2}\dot{m}v^2 + \dot{m}gh + \dot{H})_{i,in} - \Sigma(\frac{1}{2}\dot{m}v^2 + \dot{m}gh + \dot{H})_{i,out} + \Sigma\dot{Q}_j + \dot{W}_s = Accumulation$$

Following the logic from the mass balance, we obtain for the accumulation :

$$\frac{M_{sys,t+\Delta t} - E_{sys,t}}{\Delta t}$$

Therefore, we have :

$$\Sigma(\frac{1}{2}\dot{m}v^2 + \dot{m}gh + \dot{H})_{i,in} - \Sigma(\frac{1}{2}\dot{m}v^2 + \dot{m}gh + \dot{H})_{i,out} + \Sigma\dot{Q}_j + \dot{W}_s = \frac{M_{sys,t+\Delta t} - E_{sys,t}}{\Delta t}$$

Like in the case of the mass balance, we can only consider the total energy change over a total amount of time using this equation. To do this, we multiply the entire equation above by the time change from some starting point to the point of interest.

Now we need some definitions :

1. $Q = \dot{Q} * \Delta t$ is the TOTAL heat flow over the time period.

2. $W_s = \dot{W}_s * \Delta t$ is the TOTAL shaft work over the time period.

3. $\dot{m}_i * \Delta t = \Delta m_i$ is the TOTAL mass flow into (or out of) the system due to stream i during the time period.

4. $\dot{H}_i * \Delta t = H$ is the TOTAL enthalpy carried into (or out of) the system due to stream i during the time period.

The major assumption here is that the enthalpies, heat flow rates, and shaft work on the left hand side of the equals sign must either be constant, or the average value over the whole time period must be used, in order for this equation to be valid. Whether this assumption is valid or not depends on the situation (for example, it depends on whether the process feeding mass to your process is itself at steady state or not).

With these in mind, we multiply by delta t in order to obtain the following, unsteady state energy balance.

Unsteady State Energy Balance

$$\Sigma(\frac{1}{2}\Delta mv^2 + \Delta mgh + H)_{i,in} - \Sigma(\frac{1}{2}\Delta mv^2 + \Delta mgh + H)_{i,out} + \Sigma Q_j + W_s = M_{sys,t+\Delta t} - E_{sys,t}$$

ENERGY BALANCES

As with mass balances it is useful to start with our initial definition of a balance equation :

IN - OUT + GENERATION - CONSUMPTION = ACCUMULATION

Where IN and OUT correspond to energy flowing into and out of the system, respectively. In contrast to mass balances, however, energy is a *conserved* quantity (First Law of Thermodynamics) and therefore GENERATION=CONSUMPTION=0! (We will see as we go that this is *also* true for total mass, and even for atoms, but *not* for specific molecular species). This gives us :

IN - OUT = ACCUMULATION

In order to finish this balance we need to look at exactly *how* energy can move IN and OUT of a system...

Also, as we did with mass balances, we need to start out with some definitions.

TYPES OF ENERGY

Definition :

Kinetic energy is energy due to the translational (or rotational) motion relative to some frame of reference.

Definition :

Potential energy is energy due to something's position in a potential field (electromagnetic or gravitational, for example).

Definition :

Internal energy is where we lump everyone else! (Thermal energy, chemical energy, *etc.*)

Each bit of mass within our balances will contain some amount of each of these forms of energy, so that we can write the *total* energy of some bit of mass as :

$$E_{total} = U + E_k + E_p$$

where U is the internal energy, E_k is the kinetic energy, and E_p is the potential energy.

HOW ENERGY MOVES

Obviously, we can imagine that STUFF has energy (a hot potato, for example).

One way for the energy to move would be to move the entire potato! Therefore, like we had in mass balances we can have an IN-flow and OUT-flow in the system through moving mass. (Some potatoes may be hotter or colder, moving faster or slower, *etc.*)

In contrast to mass balances, however, we can also *change the energy that is in the potatoes already in the system*. Imagine putting a bucket of potatoes in the oven (we use heat to change the internal energy of the potatoes already in the bucket). Or perhaps we vigorously shake the bucket of potatoes; here, we convey mechanical (kinetic) energy right through the walls!

Energy Balance Equations

As with mass balances, energy balances will have an integral and differential form.

Recalling that our general energy balance equation looks like this :

$$IN - OUT = ACCUMULATION$$

We can start with our differential form by writing :

$dtdEtotal = Etotalin. - Etotalout.$

Recalling that total energy is given by :

$$E_{total} = U + E_k + E_p$$

we now simply need to account for the possibility of sticking stuff in an oven, or "shaking" it. We will denote the net flow of heat into the system as Q, and the net transfer of mechanical energy as W. The only catch is that we will consider a negative W to be going *into* the system, while a negative Q will be *coming out* of the system. This gives us :

General Differential Energy Balance

$dtdU + dtdEk + dtdEp = Uin. - Uout. + Ekin. - Ekout. + Epin. - Epout. + Q - W$

For a steady state problem this reduces to :

$Uout. - Uin. + Ekout. - Ekin. + Epout. - Epin. = Q - W$

As we did with mass, we can integrate this to yield this integral form of the balance :

General Integral Energy Balance

$Ufinal. - Uinitial. + Ekfinal. - Ekinitial. + Epfinal. - Epinitial. = \int tfinaltinitialUin.dt - \int tfinaltinitialUout.dt + \int tfinaltinitialEkin.dt - \int tfinaltinitialEkout.dt + \int tfinaltinitialEpin.dt - \int tfinaltinitialEpout.dt + \int tfinaltinitialQdt - \int tfinaltinitialWdt$

which for a closed system simplifies to :

$Ufinal - Uinitial + Ekfinal - Ekinitial + Epfinal - Epinitial = \int tfinaltinitialQdt - \int tfinaltinitialWdt.$

Explanation

There are really three types of energy that we'll deal with often in chemical engineering: Kinetic Energy, *Ek*, Potential Energy, *Ep*, and Internal Energy, U. Closely related to Internal Energy is Enthalpy, H. Now, lets introduce the definitions, mathematical descriptions, and the concepts for each type of energy.

Kinetic Energy

Definition: Energy due to motion

Relation:

$$E_k = \frac{1}{2}mv^2$$

$$\Delta E_k = \frac{1}{2}m(v_{final}{}^2 - v_{initial}{}^2)$$

$$\dot{E}_k = \frac{1}{2}\dot{m}v^2$$

$$\Delta \dot{E}_k = \frac{1}{2}\dot{m}(v_{final}{}^2 - v_{initial}{}^2)$$

Where m is for mass, v is for velocity, and the dot signifies a rate.

Examples: A moving car, a gas molecule, and a flowing stream all have kinetic energy. A parked car, a molecule in a solid lattice, and a stationary glass of water are all systems that do not possess kinetic energy (because they do not have a non-zero velocity).

Note: If the system is not accelerating or decelerating (changing velocity), DEK = 0. That is, the system has kinetic energy, but that energy doesn't change as you move between two point unless the system has undergone a change in velocity: the **change** in kinetic energy is zero. In the bulk of energy problems that you will deal with, that is, those in which there is a chemical reaction, a change of state, or substantial temperature change, the change in kinetic energy is very small (DEk = 0).

Potential Energy

Definition: Energy due to position

Relation:

$$E_p = mgh$$

$$\Delta E_p = mg\,(h_{final} - h_{initial})$$

$$\dot{E}_p = \dot{m}gh$$

$$\Delta \dot{E}_p = \dot{m}g\,(h_{final} - h_{initial})$$

Where m is for mass, g is gravity, h is a height, and the dot signifies a rate

Examples: A bungee jumper standing on the edge of a cliff, a book resting on a table, a jacket hung in a closet, or anything that is above the ground possesses potential energy due to gravity. So, someone standing on the ground, and shoes that rest on the ground are both systems that possess no potential energy due to gravity.

Note: If the system neither rises nor falls, the change in potential energy is zero (DEp = 0). Additionally, rarely does the system rise or fall substantially in the energy balance problems that you will perform for chemical engineering problems in this class, so you will find that the potential energy change is often negligable and set to zero. Finally, you will notice that you will only deal with potential energy due to gravity in this course. If any other source of potential energy relavent, such as electrical potential, or some other type, that will be clearly stated in the problem statement.

Internal Energy

Definition: "molecular energy", due to molecular interactions

Use: in energy balances on **closed systems**

Relation:

$$U = m \cdot \hat{U}$$

$$\Delta U = m \cdot (\hat{U}_{final} - \hat{U}_{initial})$$

$$\Delta \hat{U} = \int_{T_{initial}}^{T_{final}} C_v dT$$

$$\Delta \hat{U} = C_v \cdot (T_{final} - T_{initial})$$

CV = the heat capacity at constant volume.

Here, there are two equations to find the specific internal energy using a term with a hat over it (U-*hat*); the third equation is used if the heat capacity is a function of temperature (often for gases). The last equation is for a constant heat capacity, (often used for liquids and solids). Each of these equations involves the use of specific properties (those with hats over them, meaning that they are "per mol" or "per mass"). Once we find the specific internal energy change with the third or fourth equations, we can find the internal energy change by multiplying it by the amount of substance undergoing that change. Additionally, we can find the rate of internal energy change (DU-dot) by multiplying by the mass or molar flow rate.

It is assumed that U-hat has units of Energy/mass in the above equations, when the first formulas could be used. If U-hat has units of Energy/mole, of course, we would multiply the specific internal energy by moles (and not mass) to get the internal energy (using, for example, U = n * U-hat). It may be more correct to write U = (n or m) * U-hat. The key to getting this write is to pay attention to the units of the specific property. This of course applies to all specific properties (U-hat, H-hat, V-hat, *etc.*) and is probably a point that you are already familiar with.

Example:

Temperature Change of a Solid

Imagine a molecule, say a water molecule, is stuck in a lattice (an ice cube). Here, it is able to vibrate a little, but it is fixed in position with respect to its neighboring water molecules. If we neither increase nor decrease the temperature of the freezer (no temperature change) its internal energy ("molecular energy") doesn't change; it simply continues to vibrate to the same extent. INow, if we make the system colder, thes molecule vibrates less, and there is a change in internal energy (a decrease). If we take the ice cube out of the freezer and place it where it is warmer, the molecule vibrates more (gaining energy, an increase in internal energy). If it melts, it further gains internal energy. In this example there is a temperature change to accompany the molecule's change in internal energy.

Note: Changes in internal energy occur with processes that change its molecular energy. Thus, internal energy changes occur with a change in temperature (effects a molecule's motion), a chemical reaction (breaking and making bonds), a change of state (change in intermolecular energy), or a substantial pressure change(under high pressures, our H20 molecule in our ice cube vibrates less).

Enthalpy

Definition: "molecular energy", due to molecular interactions

Use: in energy balances on **open systems**

Relation:

$$H = m \cdot \hat{H}$$

$$\Delta H = m \cdot (\hat{H}_{final} - \hat{H}_{initial})$$

$$\Delta \hat{H} = \int_{T_{initial}}^{T_{final}} C_p dT$$

$$\Delta \hat{H} = C_p \cdot (T_{final} - T_{initial})$$

$$\hat{H} = \hat{U} + p\hat{V}$$

Cp = the heat capacity at constant pressure.

Here, the usage of the first -> fourth equations is analogous to scenarios for those equations for internal energy already explained above. However, the last equations describe the relationship between enthalpy and internal energy. Similar discussions given above in the internal energy section about flow rates and specific properties are applicable here.

Example: Enthalpy can be thought of as the "molecular energy," or the internal energy for moving fluids. Thus, if a flow of water in a pipe experiences a change in temperature, it is accompanied by a change in enthalpy.

Note: A system experiences a change in enthalpy if there is a change in temperature, a substantial change in pressure, a chemical reaction, or if there is a change in state.

Note: Additionally, there is a relationship between the heat capacity at constant volume and the heat capacity at constant pressure, this relation is deferred until later.

TYPES OF ENERGY TRANSFER

Explanation:

There are two types of energy transfer, heat and work. There should be a clear distinction between these two types of energy transfer and the forms of energy just mentioned. Where kinetic, potential, and internal energy are all forms of energy, heat and work are energy vehicles that move energy from one place to another. They act to transport energy by carrying energy from the surroundings into the system or from the system to the surroundings. Further examples follow, but let me give you one now: First, we need to name the system and the surrounding. The system could be a pot of water, and let's say that it experiences an increase in temperature (an increase in internal energy). The system gets energy from the surroundings and we say that this energy is transported by heat.

HEAT

Definition: Energy transported due to a difference in temperature between the system and the surrounding.

Notation:

$$W \quad \text{or} \quad \dot{Q}$$

where the first notation designates some quantity of heat transported, and the second tells how fast heat is being transported per time.

Example: Transport of energy from a stove burner to the pot, from the human body to the air, or from a chemical reaction to a flask is done by heat.

Sign convention: "+Q for heat into the system". In otherwords, heat into the system is positive. If a pot of water is the system and the temperature is being increased due to heating, the heat is positive (+Q). If we then let the pot sit and cool, it looses heat to the surrounding, thus a negative heat (-Q).

Note: The words adiabatic and insulated are used to describe a system in which heat transfer is zero (Q = 0).

WORK

Definition: Energy transport due to any mechanism other than heat (by a piston, propeller, electric current, electromagnetic radiation, etc).

Notation:

$$W \quad \text{or} \quad \dot{W}$$
$$\dot{W}$$

where the first expression designates the general work expression (units, Joule), the second designates work per time (units, Joule/second), and the third is the modified work expression that we will use for open systems, where the s subscript designates shaft work, such as work by a propeller (units, Joule/second).

Example: Energy transport due to work when stirring a cup of water increases its temperature, when irradiating a cup of water with microwaves, thus increasing its temperature, or the hammering of a nail, resulting in a warmed metal nail.

Sign Convention: Work done on the system is positive (stirring our cup of water), work done by the system is negative (if our system is the computer screen, it releases work to the surroundings in the form of electromagnetic waves).

The 1st Law of Thermodynamics

Explanation :

"Energy is neither created nor destroyed." There are several other ways to say this, such as the energy of the universe is constant, or that energy doesn't appear spontaneously. That is, our cup of water doesn't increase its temperature sponaneously, but due to energy brought in from the surroundings via heat or work.

This statement is what allows us to make the energy balances. Energy, like mass, is conserved.

Getting to Specific Properties

The last topic has to do with the way properties are given. In earlier sections, we determined that m stood for mass, and m*dot* stood for mass flow rate. Here, we can determine that V stands for volume, or V*hat*, stands for specific volume, volume per amount (mass or mole). The following discussion leads to ways of differentiating between quantities such as volume and specific volume. First we introduce the two following definitions.

Intensive properties : properties independent of the amount of material.

Extensive properties : properties dependent on the amount of material.

Explanation :

Let me offer you a way to determine whether a property is an external or internal property - If the answer to the following question is yes, then the property is external (otherwise, it is an internal property) : Does the property change if we go from one gram to a 100 grams for the substance?

for example, determine whether the following properties are external or internal. mass (kg), temperature (K), volume (gallons), energy (joules), density (g/mL). The extensive properties listed here (those that change when we change the amount of substance) are : mass, volume, and energy.

Now we understand how to categorize a certain property as extensive or intensive. Finally, note that any extensive property can be turned into an intensive property if it is divided by an amount. The resulting intensive property is called a specific property.

For example, mass is an extensive property. If we divide mass by moles, we get an intensive property, namely, the molecular weight of a substance. We can divide an extensive property by either mass or an amount (such as moles) to turn it into an intensive property.

The reason intensive and extensive properties are covered is to introduce specific properties because we will use them so much in energy balances. That is, many of the physical properties taken from the tables in energy balance problems are listed as specific properties.

HEAT TRANSFER

Heat transfer describes the exchange of thermal energy, between physical systems depending on the temperature and pressure, by dissipating heat. Systems which are not isolated may decrease in entropy. Most objects emit infrared thermal radiation near room temperature. The fundamental modes of heat transfer are *conduction* or *diffusion, convection, advection* and *radiation*.

The exchange of kinetic energy of particles through the boundary between two systems is at a different temperature from another body or its surroundings. Heat transfer always occurs from a region of high temperature to another region of lower temperature. Heat transfer changes the internal energy of both systems involved according to the First Law of Thermodynamics. The Second Law of Thermodynamics defines the concept of thermodynamic entropy, by measurable heat transfer.

Thermal equilibrium is reached when all involved bodies and the surroundings reach the same temperature. Thermal expansion is the tendency of matter to change in volume in response to a change in temperature.

Overview

Heat is defined in physics as the transfer of thermal energy across a well-defined boundary around a thermodynamic system. The thermodynamic free energy is the amount of work that a thermodynamic system can perform. Enthalpy is a thermodynamic potential, designated by the letter "H", that is the sum of the internal energy of the system (U) plus the product of pressure (P) and volume (V). Joule is a unit to quantify energy, work, or the amount of heat.

Heat transfer is a process function (or path function), as opposed to functions of state; therefore, the amount of heat transferred in a thermodynamic process that changes the state of a system depends on how that process occurs, not only the net difference between the initial and final states of the process.

Thermodynamic and mechanical heat transfer is calculated with the heat transfer coefficient, the proportionality between the heat flux and the thermodynamic driving force for the flow of heat. Heat flux is a quantitative, vectorial representation of the heat flow through a surface.

In engineering contexts, the term *heat* is taken as synonymous to thermal energy. This usage has its origin in the historical interpretation of heat as a fluid (*caloric*) that can be transferred by various causes, and that is also common in the language of laymen and everyday life.

The transport equations for thermal energy (Fourier's law), mechanical momentum (Newton's law for fluids), and mass transfer (Fick's laws of diffusion) are similar, and analogies among these three transport processes have been developed to facilitate prediction of conversion from any one to the others.

Thermal engineering concerns the generation, use, conversion, and exchange of heat transfer. As such, heat transfer is involved in almost every sector of the economy. Heat transfer is classified into various mechanisms, such as thermal conduction, thermal convection, thermal radiation, and transfer of energy by phase changes.

Mechanisms

The fundamental modes of heat transfer are :

Advection

Advection is the transport mechanism of a fluid substance or conserved property from one location to another, depending on motion and momentum.

Conduction or diffusion

The transfer of energy between objects that are in physical contact. Thermal conductivity is the property of a material to conduct heat and evaluated primarily in terms of Fourier's Law for heat conduction.

Convection

The transfer of energy between an object and its environment, due to fluid motion. The average temperature, is a reference for evaluating properties related to convective heat transfer.

Radiation

The transfer of energy from the movement of charged particles within atoms is converted to electromagnetic radiation.

Advection

By transferring matter, energy — including thermal energy — is moved by the physical transfer of a hot or cold object from one place to another. This can be as simple as placing hot water in a bottle and heating a bed, or the movement of an iceberg in changing ocean currents. A practical example is thermal hydraulics. This can be described by the formula :

$$Q = v \cdot \rho \cdot c_p \cdot \Delta T$$

where Q is heat flux (W/m²), ρ is density (kg/m³), c_p is heat capacity at constant pressure (J/(kg*K)), Δ T is the change in temperature (K), v is velocity (m/s).

Conduction

Main Article : Thermal Conduction

On a microscopic scale, heat conduction occurs as hot, rapidly moving or vibrating atoms and molecules interact with neighboring atoms and molecules, transferring some of their energy (heat) to these neighboring particles. In other words, heat is transferred by conduction when adjacent atoms vibrate against one another, or as electrons move from one atom to another. Conduction is the most significant means of heat transfer within a solid or between solid objects in thermal contact. Fluids — especially gases — are less conductive. Thermal contact conductance is the study of heat conduction between solid bodies in contact.

Steady state conduction (see Fourier's law) is a form of conduction that happens when the temperature difference driving the conduction is constant, so that after an equilibration time, the spatial distribution of temperatures in the conducting object does not change any further. In steady state conduction, the amount of heat entering a section is equal to amount of heat coming out.

Transient conduction (see Heat equation) occurs when the temperature within an object changes as a function of time. Analysis of transient systems is more complex and often calls for the application of approximation theories or numerical analysis by computer.

Convection

Main Article : Convection

The flow of fluid may be forced by external processes, or sometimes (in gravitational fields) by buoyancy forces caused when thermal energy expands the fluid (for example in a fire plume), thus influencing its own transfer. The latter process is often called "natural convection". All convective processes also move heat partly by diffusion, as well. Another form of convection is forced convection. In this case the fluid is forced to flow by use of a pump, fan or other mechanical means.

Convective heat transfer, or convection, is the transfer of heat from one place to another by the movement of fluids, a process that is essentially the transfer of heat via mass transfer. Bulk motion of fluid enhances heat transfer in many physical situations, such as (for example) between a solid surface and the fluid. Convection is usually the dominant form of heat transfer in liquids and gases. Although sometimes discussed as a third method of heat transfer, convection is usually used to describe the combined effects of heat conduction within the fluid (diffusion) and heat transference by bulk fluid flow streaming. The process of transport by fluid streaming is known as advection, but pure advection is a term that is generally associated only with mass transport in fluids, such as advection of pebbles in a river. In the case of heat transfer in fluids, where transport by advection in a fluid is always also accompanied by transport via heat diffusion (also known as heat conduction) the process of heat convection is understood to refer to the sum of heat transport by advection and diffusion/conduction.

Free, or natural, convection occurs when bulk fluid motions (steams and currents) are caused by buoyancy forces that result from density variations due to variations of temperature in the fluid. *Forced* convection is a term used when the streams and currents in the fluid are induced by external means—such as fans, stirrers, and pumps—creating an artificially induced convection current.

Convection-Cooling

See Also : Nusselt Number

Convective cooling is sometimes described as Newton's law of cooling :

The rate of heat loss of a body is proportional to the temperature difference between the body and its surroundings.

However, by definition, the validity of Newton's law of cooling requires that the rate of heat loss from convection be a linear function of ("proportional to") the temperature difference that drives heat transfer, and in convective cooling this is sometimes not the case. In general, convection is not linearly dependent on temperature gradients, and in some cases is strongly nonlinear. In these cases, Newton's law does not apply.

Convection vs. Conduction

In a body of fluid that is heated from underneath its container, conduction and convection can be considered to compete for dominance. If heat conduction is too great, fluid moving down by convection is heated by conduction so fast that its downward movement will be stopped due to its buoyancy, while fluid moving up by convection is cooled by conduction so fast that its driving buoyancy will diminish. On the other hand, if heat conduction is very low, a large temperature gradient may be formed and convection might be very strong.

The Rayleigh number (Ra) is a measure determining the relative strength of conduction and convection.

$$Ra = \frac{g \Delta \rho L^3}{\mu \alpha} = \frac{g \beta \Delta T L^3}{\nu \alpha}$$

where

- g is acceleration due to gravity,
- ρ is the density with $\Delta \rho$ being the density difference between the lower and upper ends,
- μ is the dynamic viscosity,
- α is the Thermal diffusivity,
- β is the volume thermal expansivity (sometimes denoted a elsewhere),

- T is the temperature,
- v is the kinematic viscosity, and
- L is characteristic length.

The Rayleigh number can be understood as the ratio between the rate of heat transfer by convection to the rate of heat transfer by conduction; or, equivalently, the ratio between the corresponding time scales (*i.e.* conduction timescale divided by convection timescale), up to a numerical factor. This can be seen as follows, where all calculations are up to numerical factors depending on the geometry of the system.

The buoyancy force driving the convection is roughly $g\Delta\rho L^3$, so the corresponding pressure is roughly $g\Delta\rho L$. In steady state, this is canceled by the shear stress due to viscosity, and therefore roughly equals $\mu V/L = \mu/T_{conv}$, where V is the typical fluid velocity due to convection and T_{conv} the order of its timescale. The conduction timescale, on the other hand, is of the order of $T_{conv} = L^2/\alpha$.

Convection occurs when the Rayleigh number is above 1,000–2,000. Radiation

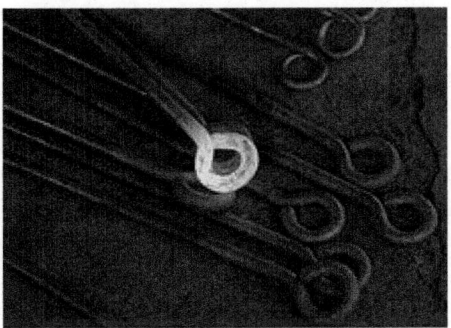

Red-hot iron object, transferring heat to the surrounding environment primarily through thermal radiation.

Thermal radiation occurs through a vacuum or any transparent medium (solid or fluid). It is the transfer of energy by means of photons in electromagnetic waves governed by the same laws. Earth's radiation balance depends on the incoming and the outgoing thermal radiation,Earth's energy budget. Anthropogenic perturbations in the climate system, are responsible for a positive radiative forcing which reduces the net long wave radiation loss out to Space.

Thermal radiation is energy emitted by matter as electromagnetic waves, due to the pool of thermal energy in all matter with a temperature above absolute zero. Thermal radiation propagates without the presence of matter through the vacuum of space.

Thermal radiation is a direct result of the random movements of atoms and molecules in matter. Since these atoms and molecules are composed of charged particles (protons and electrons), their movement results in the emission of electromagnetic radiation, which carries energy away from the surface.

The Stefan-Boltzmann equation, which describes the rate of transfer of radiant energy, is as follows for an object in a vacuum :

$$Q = \epsilon \sigma T^4$$

For radiative transfer between two objects, the equation is as follows :

where Q is the rate of heat transfer, ε is the emissivity (unity for a black body), σ is the Stefan-Boltzmann constant, and T is the absolute temperature (in Kelvin or Rankine). Radiation is typically only important for very hot objects, or for objects with a large temperature difference.

Radiation from the sun, or solar radiation, can be harvested for heat and power. Unlike conductive and convective forms of heat transfer, thermal radiation can be concentrated in a small spot by using reflecting mirrors, which is exploited in concentrating solar power generation. For example, the sunlight reflected from mirrors heats the PS10 solar power tower and during the day it can heat water to 285 °C (545 °F).

Phase Transition

Lightning is a highly visible form of energy transfer and is an example of plasma present at Earth's surface. Typically, lightning discharges 30,000 amperes at up to 100 million volts, and emits light, radio waves, X-rays and even gamma rays. Plasma temperatures in lightning can approach 28,000 Kelvin (27,726.85 °C) (49,940.33 °F) and electron densities may exceed 10^{24} m^{-3}.

Phase transition or phase change, takes place in a thermodynamic system from one phase or state of matter to another one by heat transfer. Phase change examples are the melting of ice or the boiling of water. The Mason equation explains the growth of a water droplet based on the effects of heat transport one vaporation and condensation.

Types of phase transition occurring in the four fundamental states of matter, include :

- **Solid** - Deposition, freezing and solid to solid transformation.
- **Gas** - Boiling/evaporation, recombination/deionization, and sublimation.
- **Liquid** - Condensation and melting/fusion.
- **Plasma** - Ionization.

Boiling

Main Article : Boiling

Nucleate boiling of water.

The boiling point of a substance is the temperature at which the vapor pressure of the liquid equals the pressure surrounding the liquid and the liquid evaporates resulting in an abrupt change in vapor volume.

Saturation temperature means boiling point. The saturation temperature is the temperature for a corresponding saturation pressure at which a liquid boils into its vapor phase. The liquid can be said to be saturated with thermal energy. Any addition of thermal energy results in a phase transition.

At low temperatures, no boiling occurs and the heat transfer rate is controlled by the usual single-phase mechanisms. As the surface temperature is increased, local boiling occurs and vapor bubbles nucleate, grow into the surrounding cooler fluid, and collapse. This is *sub-cooled nucleate boiling*, and is a very efficient heat transfer mechanism. At high bubble generation rates, the bubbles begin to interfere and the heat flux no longer increases rapidly with surface temperature (this is the departure from nucleate boiling, or DNB).

At high temperatures, the hydrodynamically-quieter regime of film boiling is reached. Heat fluxes across the stable vapor layers are low, but rise slowly with temperature. Any contact between fluid and the surface that may be seen probably leads to the extremely rapid nucleation of a fresh vapor layer ("spontaneous nucleation"). At higher temperatures still, a maximum in the heat flux is reached (the critical heat flux, or CHF).

The Leidenfrost Effect demonstrates how nucleate boiling slows heat transfer due to gas bubbles on the heater's surface. As mentioned, gas-phase thermal conductivity is much lower than liquid-phase thermal conductivity, so the outcome is a kind of "gas thermal barrier".

Condensation

Main Article : Condensation

Condensation occurs when a vapor is cooled and changes its phase to a liquid. During condensation, the latent heat of vaporizationmust be released. The amount of the heat is the same as that absorbed during vaporization at the same fluid pressure.

There are several types of condensation :

- Homogeneous condensation, as during a formation of fog.
- Condensation in direct contact with subcooled liquid.
- Condensation on direct contact with a cooling wall of a heat exchanger : This is the most common mode used in industry :
- Filmwise condensation is when a liquid film is formed on the subcooled surface, and usually occurs when the liquid wets the surface.
- Dropwise condensation is when liquid drops are formed on the subcooled surface, and usually occurs when the liquid does not wet the surface.

Dropwise condensation is difficult to sustain reliably; therefore, industrial equipment is normally designed to operate in filmwise condensation mode.

Melting

Main Article : Melting

Ice Melting

Melting is a physical process that results in the phase transition of a substance from a solid to a liquid. The internal energy of a substance is increased, typically by the application of heat or pressure, resulting in a rise of its temperature to the melting point, at which the ordering of ionic or molecular entities in the solid breaks down to a less ordered state and the solid liquefies. An object that has melted completely is molten. Substances in the molten state generally have reduced viscosity with elevated temperature; an exception to this maxim is the element sulfur, whose viscosity increases to a point due to polymerization and then decreases with higher temperatures in its molten state.

MODELING APPROACHES

Heat transfer can be modeled in the following ways.

Climate Models

Climate models study the radiant heat transfer by using quantitative methods to simulate the interactions of the atmosphere, oceans, land surface, and ice.

Heat Equation

The heat equation is an important partial differential equation that describes the distribution of heat (or variation in temperature) in a given region over time. In some cases, exact solutions of the equation are available; in other cases the equation must be solved numerically using computational methods.

Lumped System Analysis

Lumped system analysis often reduces the complexity of the equations to one first-order linear differential equation, in which case heating and cooling are described by a simple exponential solution, often referred to as Newton's law of cooling.

System analysis by the lumped capacitance model is a common approximation in transient conduction that may be used whenever heat conduction within an object is much faster than heat conduction across the boundary of the object. This is a method of approximation that reduces one aspect of the transient conduction system — that within the object — to an equivalent steady state system. That is, the method assumes that the temperature within the object is completely uniform, although its value may be changing in time.

In this method, the ratio of the conductive heat resistance within the object to the convective heat transfer resistance across the object's boundary, known as the *Biot number*, is calculated. For small Biot numbers, the approximation of *spatially uniform temperature within the object* can be used : it can be presumed that heat transferred into the object has time to uniformly distribute itself, due to the lower resistance to doing so, as compared with the resistance to heat entering the object.

Engineering

Heat exposure as part of a fire test for firestop products

Heat transfer has broad application to the functioning of numerous devices and systems. Heat-transfer principles may be used to preserve, increase, or decrease temperature in a wide variety of circumstances. Heat transfer methods are used in numerous disciplines, such as automotive engineering, thermal management of electronic devices and systems,climate control, insulation, materials processing, and power station engineering.

Insulation, Radiance and Resistance

Thermal insulators are materials specifically designed to reduce the flow of heat by limiting conduction, convection, or both. Thermal resistance is a heat property and the measurement by which an object or material resists to heat flow (heat per time unit or thermal resistance) to temperature difference.

Radiance or spectral radiance are measures of the quantity of radiation that passes through or is emitted. Radiant barriers are materials that reflect radiation, and therefore reduce the flow of heat from radiation sources. Good insulators are not necessarily good radiant barriers, and vice versa. Metal, for instance, is an excellent reflector and a poor insulator.

The effectiveness of a radiant barrier is indicated by its reflectivity, which is the fraction of radiation reflected. A material with a high reflectivity (at a given wavelength) has a low emissivity (at that same wavelength), and vice versa. At any specific wavelength, reflectivity = 1 - emissivity. An ideal radiant barrier would have a reflectivity of 1, and would therefore reflect 100 percent of incoming radiation. Vacuum flasks, or Dewars, are silvered to approach this ideal. In the vacuum of space, satellites use multi-layer insulation, which consists of many layers of aluminized (shiny) Mylar to greatly reduce radiation heat transfer and control satellite temperature.

Critical Insulation Thickness

Low thermal conductivity (k) materials reduce heat fluxes. The smaller the k value, the larger the corresponding thermal resistance (R) value. Thermal conductivity is measured in watts-per-meter per kelvin (W·m^{-1}·K^{-1}), represented as k. As the thickness of insulating material increases, the thermal resistance— or R-value—also increases.

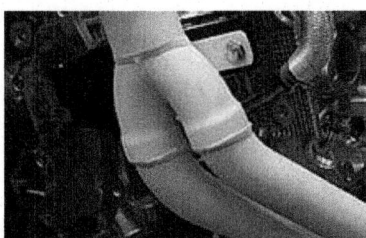

Car exhausts usually require some form of heat barrier, especially high performance exhausts where a ceramic coating is often applied.

For a cylinder, the convective thermal resistance is inversely proportional to the surface area and therefore the radius of the cylinder, while the thermal resistance of a cylindrical shell (the insulation layer) depends on the ratio between outside and inside radius, not on the radius itself. Suppose for example that we double the outside radius of a cylinder by applying insulation. We have added a fixed amount of conductive resistance (equal to $\ln(2)/(2\pi kL)$) but at the same time we have halved the value of the convective resistance. Because convective resistance tends to infinity when the radius approaches zero, at small enough radiuses the decrease in convective resistance will be larger than the added conductive resistance, resulting in lower total resistance. This implies that a critical radius exists at which the heat transfer is maximum. Above this critical radius, added insulation decreases the heat transfer. For insulated cylinders, the critical radius is given by the equation

$$r_{critical} = \frac{k}{h}$$

This equation shows that the critical radius depends only on the heat transfer coefficient and the thermal conductivity of the insulation. If the radius of the uninsulated cylinder is larger than the critical radius for insulation, the addition of any amount of insulation will decrease the heat transfer.

Devices

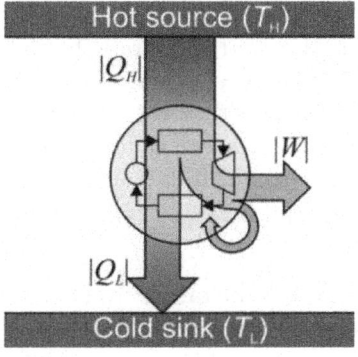

Heat Engine Diagram

- Heat engine is a system that performs the conversion of heat or thermal energyto mechanical energy which can then be used to do mechanical work.

- Thermocouple is a temperature-measuring device and widely used type of temperature sensor for measurement and control, and can also be used to convert heat into electric power.

- Thermoelectric cooler is a solid state electronic device that pumps (transfers) heat from one side of the device to the other when electrical current is passed through it. It is based on the Peltier effect.

- Thermal diode or thermal rectifier is a device that causes heat to flow preferentially in one direction.

Heat Exchangers

Main Article : Heat Exchanger

A heat exchanger is used for more efficient heat transfer or to dissipate heat. Heat exchangers are widely used in refrigeration, air conditioning, space heating, power generation, and chemical processing. One common example of a heat exchanger is a car's radiator, in which the hot coolant fluid is cooled by the flow of air over the radiator's surface.

Common types of heat exchanger flows include parallel flow, counter flow, and cross flow. In parallel flow, both fluids move in the same direction while transferring heat; in counter flow, the fluids move in opposite directions; and in cross flow, the fluids move at right angles to each other. Common constructions for heat exchanger include shell and tube, double pipe, extruded finned pipe, spiral fin pipe, u-tube, and stacked plate.

A heat sink is a component that transfers heat generated within a solid material to a fluid medium, such as air or a liquid. Examples of heat sinks are the heat exchangers used in refrigeration and air conditioning systems or the radiator in a car. A heat pipe is another heat-transfer device that combines thermal conductivity and phase transition to efficiently transfer heat between two solid interfaces.

Architecture

Efficient energy use is the goal to reduce the amount of energy required in heating or cooling. In architecture, condensation and air currents can cause cosmetic or structural damage. An energy audit, can help to assess the implementation of recommended corrective procedures. For instance, insulation improvements, air sealing of structural leaks or the addition of energy-efficient windows and doors.

- Smart meter is a device that records electric energy consumption in intervals.

- Thermal transmittance is the rate of transfer of heat through a structure divided by the difference in temperature across the structure. It is expressed in watts per square meter per kelvin, or W/m²K. Well-insulated parts of a building have a low thermal transmittance, whereas poorly-insulated parts of a building have a high thermal transmittance.

- Thermostat is a device to monitor and control temperature.

Climate Engineering

See Also : Anthropogenic Heat

An example application in climate engineering includes the creation of Biochar through the pyrolysisprocess. Thus, storing greenhouse gases in carbon reduces the radiative forcing capacity in the atmosphere, causing more long-wave (infrared) radiation out to Space.

Climate engineering consist of carbon dioxide removal and solar radiation management. Since the amount of carbon dioxide determines the radiative balance of Earth atmosphere, carbon dioxide removal techniques can be applied to reduce the radiative forcing. Solar radiation management is the attempt to absorb less solar radiation to offset the effects of greenhouse gases.

Greenhouse Effect

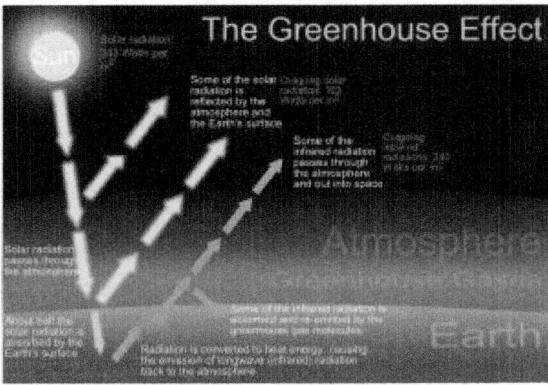

A representation of the exchanges of energy between the source (the Sun), the Earth's surface, the Earth's atmosphere, and the ultimate sink outer space.

The ability of the atmosphere to capture and recycle energy emitted by the Earth surface is the defining characteristic of the greenhouse effect.

The greenhouse effect is a process by which thermal radiation from a planetary surface is absorbed by atmospheric greenhouse gases, and is re-radiated in all directions. Since part of this re-radiation is back towards the surface and the lower atmosphere, it results in an elevation of the average surface temperature above what it would be in the absence of the gases.

Heat Transfer in the Human Body

The principles of heat transfer in engineering systems can be applied to the human body in order to determine how the body transfers heat. Heat is produced in the body by the continuous metabolism of nutrients which provides energy for the systems of the body. The human body must maintain a consistent internal temperature in order to maintain healthy bodily functions. Therefore, excess heat must be dissipated from the body to keep it from overheating. When a person engages in elevated levels of physical activity, the body requires additional fuel which increases the metabolic rate and the rate of heat production. The body must then use additional methods to remove the additional heat produced in order to keep the internal temperature at a healthy level.

Heat transfer by convection is driven by the movement of fluids over the surface of the body. This convective fluid can be either a liquid or a gas. For heat transfer from the outer surface of the body, the convection mechanism is dependent on the surface area of the body, the velocity of the air, and the temperature gradient between the surface of the skin and the ambient air. The normal temperature of the body is approximately 37°C. Heat transfer occurs more readily when the temperature of the surroundings is significantly less than the normal body temperature. This concept explains why a person feels "cold" when not enough covering is worn when exposed to a cold environment. Clothing can be considered an insulator which provides thermal resistance to heat flow over the covered portion of the body. This thermal resistance causes the temperature on the surface of the clothing to be less than the temperature on the surface of the skin. This smaller temperature gradient between the surface temperature and the ambient temperature will cause a lower rate of heat transfer than if the skin were not covered.

In order to ensure that one portion of the body is not significantly hotter than another portion, heat must be distributed evenly through the bodily tissues. Blood flowing through blood vessels acts as a convective fluid and helps to prevent any buildup of excess heat inside the tissues of the body. This flow of blood through the vessels can be modeled as pipe flow in an engineering system. The heat carried by the blood is determined by the temperature of the surrounding tissue, the diameter of the blood vessel, the thickness of the fluid, velocity of the flow, and the heat transfer coefficient of the blood. The velocity, blood vessel diameter, and the fluid thickness can all be related with the Reynolds Number, a dimensionless number used in fluid mechanics to characterize the flow of fluids.

Latent heat loss, also known as evaporative heat loss, accounts for a large fraction of heat loss from the body. When the core temperature of the body increases, the body triggers sweat glands in the skin to bring additional moisture to the surface of the skin. The liquid is then transformed into vapor which removes heat from the surface of the body. The rate of evaporation heat loss is directly related to the vapor pressure at the skin surface and the amount of moisture present on the skin. Therefore, the maximum of heat transfer will occur when the skin is completely wet. The body continuously loses water by evaporation but the most significant amount of heat loss occurs during periods of increased physical activity.

Cooling Techniques

Evaporative Cooling

A traditional air cooler in Mirzapur,Uttar Pradesh, India

Evaporative cooling happens when water vapor is added to the surrounding air. The energy needed to evaporate the water is taken from the air in the form of sensible heat and converted into latent heat, while the air remains at a constant enthalpy. Latent heat describes the amount of heat that is needed to evaporate the liquid; this heat comes from the liquid itself and the surrounding gas and surfaces. The greater the difference between the two temperatures, the greater the evaporative cooling effect. When the temperatures are the same, no net evaporation of water in air occurs; thus, there is no cooling effect.

Laser Cooling

In Quantum Physics laser cooling is used to achieve temperatures of near absolute zero(−273.15°C, −459.67°F) of atomic and molecular samples, to observe unique quantum effects that can only occur at this heat level.

- Doppler cooling is the most common method of laser cooling.

- Sympathetic cooling is a process in which particles of one type cool particles of another type. Typically, atomic ions that can be directly laser-cooled are used to cool nearby ions or atoms. This technique allows cooling of ions and atoms that cannot be laser cooled directly.

Magnetic Cooling

Magnetic evaporative cooling is a process for lowering the temperature of a group of atoms, after pre-cooled by methods such as laser cooling. Magnetic refrigeration cools below 0.3K, by making use of the magnetocaloric effect.

Radiative Cooling

Radiative cooling is the process by which a body loses heat by radiation. Outgoing energy is an important effect in the Earth's energy budget. In the case of the Earth-atmosphere system, it refers to the process by which long-wave (infrared) radiation is emitted to balance the absorption of short-wave (visible) energy from the Sun. Convective transport of heat and evaporative transport of latent heat both remove heat from the surface and redistribute it in the atmosphere.

Thermal Energy Storage

Thermal energy storage refers to technologies used to collect and store energy for later use. They can be employed to balance energy demand between day and nighttime. The thermal reservoir may be maintained at a temperature above (hotter) or below (colder) than that of the ambient environment. Applications include later use in space heating, domestic or process hot water, or to generate electricity.

Heat Exchanger in Bio-Chemical Process

The main purpose of the heat exchanger in a bio-process is **sterilization**.

There are other ways to kill unwanted organisms (contaminants), such as using chemicals and filtration. However, using heat energy seems to be the best way to sterilize feed before entering to the reactor.

Here is a simple flow diagram showing how heat transfers in a heat exchanger.

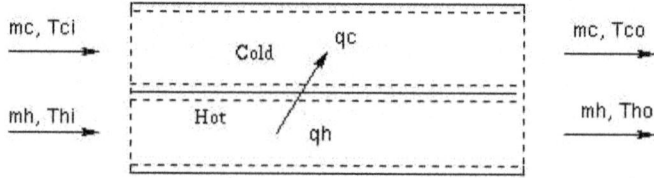

This flow arrangement is called co-current. If the direction of one of the stream is reversed, the arrangement is called counter-current flow.

Here are the temperature profiles along the heat exchanger. Note that the temperature profiles are different for co-current flow and for counter-current flow.

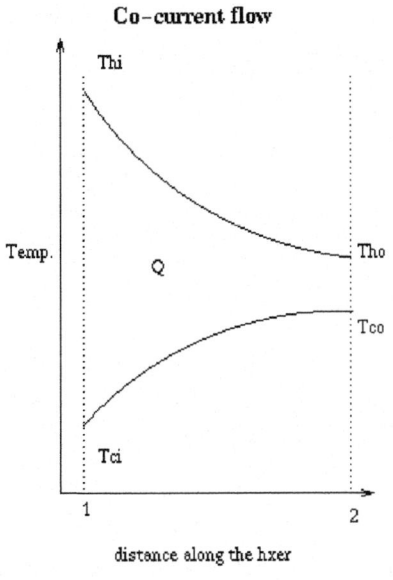

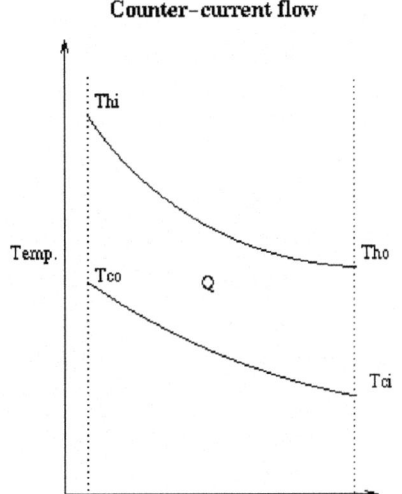

T_{ci} = Cold fluid inlet temperature
T_{co} = Cold fluid outlet temperature
T_{hi} = Hot fluid inlet temperature
T_{ho} = Hot fluid outlet temperature

The area between the curve is the heat transfer rate (Q). We can see that the heat transfer rate for counter-current flow is larger than the rate for co-current flow.

Counter flow heat exchanger provides more effective heat transfer. Most of the industrial heat exchangers are counter-current flow design.

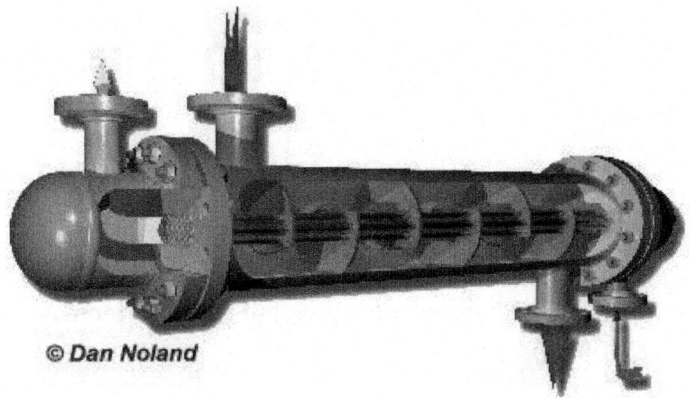

© Dan Noland

Design of a heat exchanger varies with needs. Click on the names to see the design of the heat exchanger. These pictures are taken from the "Continuous Sterilization" exercise from RPI's Chemical Engineering department.

Shell and Tube Heat Exchanger

Image with permission from Dan Nolan of Southern Heat Exchanger Corp.

Plate-and-Frame Heat Exchanger

Cooling Coil Heat Exchanger : There used to be a nice picture at this site for Owen's Publications. Now the link goes to their home page in case you would like to browse to try and find it.

2. On what does the performance of a heat exchanger depend?

To maximize the performance of a heat exchanger means saving money, especially if the process is built for a long-term project. Here are some ways to improve the performance of a heat exchanger :

- Heat transfer area
- Fluid flow velocity
- Temperature gradient

These suggested ways of improvements are based on the **equation for heat transfer rate of a heat exchanger**, which is :

$$Q=U*A*dTlm$$

where

Q = Heat transfer rate between the fluids

U = Overall heat transfer coefficient

A = Heat transfer area

dTlm = Log mean temperature difference of the system

The suggested improvements are also based on the data obtained from **the heat transfer lab experiment**, which is a required experiment for all seniors in Rensselaer Polytechnic Institute Chemical Engineering Department.

Heat Transfer Area

As the equation shown above, the heat transfer area (or contact area) is directly proportional to the heat transfer rate. If the heat transfer area increases, heat transfer rate increases as well.

A common way to increase heat transfer area is adding fins to the surface. It is cheap to put fins to the heat transfer area but fins also increase fouling, especially in bio-process.

Fluid Flow Rate

The importance of the fluid flow in a heat exchanger is that it changes the overall heat transfer coefficient, U. The data obtained from the heat transfer experiment shows that the velocity of the cooling fluid is directly proportional to the overall transfer coefficient. The following is a plot of $1/U$ vs. $1/V^{0.8}$ during one of the runs during the lab experiment.

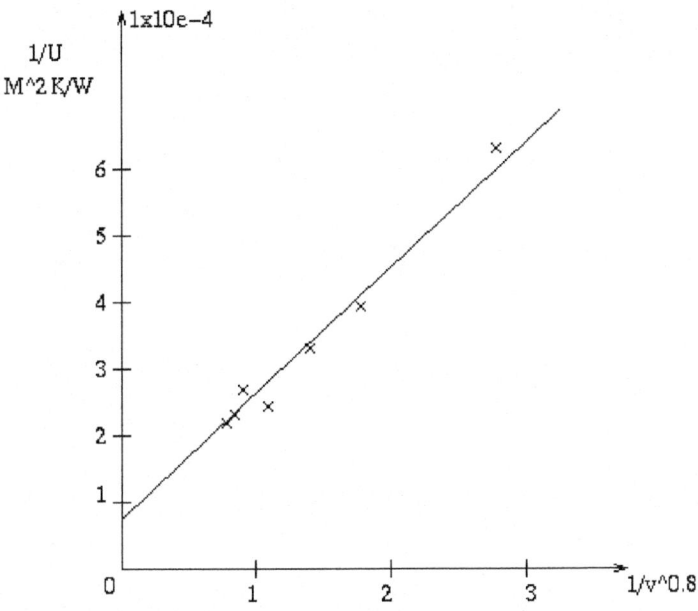

Wilson Plot for hot water counter-current flow

Data points:

$1/V^{0.8}$	$1/U$ (x10e-4)
0.84	2.15
0.86	2.22
0.99	2.78
1.38	3.37
1.14	2.44
1.77	3.94
2.85	6.37

As the cooling fluid velocity increases, the cooling fluid is able to dissipate heat more effectively. The data have shown that it is the case.

Although increasing flow velocity can give more effective heat transfer, it may not be a good idea in some bio-process. High velocity creates high shear stress in flow. Some proteins' or cells' structures are very delicate. They cannot withhold such force and they will be destroyed. The whole batch can be ruined.

Temperature Gradient

Temperature gradient is certainly a important part of heat transfer. It is the driving force for heat transfer. If we can introduce fluids with greater temperature difference into the heat exchanger, the heat transfer rate (Q) will be greater. If we go back to the temperature profiles of the co-current and counterflow, we can see that the driving force is great for co-current at the beginning but decreases drastically as it moves along the heat exchanger. The counter-current flow provides relatively consistent driving force and therefore performs better than co-current flow.

Chapter 11

DELAYED COKING

Delayed coking is a thermal cracking process used in petroleum refineries to upgrade and convert petroleum residuum (bottoms from atmospheric and vacuum distillation of crude oil) into liquid and gas product streams leaving behind a solid concentrated carbon material, petroleum coke. A fired heater with horizontal tubesis used in the process to reach thermal cracking temperatures of 485 to 505°C (905 to 941°F).

With short residence time in the furnace tubes, coking of the feed material is thereby "delayed" until it reaches large coking drums downstream of the heater. Three physical structures of petroleum coke : shot, sponge, or needle coke can be produced by delayed coking. These physical structures and chemical properties of the petroleum coke determine the end use of the material which can be burned as fuel, calcined for use in the aluminum, chemical, or steel industries, or gasified to produce steam, electricity, or gas feedstocks for the petrochemicals industry.

CRUDE OIL REFINING WITH DELAYED COKING

To understand the delayed coking process, one must understand how the delayed coker is integrated with the rest of the refinery. Delayed coker feed originates from the crude oil supplied to the refinery. Therefore,brief descriptions of each of the processing steps preceding the delayed coking unit are provided below.

Crude Oil Desalting

Crude oil contains around 0.2% water in which is mixed soluble salts such as sodium chloride and other metals which are on the edge of the sphere of water. In desalting, crude oil is washed with around 5% water to remove the salts and dirt from the crude oil. The water, being heavier than the oil, drops out of the bottom, and the cleaned oil flows overhead with around 0.1% water.

Atmospheric Distillation

The desalted crude oil is heated in a tube furnace to over 385°C (725°F), just below the temperature that cracking of the oil can occur, then flashed into a distillation column. The primary products are straight run gasoline, kerosene, jet fuel, diesel, atmospheric gas oil (AGO) and atmospheric reduced crude.

Vacuum Distillation

The atmospheric reduced crude (ARC) is then heated to around 395°C (743°F) and flashed into a vacuum distillation column that is operated at low pressures, 10 mm Hg absolute desired but more common 25 to 100 mm Hg absolute. The desired aim is to lift the maximum amount of oil boiling below 565°C into heavy vacuum gas oil (HVGO) reducing the production of vacuum reduced crude (VRC), the main feedstock to the delayed coker. The HVGO and the AGO are the principal feedstocks to a fluid catalytic cracking unit (FCCU) for the production of gasoline and diesel. Improving vacuum distillation is one of the best methods for increasing gas oil yield in a refinery while at the same time reducing the amount of vacuum reduced crude (coker feed). This enables higher refinery throughput rates to be achieved.

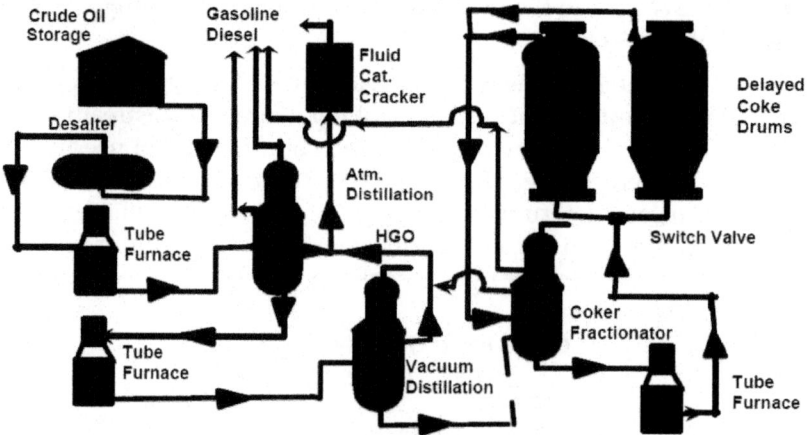

Fig. Basic Refinery.

Vacuum Reduced Crude Processing Options

- **Visbreaking:** Primary function is to reduce viscosity of the oil with some production of heavy gas oil.
- **Resid FCC:** Residuum Fluid Catalytic Cracking, metals deactivate catalyst, must use passivating chemicals to reduce unwanted reactions
- **Resid Hydrocracking:** Feed is contacted with a catalyst and hydrogen at high temperature and pressure to remove sulfur, nitrogen, and some aromatic compounds with some conversion to lighter liquid products.

- **ROSE :** Residual Oil Super critical Extraction for production of metal free gas oil, as phaltenes and resins
- **Propane Deasphalting / Bright Stock :** Solvent extraction of heavy lubrication oils
- **Roofing Asphalt :** May require air blowing to increase hardness
- **Fuel Oil :** Burner and slow RPM marine diesel

HISTORY OF THE DELAYED COKING PROCESS

"Petroleum coke was first made by the pioneer oil refineries in Northwestern Pennsylvania in the 1860's. These primitive refineries boiled oil in small, iron stills to recover kerosene, a valuable and much needed luminescent. The stills were heated by wood or coal fires built underneath which over-heated and coked the oil near the bottom. After the distillation was completed, the still was allowed to cool so the workmen could dig out the coke and tar before the next run." The use of single horizontal shell stills for distillation of the crude was used until the 1880's, with the process sometimes stopped before bottoms coked to produce a heavy lubricating oil.

Multiple stills were used to process more fractions by running the stills in series with the first still producing the coke. In the 1920's the tube furnace with distillation columns (bubble cap distillation trays patented by Koch ushered in the modern distillation column) were being built with the bottoms from the distillation column going to wrought iron stills in which the total outside of the horizontal still was in direct contact with the flue gases. This produced the maximum amount of heavy gas oil.

Some of these units were still in operation after World War II. Operators assigned as decokers used picks, shovels,and wheelbarrows and had rags wrapped around their heads to protect against the heat. The coke that was produced in the horizontal stills had a high density, low volatile matter (VM) content of around 8wt%, and less than 1wt% moisture. One problem was that ash content was high, around 1wt%compared to under 0.2wt% in most modern delayed cokers. Conners thought that this was due to the lack of desalting and washing of the crude oils processed at that time. The origin of the vertical coke drum was probably from thermal cracking of gas oil for the production of gasoline and diesel fuel.

From 1912 to 1935 the Burton process developed by Standard Oil at Whiting, Indiana converted gas oil to gasoline with the production of petroleum coke. Dubbs and other thermal cracking processes also produced petroleum coke. Lack of an adequate supply of crude oil and the lack of a heavy oil market caused land-locked middle American refineries to process the heavy fuel oil (atmospheric distillation bottoms and vacuum distillation bottoms) in a delayed coker to produce more gasoline and diesel fuel. Decoking the drums was difficult. "Manual decoking was a hot and dirty job. ...various mechanical devices were tried.

One of the common systems employed was to wind several thousand feet of steel cable on holding devices in the drum. The cable was pulled by a winch, to loosen the coke. Coke was also removed by drilling a small hole, then a large hole, after which beater balls on a rotating stem knocked out the remaining coke."The first delayed coker was built by Standard Oil of Indiana at Whiting, Indiana in 1929. The development of hydraulic decoking came in the late 1930's. Shell Oil at Wood River, Illinois presented apaper on hydraulic decoking 4.0 m (13ft) diameter Dubbs units and stated that they had patents along with Worthington Pump Company on hydraulic decoking bits and nozzles.

Standard Oil of Indiana had patents on the original cutting nozzles used by Pacific Pump. A very similar nozzle is currently used in the new compact combination coke cutting unit. A pilot hole is drilled down through the coke in the drum using high pressure water, and then the coke is cut out with a drilling bit with horizontal water nozzles. Roy Diwoky while at Standard Oil Whiting was one of the key people in developing the hydraulic decoking in the 1930's.Diwoky in May 1952, while Executive Vice President of Pan Am Southern Corp. (Owned by Standard Oil of Indiana), worked with Great Lakes Carbon Corporation to produce the first needle coke in a delayed coker.

Bernard Gams on, the Director of Research and Development for Great Lakes Carbon at the time,stated in a report that Diwoky was "the father of delayed coking." Delayed coking combined a number of the features and improvements from the development of the thermal cracking process. The use of pressure as well as heat for cracking and separating the heater from the coker and the use of two drums enabled the delayed coker to operate on a continuous basis. The number of cokers built before 1955 was small, with a surge in delayed coker construction between 1955 to 1975 at 6% per year and an 11% growth rate during the 1965 to 1970 period.

The growth of delayed cokers was in step with the growth of fluid catalytic cracking and rapid decline in thermal cracking. A fluid coker, similar to a fluid catalytic cracker except that fluid coke is circulated instead of catalyst, was first built in 1954 at Billings,Montana. Five more fluid cokers were built in the late fifties, and one in 1970. In 1958, the head of petroleum refining engineering at Colorado School of Mines, J.O. Ball, stated that there would not be anymore delayed cokers built. Ball thought all new cokers would be fluid cokers, and that a delayed coker was just a garbage can in the refinery. Today there are 49 operating delayed cokers in the U.S. and only six fluid cokers / flexicokers.

MODERN DELAYED COKING PROCESS

The delayed coker is the only main process in a modern petroleum refinery that is a batch-continuous process. The flow through the tube furnace is continuous. The feed stream is switched between two drums. One drum is on-line filling with coke while the other drum is being steam-stripped, cooled, decoked, pressure checked, and warmed up. The overhead vapors from the coke drums flow to a fractionator, usually called a combination tower. This fractionator tower has a reservoir in the bottom where the fresh feed is combined with condensed product vapors (recycle) to make up the feed to the coker heater.

Delayed Coking Drum Cycle

Since the feed stream is regularly switched between drums, a cycle of events will occur on a regular interval depending on the delayed coking unit feed rate, drum size, and throughput capacity. Most typical delayed cokers currently run drum cycle times of about 16 hours with one drum filling on-line while its counter-partis off-line for stripping, cooling, and decoking. Drum cycle event approximate time requirements for such a cycle are shown below in Table.

Shortening the cycle time is one method of increasing throughput on delayed coking units. One refinery regularly runs 12 hour drum cycles and has attempted 10 and 11 hour cycles, but cycles this short are extremely difficult due to minimum time requirements for each of the steps of the drum cycle. Some of the more important drum cycle steps are described in detail in the following sections.

Table. Typical Short Cycle Coking Operations.

Drum Cycle	Hours
Steam to Fractionator	0.5
Steam to Blow Down	0.5
Depressure, Water Quench and Fill	4.5
Drain	2.0
Unhead Top and Bottom	0.5
Cutting Coke	3.0
Rehead / Steam Test / Purge	1.0
Drum Warm-Up (Vapor Heat)	4.0
Total Time	16.0

Drum Warm-Up (Vapor Heat)

To prepare the cold empty coke drum to be put back on-line to receive the hot feed, hot vapors from the on-line drum are circulated into the cold empty drum. The hot 415°C (780°F) vapors condense in the cold drum, heating the drum to a target temperature of around 340°C(650°F). While the drum is heating, the condensed vapors are continuously drained out of the drum.

On-line Filling.

After the cold drum has been vapor heated for a few hours, hot oil from the tube furnace at about 485°C (905°F) is switched into the drum. Most of the hot vapors condense on the colder walls of the drum, and a large amount of liquid runs down the sides of the drum into a boiling turbulent pool at the bottom of the drum. The drum walls are heated up by the condensing vapors, so less and less vapors are condensing and the liquid at the bottom of the drum starts to heat up to coking temperatures. A main channel is formed similar to the trunk of a tree.

As time goes on the liquid pool above the coke decreases and the liquid turns to a more viscous type tar. This tar keeps trying to run back down the main channel which can coke at the top causing the channel to branch. So the limbs of the "tree in the drum" appear. This progresses up though the coke drum. Sponge coke, which includes needle coke, is formed from this liquid which remains in a quiescent zone between the main branches or channels up through the coker. The liquid pools in the quiescent zones slowly turn to solid coke.

Shot coke has a different type of coke structure indicating that it is produced while suspended in the vapor phase in the drum. This will be discussed in detail-later in the paper. On top of the liquid layer is foam or froth. Paraffinic type feed-stocks with some sodium present foam readily compared to aromatic feedstocks which tend to have smaller foam heights. Higher temperatures greatly decrease the height of the foam. At high temperature, needle coke has very small or no foam present. After the coke drum is filled, the hot oil is switched to the new drum.

Steam-Stripping / "Hot Spots."

Steam must be flowing before the switch and immediately after the switch; otherwise, the yet unconverted liquid feed on top of the coke bed will run down the channels which will coke or solidify and plug the channels. The plugging of the channels causes problems in cooling the coke since sections of the coke bed will be isolated from the steam and cooling water by the plugged channels. This is the cause for "hot spots" and "steam eruptions" when cutting the coke. Cold water from the cutting nozzle hits the exposed hot coke which results in a steam explosion.

This is particularly hazardous when the pilot hole is being cut, since the drum is filled with a large quantity of hot water. A steam explosion during pilot hole cutting can cause the hot water to erupt out of the top of the drum and has caused fatalities in the past. Steam stripping also serves to transfer heat from the hot bottom section of the coke bed to the unconverted liquid present at the top of the coke drum. Adequate steam stripping increases the amount of recovered gas oil yield while at the same time reduces the amount of volatile matter and pitch left in the to section of the coke drum.

After the steam has been flowing up through the coke bed for about thirty minutes with the vapors going to the fractionator, the vapor line is vented to blow down system. Steam is increased for a short time or in some cases water is immediately introduced at the bottom of the drum which instantly flashes to steam. The steam is backed out and the flow of cooling water is gradually increased. The top vapor temperature in the drum may increase slightly at first before cooling due to the increased flow of steam up through the coker.

Water Cooling / "Drum Bulging"

The rate of cooling water injection is critical. Increasing the flow of water too rapidly can "case harden" the main channels up through the coker without cooling all of the coke radially across the coke bed. The coke has low porosity (the porosity comes from the thermal cracking) which then allows the water to flow away from the main channels in the coke drum. Porosity of delayed coke has been measured experimentally in the past by measuring water flow through cores about the size of hockey pucks cut from large chunks of needle coke from different areas of a commercial coke drum.

Most of the coke cores were found to have no porosity except the coke right at the wall which had some porosity . This explains problems that have been found to occur with drums bulging during cool down. Ifthe rate of water is too high, the high pressure causes the water to flow up the outside of the coke bed cooling the wall of the coke drum. Coke has a higher coefficient of thermal expansion than does steel (154 for coke versus 120 for steel, cm/cm/°C x 10^{-7}). This was measured in the transverse direction from a chunk of needle coke. The coefficient of thermal expansion for raw sponge coke is probably even greater than that of the needle coke tested.

Delayed Coking Unit Hardware

A basic coker operation flow diagram is shown below in Figure to illustrate some of the delayed cokingun it hardware.

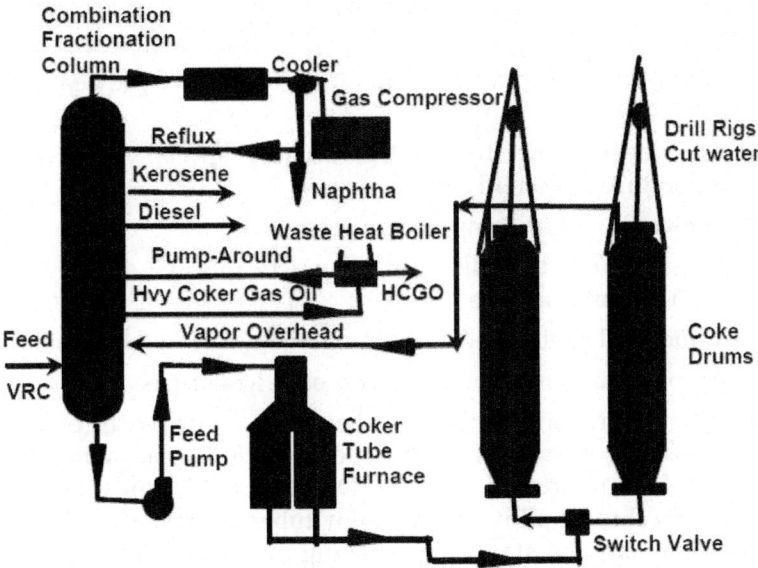

Fig. Basic Coker Operation.

Feed Preheat

In some refineries, delayed coker feed which is usually vacuum reduced crude (VRC) arrives at the coker hot,straight from the vacuum distillation unit, but in most cases, delayed coker feed is relatively cold coming from tankage. The feed is preheated by heat exchangers with gas oil products or in some rare cases by a fired coker pre-heater (tube furnace). In some refineries, the convection section of the main coker furnace is used to preheat the cold feed. The hot coker feed, ranging from 360 to 400°C (680 to 750°F), then enters the bottom of the fractionator / combination tower where the fresh feed is combined with some condensed product vapors (recycle) to make up the feed to the coker heater.

The fractionator bottom provides some surge storage capacity for the incoming fresh feed, and in some units, heat is transferred to the fresh feed by flowing a split of the fresh feed above the drum overhead vapor entrance to the fractionator. This practice usually results in increased amounts of heavy coker gas oil recycle in the furnace charge.

Coker Charge Pumps : The coker charge pumps located between the fractionator bottom and the coker heater are normally driven by an electric motor with a steam-driven turbine pump as a backup. The pressure is in excess of 35 bars (500 psig) with a mechanical seal operating up to 382°C (720°F).

Coker Tube Furnace : The coker tube furnace is the heart of the delayed coking process. The heater furnishes all of the heat in the process. The outlet temperature of a coker furnace is typically around 500°C (930°F) with a pressure of 4 bars (60 psig).

Coker Furnace Design : Delayed coker furnace design objectives according to Elliott are :

- High in-tube velocities resulting in maximum inside heat transfer coefficient
- Minimum residence time in the furnace, especially above the cracking temperature threshold
- A constantly rising temperature gradient
- Optimum flux rate with minimum practicable mal-distribution based on peripheral tube surface
- Symmetrical piping and coil arrangement within the furnace enclosure
- Multiple steam injection points for each heater pass.

Normally the modern-day furnace has two to four passes per furnace. The tubes are mounted horizontally on the side and held in place with alloy hangers. The furnace tubes are around 100 mm ID with 6 to 12 mm wall thickness and are at least a 9% chrome alloy. Higher alloy tubes are being used with the more rapid steam spalling and steam-air decoking methods. Aluminized tubes have been tried, but offer no advantage. Multiple burners are along the bottom of the radiant wall opposite from the tubes and are fired vertically upward.

The burners for each firebox are controlled by the temperatures of tubes in that firebox only. The control thermocouple for the firebox should be three or more tubes back from the outlet to prevent coke forming on the thermocouple. The outlet thermocouple is initially read and an off-set from the control thermocouple is then used to control the furnace. Tall furnaces are advantageous since the roof tubes are less likely to have flame impingement and overheating by both radiation and convection. Normally just the radiant section of the heater is used to heat the oil for a delayed coker.

The upper convection section of the coker heater is used in some refineries to preheat the oil going to the fractionator or for other uses such as steam generation. The typical gas burners in a delayed coker furnace are 3 MM BTU size. Adams stated that the burners will produce flame height of around 0.33 meter per 1 MM BTU. Elliott and others state that the average radiant flux rate should be below 9000 BTU/HR/FT^2 with cold oil velocity of 2 meters/sec (6ft/sec)or mass velocity of 1800 kg/sec/meter 2(400lb/sec/ft^2) or greater. Velocity steam is added at around 1wt%of the feed.

This helps increase the velocity in the tube furnace, and reduces the partial pressure in the drum so that more gas oil product is carried out of the drum. The specific heat of the steam is less than the oil, so steam is not a good source of heat in the drum. The main use for the steam is that it keeps the velocity flowing in the tube furnace if the oil flow is momentarily is lost or decreased which reduces the chance of coking up the furnace tubes.

Heater Tube Decoking

When coke forms in the heater tubes, it insulates the inside of the tube which results in elevated temperatures on the outside of the tube. With good operational practices, coker furnace run lengths of 18 months are possible before decoking of the tubes is needed. When temperatures approach 677°C (1250°F) on the exterior skin thermocouple, the furnace must be steam spalled and/or steam-air decoked or cooled down and cleaned by hydraulic pigging.

Steam Spalling

Steam spalling was probably first practiced by Exxon but was perfected byLloyd Langseth while operating the cokers at Arco in Houston, Texas in the 1970's. He was able to operate a coker furnace over four years without shutting down by practicing on-line steam spalling. The only reason he had to shut down was that Texas had a law that required steam boilers to be inspected every five years.On-line steam spalling requires replacing the oil with steam in the pass and then heating and cooling the tubes to snap or spall off the coke inside the tube.

The steam and coke go into the drum. The main problem is in controlling the velocity and speed of spalling off the coke. Too rapid spalling can plug the tube outlet, and too high steam velocity can erode the metal in the elbows. In one refinery, return bends failed after the second steam spalling. Steam spalling requires that the delayed coker be supplied with four passes or more. Attempts to steam spall a two-pass furnace has been tried, but the large amount of steam being handled caused problems in the fractionator.

Steam-Air Decoking and Pigging

The usual method of decoking the tubes in a coker furnace is to take the furnace off-line, steam spall, then burn the coke out of the tubes by steam-air decoking. After steam-air decoking, the tubes need to be water washed since the salts still remain in the tubes and will cause rapid coking of the tubes. A new method of decoking the tubes is to steam spall, and then use water pressure to push Styrofoam pigs with studs and grit on the exterior through the tubes and around u-bends(even u-bends with clean-out plugs).

The pigs scrape out the coke without scratching the tube walls. Early methods of pigging coker heaters left scratches on the tube walls, but with the grit-coated pigs, pigging just polishes the inside of the tube wall. Pigging is faster than steam-air decoking, and refiners generally have longer campaigns on the heater compared to steam-air decoking.

Heater Tube Deposits

Iron sulfide is probably not totally removed in steam-air decoking. Coke deposits have very high content of iron, silica and sodium. Deposits recovered from return bend clean-out plugs are sometimes long cylindrical shapes and in another case looked like a thick scallop shell. These deposits were mostly sodium and calcium.

TRANSFER LINE AND SWITCH VALVE

Transfer Line

The line from the furnace to the switch valve and on to the drum is referred to as the transfer line. The transfer line must be very well insulated to prevent coking and plugging. The shorter the line the better. Long transfer lines with many crosses and tee's used for clean outs will rapidly coke and increase the pressure on the furnace which usually results in increased fouling of the tubes in the furnace.

Flanges near the drums are difficult to insulate without causing the joints to leak. Some transfer lines have a pressure relief valve in the line, but most furnaces and transfer lines are designed to withstand the maximum pressure the charge pump can produce in case of an accidental switch into a blinded valve.

Switch Valve

The switch valve is a three-way valve with ports to the two drums and a port(recirculation line) back to the fractionator which is used in startup and shutdown. Older cokers used a manually operated Wilson-Snyder valve which was a tapered plug valve that required unseating before rotation. The newer units and retrofits are using ball valves which are usually motorized. One problem with the ball valves is that many separate steam purge lines are required to keep coke from forming on the seal bellows. If the steam purges are not monitored they can decrease the temperature of the oil going to the coke drum resulting in high volatile matter coke being produced.

Coke Drums

The coke drum diameters range from 4 to 9 meters (13 to 30ft) with the straight side being around 25 meters (82ft) with a 1.5 meter diameter top blind flange closure and a two meter diameter bottom blind flange in which the 15 to 30 cm diameter inlet nozzle is attached. Both the top blind flange and the bottom must be removed when decoking the drum. Usually the drum is constructed from 25 mm of carbon steel and is clad internally with 2.8 mm of stainless steel for protection against sulfur corrosion.

The pressure ranges from 1 to 5.9 bars, typically around 2 to 3 bars. The vapor outlet nozzles, 30 to 60 cm diameter, are located at the top of the drum. Pressure relief valves are also located on the top of the drum on modern cokers. The outside of the drum is insulated with around 10 cm (4 in.) of fiber glass insulation with an aluminum or stainless steel covering. The coke level in the drum is usually determined with three nuclear backs catter devices mounted on the outside of the drum.

Overhead Vapor Lines

The vapor overhead line runs from the top of the coke drum to the fractionator. The temperature in the line is around 443°C (830°F). The temperature is decreased by about 28°C (50°F) by injecting hot heavy cokergas oil into the line as quench oil. This prevents coking in the line. The heavy coker gas oil is a wash oil coating the inside of the pipe. If the liquid layer dries out, coke starts to form. Some refineries leave the insulation off the overhead lines to help drop the temperature and keep the inside wetted.

Prevention of coke in the line is important since this will increase the pressure in the coke drum thus increasing reflux of gas oil in the drum. Decreasing coke drum pressure increases liquid yield (decreases coke yield). Also, high pressure drops in overhead lines can cause foaming in the coke drum during the drum switch. Vapor line sizes are very large in order to obtain the minimum amount of pressure drop. One refinery used two 760 mm (30 inch) vapor lines in parallel.

Antifoam Injection System

Injection of silicon anti foam should always be furthest away from the vapor overhead line outlet at the top of the drum to prevent silicon from being carried overhead into the vapor lines to the fractionator. The heaviest possible antifoam that can be handled in the refinery should be used. Lower viscosity antifoamsappear to break down at lower temperatures and are not as effective. Usually a carrier stream is used to anti foam into the drum, heavier carrier material would not be as easily flashed off in the drum.

Several refineries are using less antifoam and having less problems with foam since starting continuous injection of antifoam. A Dow Chemical Company representative stated in 1981 that it is easier to prevent a foam than it is to kill a foam. Also, when a foam is broken down, it still leaves a mist which can cause coking in the bottom of the fractionator. A rule of thumb is that antifoam should cost around $0.10 per ton of coke produced. Costs different than this may indicate that too much or too little antifoam is being used.

Coker Fractionator

The fractionator or combination distillation tower separates the coker overheads into gases, gasoline, diesel,heavy coker gas oil (HCGO), and recycle. An oversized fractionator can be used to maximize the amount of diesel product and minimize the heavy coker gas oil to the FCCU. Hot overhead vapors can cause coking in the lower section of the fractionator if trays are not kept washed (wet). The major amount of heat is removed in the heavy coker gas oil section by trapping out the oil and then extracting the heat with heat exchangers or steam boilers. This pump-around HCGO is then pumped back into the tray above the trap-out tray.

Some of the HCGO is sprayed below the trap-out tray to wash and cool the hot vapors. Trap-out trayscan be used to catch some of this oil and reduce the amount of recycle oil going back to the furnace. Packing can be used in fractionators to reduce the pressure drop, but it is critical to keep the packing wet to prevent-coking in the packing. The pressure in the fractionator and also the coke drums is controlled by the gas compressor at the top of the fractionator. The fresh feed from the vacuum distillation (VRC) should go directly to the bottom of the tower since the effective temperature of distillation is higher than in the fractionator.

Originally when some cokers were designed to coke atmospheric reduced crude, the feed was sprayed into the fractionator above the vapor in let to fractionate out more light ends in the feed. If VRC is injected above the vapor it condenses out part of the HCGO into the bottom of the fractionator increasing the recycle to the coker furnace. The bottom of the fractionator should be operated at as high a temperature as possible without causing coking in the bottom in order to keep the tube furnace duty low.

Normally the temperature in the bottom ranges from 343°C(650°F) to 382°C (720°F) without coke formation in the bottom of the fractionator. A slotted stand pipe in the bottom of the fractionator feeds the furnace charge pump.

HYDRAULIC COKE CUTTING SYSTEM

Cut Water Pump

High pressure water is used to cut the coke out of the drum. Water pressures range from 86 bars (1250 psig) to 275 bars (4000 psig) and flow rates range from 2.8 cubic meters per minute (750 GPM) to 4.7 cubic meters per minute (1250 GPM).

Cut water pumps are multistage barrel type or split case multistage pumps which were originally developed for feed water pumps for steam boilers. The pumps are usually powered with an electric motor, but some older units use steam-driven turbines.

Cutting Equipment

Derricks are built on top of the drum so that the drill stem (5 to 6 inch extra-heavy pipe) can be moved with a winch and cable. The high pressure water flows through an API 10,000 psi drilling hose to the top of the drill stem. The drill stem is rotated with an air motor at the top through a rotary joint. The cutting nozzles are the pilot bit with down facing nozzles and the cutting bit with nozzles facing outward. New units have both nozzles incorporated into a single drilling head.

Coke Cutting Technique

A pilot hole approximately one meter in diameter is drilled from the topof the drum to the bottom. The pilot hole must be cut down through the coke with minimum weight on the bit, since if pushed, the bit can follow the main channel in the coke drum, bend, and stick the drill stem in the coke. After completing the pilot hole, the pilot bit is changed to the cutting bit, and the bottom of the hole is belled out and opened up to around two meters in diameter to prevent plugging.

The bit is then pulled to the top of the drum and cutting begins by spiraling downward at four to six RPM with vertical movement of one-half meter per revolution of the drill stem. Usually a vertical four meter section will be cut by moving the drill stem up and down until the coke is all cut out of the section. Normally around 15 to 20 minutes are required to drill out the pilot hole and three to four hours to cut the coke. The coke can be cut directly into rail cars, cut into a crusher car and the coke pumped hydraulically, or cut into a pit or pad with cranes or end loaders moving the coke.

COKE FORMATION, PROPERTIES, AND STRUCTURE

Crude Oil Origin

In order to understand the components of petroleum coke, we must review the origin of the vacuum reduced crude (coker feed) and crude oil from which it originates. The formation of crude oil is thought to be derived from ancient remains of animals and plants. The organic matter was squeezed out of the strata probably by the connate water or water that was originally in formation. Ancient stream beds, reefs, and sand beaches had the porosity to allow migration of the oil water and provided a conduit for this oil and water material to flow. The final requirement is a trap for the oil.

Oil is not in an open pool but is trapped between layers of sand or in cracks of limestone. The basic trap is the anticline where the formations were pushed up into a dome shaped area where the oil and gas accumulated. Also, faulting of geologic formations pushed oil up against some impervious formations forming fault traps. The salt domes in the Gulf Coast of the United States have been big producers of oil.

Spindle top near Beaumont, Texas was a prime example. Salt laid down in the bottom of a drying sea, and these formations were covered by over five miles of sediment. The resultant pressure liquefied the salt which then started to migrate toward the surface. The geologic formations are pushed up as the salt plug punches through the formations. The salt, being impervious to the oil, forms an excellent trap. The vanadium and nickel are in the crude oil as porphyrins or metal chelates. Originally the metals were probably magnesium (chlorophyll) and iron (hemoglobin).

The ratio of the metals to each other is due to when and how they were buried. Some of the vanadium and nickel can be loosely held between the asphaltene molecules (intercalation). The other metals are complexed onto the water droplets and probably were due to the structure that the oil migrated through. Crude oils such as the Paraffinic Pennsylvanian crude contain very small amounts of asphaltenes. It is possible that the asphaltenes dropped out of the oil phase but are still down in the formation.

Parts of Crude Oil

Crude oil contains three different fractions. The "Oil" is the hydrocarbon : paraffinic, naphthenic, andaromatic which also contain sulfur and nitrogen. The second part of the crude oil, the resins, coat the asphaltene fraction so that it can be peptized into the crude oil. The resins are a brown, sticky hydrocarbon which contain nitrogen, oxygen, and sulfur, are soluble in n-pentane but insoluble in propane, and have molecular weights greater than 3000.

The asphaltenes contain the chelated metals, vanadium, nickel, and possibly some calcium along with sulfur, oxygen, and nitrogen. During crude oil distillation, the asphaltenes are not volatilized and remain in the vacuum reduced crude along with most of the resin fraction. Jakob thought that all the resins and asphaltenes dropped out in the coker and the remaining coke was made from the oil fraction.

With higher temperatures and lower pressures, the hydrocarbon part of the coke could bereduced but not the resin and asphaltene fraction. The amount of coke produced in a delayed coker is always more than the Conradson or Ramsbottom carbon residue percentage by a factor of about 1.6.

Desalter's Influence on Coke Properties

The crude oil desalter is one of the most critical pieces of equipment in the refinery for producing good quality anode grade coke (coke low in metals suitable for calcination and use in the aluminum industry) and keeping a coker furnace on line. Crude oil contains around 0.2% water in which is mixed the soluble salts such as sodium chloride and other metals which are on the edge of the sphere of water. In desalting, the crude oil is washed with 5% water to remove the salts and dirt from the crude oil. The water, being heavier than oil, drops out of the bottom, and the cleaned oil flows overhead with around 0.1% water.

Without good desalting, caustic (sodium hydroxide) or filming amines must be added to eliminate the chloride corrosion in the overheads of the distillation and vacuum distillation lines. The chlorides are usually in the form of salt (sodium, calcium, and magnesium chloride)The salt content can vary from 50 to 300lbs per 1000 barrels of crude. Since the number of droplets are high, around 9×10^{11}, the amount of dirt and other metals on the outside of these water droplets is appreciable. The magnesium chloride causes most of the corrosion since it breaks down at low temperatures in the distillation column liberating chlorine which forms hydrochloric acid that attacks the overhead lines in both the vacuum and atmospheric distillation units.

Sodium is a catalyst for burning of carbon (air and carboxyl reactivity in baked anodes) and also causes rapid tube fouling in the coker tube furnace. The mechanism for rapid fouling of tubes due to sodium is not fully understood, but it is known that if the tubes are not water washed after steam-air decoking to remove the crystals of salt, the unit will rapidly foul. Iron in fine particles, probably iron sulfide, is very difficult to remove in a desalter, but some chemical companies can do a better job than others. Metals that the desalter does not remove will end up in the coker feed and ultimately in the delayed coke.

COKE PHYSICAL STRUCTURE

Coefficient of Thermal Expansion

To determine a quantitative value describing coke structure,the coke is calcined, ground to a flour, mixed with coal tar pitch, extruded to orientate particles into 13 mm rods, baked to 850°C, and graphitized to 2900°C, and then the difference in expansion at 0°C and 50°C is measured for Coefficient of Thermal Expansion (CTE) determination. Typical values of CTE corresponding to coke structure are : needle coke (acicular), 0 to 4; sponge coke, 8 to 18; and shot coke (isotropic),>20(cm/cm/°C x 10^{-7}).

Shot Coke

The production of shot coke in a delayed coker requires high concentrations ofasphaltenes in the feedstock, dynamics (velocity and/or turbulence) in the coke drum, and high coke drum temperatures. A coker feedstock high in oxygen content can also produce shot coke. When asphaltene content compared to the Conradson carbon residue content of the coker feed is high, the production of shotcoke is very likely.

The present trend in refineries is to run heavier crudes with higher as phaltene contents and to improve operation of the vacuum distillation unit to produce a heavier VRC with a higher asphaltene content. This trend toward increased production of shot coke has been observed in refineries which originally ran atmospheric reduced crude in the delayed coker, never making shot coke, that started producing shot coke after a vacuum distillation unit was installed. Shot coke is produced as the oil flows into the coke drum. With the light ends flashing off, small globules of heavy tar are suspended in the flow.

These tar balls rapidly coke due to the exothermic heat produced by asphaltene polymerization. (Cokers going from sponge coke production to shot coke production have seen the drum overhead temperature increase by as much as 3°C.) The balls then fall back into the drum as discrete little spheres two to five millimeters in size. In the main channel up through the drum, some of the spheres will roll around and stick together forming large balls as large as 25 centimeters. When these large balls are broken, they are found to be composed of many of the two to five millimeter size balls.

Normally, small shot coke balls from different delayed cokers will be nearly the same size; however, Mexican Mayan VRC has been found to produce larger shot coke balls upon delayed coking than does Venezuelan VRC. It is thought that smaller balls are made when very high feed rates are used in the coker. Aromatic feeds, such as decant oil from the FCCU, can help eliminate shot coke formation. All other methods of eliminating shot coke such as decreasing temperature, increasing drum pressure, and increasing recycle ratio, will all increase coke yield (decrease more valuable liquid yields) which is not desired. It is very difficult to produce shot coke spheres in a pilot delayed coker.

Spherical shot coke can only be produced in pilot delayed cokers if the velocity in the drum and the temperature in the drum are both very high. In a batch (pot) coker, the typical spherical form of shot coke cannot be produced at all; but the shotcoke micro-structure in the batch-produced coke can be seen with a microscope, and the batch-produced coke does have a high CTE value similar to the spherical form. Shot coke is unique in that the small spheres two to five millimeters in diameter each have a slick shiny exterior coating of needle or acicular type carbon. The inside of each sphere contains isotropic or amorphous type coke as originally described by Marsh and Bacha.

Shot coke cannot be used in making aluminum anodes because the outer needle coke layer of the shot sphere has a very low coefficient of thermal expansion while the inside of the sphere, being isotropic, has a very high coefficient of thermal expansion. When rapidly heated in a calcining kiln, the outer layer is cracked and pulled away from the center; thus when used in an anode with a coal tar binder, the binder adheres to the outer layer (egg shell). This results in many cracks between the ball and the skin causing the anode to crack and dust in an aluminum smelter cell.

Sponge Coke

Sponge coke is named for its sponge-like appearance and is produced from VRC with a low to moderate asphaltene concentration. If sponge coke meets strict property specifications, it is considered anode grade sponge coke suitable for calcination for use in making carbon anodes for the aluminum industry. Otherwise, if sponge coke meets the more lenient fuel grade specifications, it can be used in its raw form for fuel. The biggest problem for refineries producing anode grade sponge coke is obtaining the low volatile matter (VM) required.

The metals and sulfur are strictly controlled by the crudes being processed, but the VM is in the control of the delayed coker operators. Temperature in the drum is the most critical item, along with cycle time and drum pressure. Longer residence time at temperature helps to decrease the VM. Increased recycle can increase the temperature in the drum. Insulation of the transfer line and coke drum, especially the upper sections of the coke drum, are critical for obtaining low VM coke.

Poor insulation and other bad practices on the delayed coker require higher temperatures in the tube furnace, which results in shorter campaigns and more downtime for decoking of the furnace. Monitoring the seal steam to prevent decreasing the temperature in the transfer line, elimination of seal oil on the pressure relief devices on the transfer line, and minimizing the amount of carrier oil for the antifoam all help in increasing the temperature in the drum in order to decrease VM of the resultant coke.

Raw or "green" sponge coke must be calcined before it can be used in making anodes. The density of the calcined coke is critical for producing good carbon anodes. The higher the density, the more carbon can be incorporated into the anode, and the longer the anode will last. Vibrated bulk density (VBD) of the calcined coke must be greater than 86(grams/100 cc). The best single property that correlates from the raw coke is the Hardgrove Grindability Index (HGI). Raw coke with lower than 70 HGI usually can be calcined to produce an 86 VBD.

Volatile matter is another good property used to correlate how well the raw coke will calcine. Structure is a strong factor in calcinability also, since cokes with low CTE must have volatile matter much lower than a more isotropic type coke to produce the same density. Porosity of the calcined coke should be low and is also a function of the raw coke volatile matter. The ash in the calcined coke is normally around 0.2 % with vanadium and nickel combination under 500 ppm.

Sodium and calcium are very strong catalysts for air burn of an anode. Vanadium, nickel and iron and other metals causes increased carboxyl reaction in the bottom of the anode. The sulfur in the anode must be below 3.5% to prevent the sulfur from increasing the electrical resistance of the cast iron connection between the anode and the power rod. Normally, sulfur is more of an environmental and scrubbing problem. Sulfur can cause the real density of calcined coke to decrease due to an increase of the porosity and micro cracking of the calcined coke. Sulfur does help reduce reactivity (air and carboxyl) by reacting with the caustics which are strong catalysts. Aluminum production requires around one-half kilogram of carbon per kilogram of aluminum produced.

Anode grade coke must be low in metals concentration since the exhaust from the aluminum cell is being scrubbed with the alumina used as feed to the aluminum cells. Therefore, any metals in the coke would get into the alumina and into the aluminum metal produced. The carbon is used in the aluminum smelter as a means of carrying electrical power into the cell. It takes around 15 KW of power per kg of aluminum produced.

A carbon with some porosity must be used since gases coming off the cell would block the power going into the cell if the anode was not porous. The high temperature along with the very corrosive fluoride salts used in the aluminum cell and the problem with the evolution of the gases makes the discovery of a non-consumable anode difficult.

Needle Coke

Needle coke, named for its needle-like structure, is produced from feedstocks without asphaltenes present, normally FCCU decant oils. Needle coke is the premier coke, used in graphite electrode manufacturing (used in steel arc furnaces) and commands a high price (calcined ultra-premium non-puffing,$500 per ton); but needle coke requires special feedstocks, special coking, and special calcination to obtain the optimum properties that it requires.

The Shea patent on needle coke gives an accurate description of the formation of needle coke, still relevant today. Most needle coke is produced from hydrodesulfurized decant oil (due to the low sulfur requirement for non-puffing coke, that can be nearly flash graphitized in the new direct current (DC) length-wise graphitization method, without splitting the electrode). The principle requirement for needle coke is that the CTE must be 2.0 or below (low CTE is required to prevent spalling due to the thermal stresses on the tip of the electrode which can be as high as 2000°C/cm).

Needle coke must have low sulfur (<0.6wt%) and nitrogen contents in order to be non-puffing during graphitization to 2900°C(measured by a special dynamic puffing test that is proprietary). Needle coke must also have a maximum amount of coarse sizing (>6 mm), a minimum amount of fines (<1 mm), good density (>78 grams/100 cc;4/6 mesh test), low ash content (<0.3%; any ash leaves a void when graphitized), and a high real density(2.13 grams/cc).

Even with all the property specifications, an electrode manufacturer will not accept a calcined needlecoke for production until they have actually run a trial lot through the plant and trials on the electric arcfurnace. Most graphite plants want a needle coke with low variability so that they can set up the optimum pitch level, extruding and baking to produce a good electrode. The most popular electrode is the 24 inch (60cm), with a demand for larger than 30 inch (76 cm) for DC single electrode furnaces.

Obtaining good needle coke is still a "black art" for excellent graphite electrodes. The principal property that the electric arc steel mill wants in a graphite electrode is a low amount of graphite per ton of steel melted. In single electrode DC furnaces, the amount of graphite per ton is below 2 kg/ton . With better practices and foamy slag, ACfurnaces (using three electrodes due to three-phase electric power) have approached this level.

Pilot Delayed Coker and Coke Formation Model

In the 1970's, Great Lakes Carbon obtained a design for a pilot delayed coker and modified and improved the design to produce needle coke identical to that which is commercially produced. Non-puffing needle coke was in short supply at that time with prices for raw coke running over $600 per short ton. The pilot unit had a 305 mm (12 in) diameter drum 2.13 meters (84 in) tall, a gas fired tube furnace (both convection and radiant sections), self-generated recycle (up to 400%), an operating pressure of up to 6.8 bars(100 psig), adiabatic drum conditions, load cells on feed tanks, overhead receivers, gas meters, and a Ranarex(for determining molecular weight of overhead gases).

A very good material balance could be obtained. The unit produced around 70 kg (>150lbs) of coke which was calcined in a gas fired pilot calciner. The coke drum was decoked by cutting the head off the drum with a metal band saw and then trepanning the coke out with a large core drill .Due to the extreme hardness of needle coke, the specified auger type cutter could not be used for decoking and a core bit was designed and built. By using the core bit, the coke came out in long sections which could then be sliced with a band saw.

From studying these slices of coke, GLC discovered how coke forms in the drum. This discovery was confirmed by sampling many railcars of commercially produced needle coke and observing a puzzling strange-looking coke found on the very large piece cut from the lower section of the drum near the knuckle. This coke was typical except the center contained a dull looking, very friable coke.

Also, looking up at the bottom of the coke bed after the bottom blind flange was removed on a double-deckers, a hole was observed going up in the bottom of the coke drum. From these observations of commercial delayed coke and from studying the slices from the pilot unit, the "tree in the drum" channel branching theory was formulated. This coke formation model is illustrated in Figure shown below.

Structure Orientation In Drum

Samples of coke cut both vertically and horizontally showed that the coke was oriented in the drum. The coefficient of thermal expansion is much lower in the vertical direction compared to horizontal, 132% lower in the raw coke, 505% lower in the calcined, and 2850%lower in the graphitized sample. Gas bubbles, formed from cracking, migrate upward during coke formation in the liquid, orientating the mesophase chain growth.

Table. Coke CTE by Orientation

CTE (cm/cm/°C x 10^{-7})	Vertical	Horizontal
Raw Coke	117	154
Calcined 850°C	1.8	9.1
Graphitized 2900°C	0.2	5.7

Chemical Property Distributions

The ash is mainly in the lower sections of the coke drum with a high percentage in and near the channels up through the coker. The ash being a particulate drops out of the oil, or the wall of the channel traps out the particle, otherwise known as the "fly paper effect." Across-section of the coke in the drum was cut into small cubes which were analyzed for ash content. The ash level was five to ten fold higher near the channels compared to the rest of the section.

In an experiment where a paint pigment sized chromium oxide was pumped through the tube furnace into the coke drum in an attempt to get good dispersion in the coke (chromium oxide is a puffing inhibitor in needle coke), all the chromium oxide dropped out in one spot in the lower section of the coke drum up about six inches from the inlet. This is normally the first spot where the main channel starts to branch.

Several runs were made with identical results. Injecting the pigment through the top of the drum distributed the material uniformly in the coke, indicating that there is some back mixing in the top of the drum either in the froth layer or in the liquid. Iron, silicon, and ash are in the coke as particulates. These metals concentrate in the lower section of the coke drum as shown in Table below. Vanadium and nickel are in crude oil as metal chelates or porphyrins in the asphaltene fraction.

It was puzzling that vanadium and nickel are not uniformly distributed in the drum until it was understood that some of the metals are intercalated in the structure and are not chemically bonded, so they drop out early in the coke drum similar to the ash and particulates. Volatile matter (VM) in the coke drum is normally high in the top of the drum due to the short residence time of the material. The sulfur is uniformly distributed in the drum unless the feedstock to the drum is changing as the drum is filled.

Table. Property Distributions in the Coke Drum (wppm)

	Top	Middle	Bottom
Iron	40	440	1300
Silicon	60	150	380
Ash	1000	2100	4500
Vanadium	267	310	380
Nickel	130	160	230
VM (wt%)	11.8	9.5	9.4

USES OF PETROLEUM COKE

Raw Petroleum Coke

Fuel Coke : Fuel grade coke (shot or sponge) is used in the production of cement and with fluidized bed boilers (using limestone for sulfur removal) for generation of steam and electricity. The important properties for pulverized fuel coke is the cost per BTU, high HGI, and sulfur content. Vanadium in petroleum coke does not cause corrosion on boiler tubes as does vanadium in heavy fuel oil.

Metallurgy Uses : Some raw petroleum coke, if the sulfur is low enough, can be blended into feed for slot ovens which produce blast furnace coke. Petroleum coke increases the physical strength and density of the coke when blended with coal.

Gasification : Partial oxidation of petroleum coke in a gasification process enables raw petroleum coke to be used to produce steam, electricity, or gas feedstocks for the petrochemicals industry.

Calcined Petroleum Coke - Other Uses : Some calcined petroleum coke is used in production of titanium dioxide (in the chloride process), production of carbon monoxide for production of plastics, as a feedstock for continuous particle thermal desulfurization for special low sulfur carbon raiser (steel ladle additive), or as carbon raiser in cast iron and steel making.

Chapter 12

DISTILLATION PROCESS

Distillation is a commonly used method for purifying liquids and separating mixtures of liquids into their individual components. Familiar examples include the distillation of crude fermentation broths into alcoholic spirits such as gin and vodka, and the fractionation of crude oil into useful products such as gasoline and heating oil. In the organic lab, distillation is used for purifying solvents and liquid reaction products. To understand distillation, first consider what happens upon heating a liquid. At any temperature,some molecules of a liquid possess enough kinetic energy to escape into the vapor phase(evaporation) and some of the molecules in the vapor phase return to the liquid (condensation). An equilibrium is set up, with molecules going back and forth between liquid and vapor. At higher temperatures, more molecules possess enough kinetic energy to escape, which results in a greater number of molecules being present in the vapor phase.

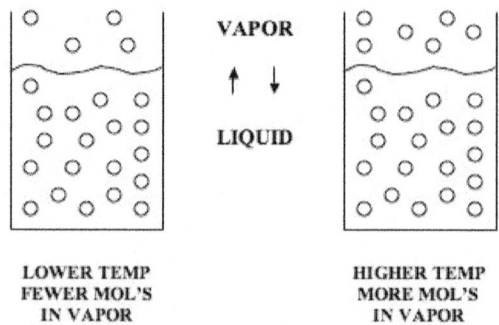

If the liquid is placed into a closed container with a pressure gauge attached, one can obtain a quantitative measure of the degree of vaporization. This pressure is defined as the vapor pressure of the compound, and can be measured at different temperatures. Consider heating cyclohexane, a liquid hydrocarbon, and measuring its vapor pressure at different temperatures. As shown in the following graph of temperature vs vapor pressure, as the temperature of cyclohexane is increased the vapor pressure also increases. This is true for allliquids. At some point, as the temperature is increased, the liquid begins to boil. This happens when the vapor pressure of the liquid equals the applied pressure [for an apparatus that is open to the atmosphere the applied pressure equals atmospheric pressure (1 atm = 760 mm Hg)].

For cyclohexane, this occurs at 81° C. The boiling point (BP) of cyclohexane therefore equals 81° C.The definition of the boiling point of a liquid in an open container then is the temperature at whichits vapor pressure equals atmospheric pressure. Note that under vacuum, the BP of a liquid will belower than the BP at atmospheric pressure. This can be exemplified by looking at the BP of water atdifferent pressures. Atmospheric pressure decreases with increasing altitude so the BP of water isfound to be about 95°C in Denver which is at about 5200' above sea level. Atop a 10,000'mountain the BP of water would be 90°C. Because liquids boil at lower temperatures undervacuum, vacuum distillation is used to distill high-boiling liquids that would decompose at theirnormal BPs.

It can also be seen from the graph that for toluene the vapor pressure equals atmospheric pressure at a temperature of 111°C. The BP of toluene is therefore 111°C. Note that at any given temperaturethe vapor pressure of cyclohexane is greater than the vapor pressure of toluene.

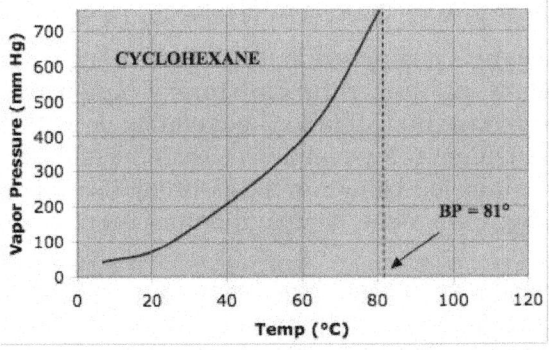

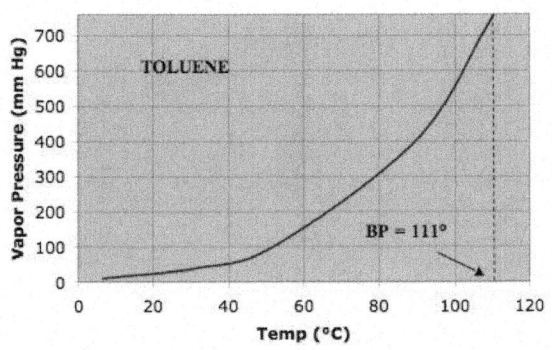

Fig. Consider next the behavior of a mixture of two liquid compounds. The example shown below is fora 1:1 mixture of cyclohexane (C) and toluene (T).

Fact: at any given temperature, the vapor pressure of the lower-boiling (lower BP) compound> the vapor pressure of the higher-boiling (higher BP) compound. Thus, the vapor above the liquid will be richer in the lower-boiling compound, compared to the relative amounts in the liquid phase.

If we were to collect the vapor above the 1:1 mixture, condense it to liquid, and analyze its composition we would find that the vapor was greater than 50% cyclohexane and less than 50% toluene. The vapor is enriched in the lower-boiling cyclohexane.

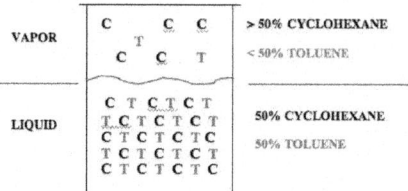

Take a look at the following simple distillation set-up. (This is not the complete experimental set-upthat will be used in this experiment. It shows only the basic pieces that exemplify the process.) If we placed the 1:1 mixture of cyclohexane and toluene into the distilling flask, heated the mixture to the BP, and allowed the cooled vapors to drip into the collection vial, we would find upon analysis that the distillate was greater than 50% cyclohexane and less than 50% toluene.

The distillate has been enriched in the lower-boiling component. This is the essence of distillation - starting with a mixture of liquids having different BPs, going through the process of distillation, and ending up with distillate that is enriched in the lower-boiling component. Because more of the lower-boiling liquid has distilled, the residue left behind in the distilling flask is necessarily enriched in the higher boilingc omponent. A separation has been accomplished.

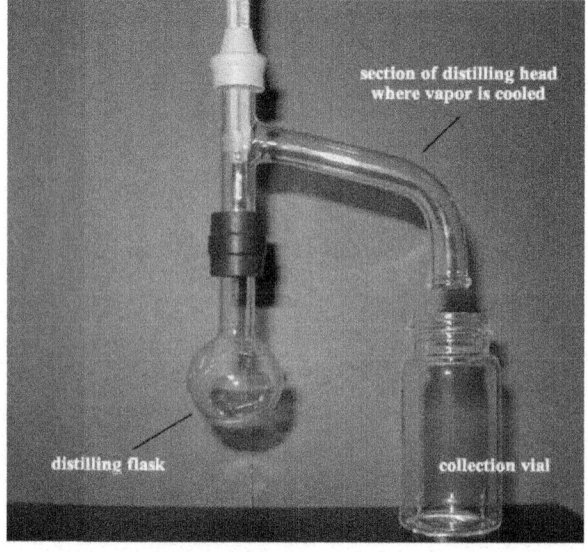

The purpose of doing a distillation is to end up with a relatively pure individual component or components. So far we have only seen that the distillate has been enriched but we have not seen by how much it has been enriched. On doing the experiment, one finds that by carrying out one vaporization - condensation step,starting with a 1:1 mixture of cyclohexane and toluene, the distillate would initially distill as a mixture of 80% cyclohexane and 20% toluene.

The distillate has been significantly enriched in cyclohexane. Generally though this would not be considered to be sufficiently pure. Our purpose is to get pure individual compounds. What if we now took the 80% cyclohexane/20%toluene mixture that we just obtained and placed it into a clean distillation set-up and distilled that? Upon analysis we would find that the distillate is now 95% cyclohexane and 5% toluene. Again this is a substantial enrichment but perhaps not yet of the desired purity.

Take that distillate and distill it again. This third distillation would produce distillate that is about 99% pure cyclohexane. This would normally be considered to be fairly"pure" cyclohexane. At the same time, as we remove cyclohexane from the mixture, the residue has been enriched in toluene. By doing three vaporization-condensation steps we have achieved 99%purity. Each vaporization-condensation step is known as a "simple distillation". Thus, for this mixture, three simple distillations have produced the desired purification.

FRACTIONAL DISTILLATION

Unfortunately, each time a distillation is run, material is lost. Some evaporates into the air and some is left behind, stuck to the apparatus. Material left behind is known as "hold-up". We would find that after doing three separate simple distillations, we have lost much material. Besides obtaining pure compounds we also want to attain high yields, with little loss. A method exists for carrying out several simple distillations in one apparatus, thereby resulting in smaller losses. This method is called "fractional distillation".

The difference between the apparatus used for simple and the apparatus used for fractional distillation is the presence of a "fractionating column" in the fractional distillation. In a distillation,liquid is converted to vapor by heating and the vapor is then condensed back to liquid by cooling.

In a simple distillation this is done one time. In a fractional distillation, as the vapor ascends the column, it encounters a cooler area and condenses. The hot ascending vapors revaporize the liquid and the vapor travels further up the column, where it encounters a cooler area and recondenses. Hot ascending vapors revaporize the liquid and it travels a bit further. Each vaporization-condensation cycle is equivalent to a simple distillation so by the time the vapor reaches the top of the column, it has undergone several simple distillations, and has thus undergone further purification than in the simple distillation apparatus. Because it was done in one apparatus, much less material is lost and the yield is greater than if several separate simple distillations had been done. Note however that even in a fractional distillation, some material is lost to evaporation and some is left behind in the apparatus ("hold-up").

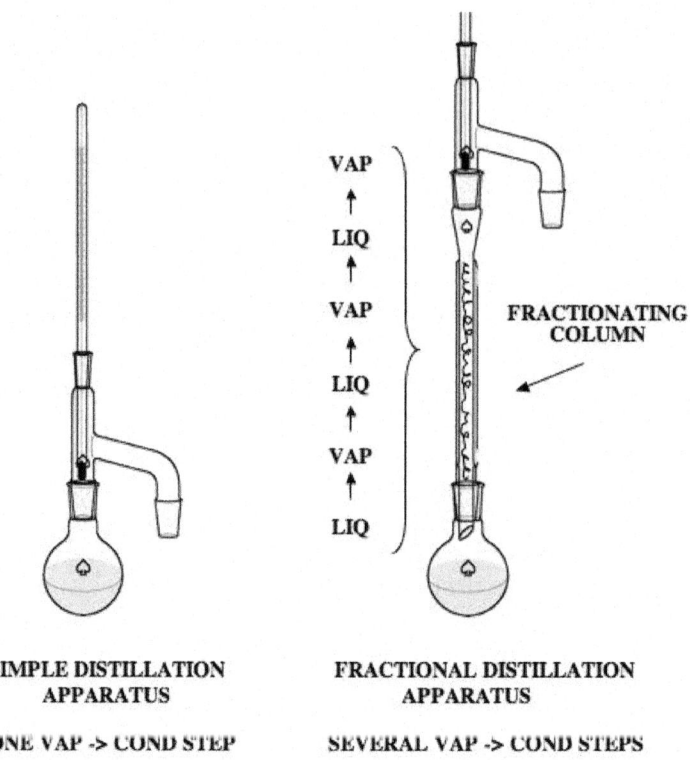

SIMPLE DISTILLATION   FRACTIONAL DISTILLATION
APPARATUS   APPARATUS

ONE VAP -> COND STEP   SEVERAL VAP -> COND STEPS

Distillation is a sacrificial process in that some material is always lost, no matter how careful or experienced the chemist. Minimizing the loss is an important goal. The number of vaporization-condensation cycles that can occur within a fractionation column determines the purity which can be attained. The efficiency of a column depends upon column length and composition. In the fractionation column used in this lab, the column is packed with copper sponge.

This increases the surface area that the ascending vapor encounters and results in more vaporization-condensation cycles compared to an empty column. A measure of efficiency of a column is known as the number of theoretical plates of that column. One theoretical plate is equivalent to one vaporization-condensation cycle, which is equivalent to one simple distillation. Thus a fractionation column that can attain the equivalent of three simple distillations would be said to have three theoretical plates. Boiling Point – Composition Curve.

A boiling point – composition curve allows us to quantify this and to predict the number of theoretical plates needed to achieve a desired separation. Such a curve would be made by taking mixtures of varying composition, heating them to the BP, measuring that temperature, analyzing the composition of the vapor above each mixture, and plotting the results. Such a curve for cyclohexane and toluene is shown below. The lower curve represents the liquid composition and the upper curve represents the vapor composition. As an example of how to read the curve, say we distill a mixture that is 20% cyclohexane and 80% toluene. Starting at the x-axis at the 20cyl / 80tol point, draw a line straight up to the liquid curve. Note that this mixture boils at about 102°.

Draw a straight horizontal line over to the vapor curve and back down to the x-axis. This gives the composition of vapor above a boiling mixture of 20% cyclohexane and 80% toluene. In other words if we were to do a simple distillation on a mixture of 20% cyclohexane and 80%toluene, the first drop of distillate would be comprised of 50% cyclohexane and 50% toluene. (There a son it is stated that "the first drop of distillate" would have this composition is that as lower boiling cyclohexane is removed, the composition of liquid in the distilling flask becomes enriched in the higher-boiling toluene, thus moving the starting point on the graph to the right. This is discussed further below.)

Reading this graph is simple. Start on the x-axis with the composition of mixture to be distilled, move straight up to the liquid curve, over to the vapor curve, and down to the x-axis, which gives the vapor composition. What is shown here is for one vaporization –condensation cycle or one simple distillation.

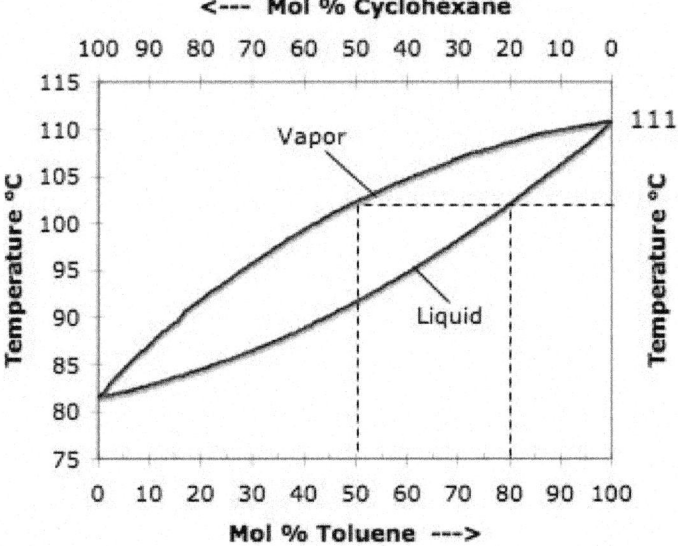

BOILING POINT – COMPOSITION CURVE

The following graph shows what would result if we carried out a second simple distillation on the distillate from the first distillation (50:50). Stating at the 50:50 point on the x-axis go up to the liquid curve, over to the vapor curve and down to the x-axis to show that the distillate for this second simple distillation would be 80% cyclohexane and 20% toluene.

BOILING POINT – COMPOSITION CURVE

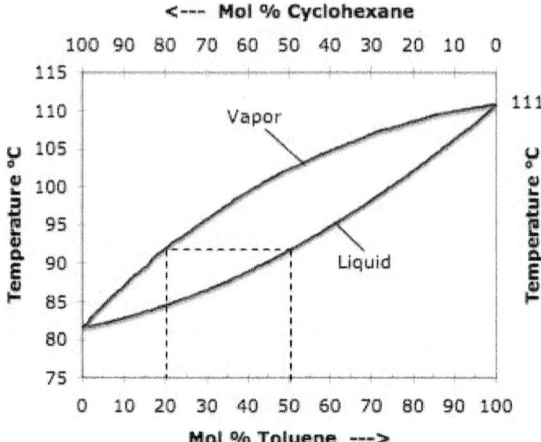

One could keep drawing additional graphs to show a third, fourth, and so on, distillation. It is easier though to combine all steps onto one graph as shown in the next figure. Starting with the original 20% cycl / 80% to l mixture, do the first vaporization – condensation cycle, but instead of going all the way back down to the x-axis, step down to the liquid curve and over to the vapor, and so on.

Each Step is Equivalent to a Simple Distillation

The graph shows that starting with a mixture that is 20% cyclohexane and 80% toluene, a fractionation column having efficiency equal to three theoretical plates would be needed to result in distillate that is 95% pure. A fourth theoretical plate would result in distillate that is about 99% pure cyclohexane.

BOILING POINT – COMPOSITION CURVE

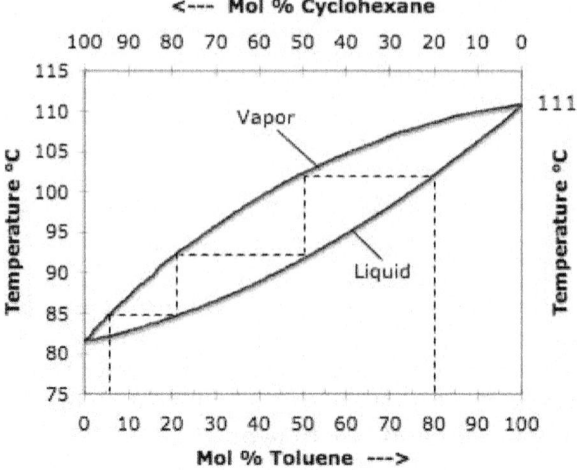

There is a catch however. As mentioned above, as a distillation proceeds and lower-boiling component is removed, the composition of the mixture in the distilling flask is becoming enriched in the higher-boiling component. This is a dynamic process. This means that after the first drop of distillate is obtained, the starting point on the x-axis has moved to the right (towards a composition that is higher in the higher-boiling toluene). For our example, let the distillation proceed for a while until the starting point on the graph is at 5% cyclohexane / 95% toluene.

How many theoretical plates would be needed now to attain distillate that is 99% cyclohexane. Work it out. The answer is about 5. As a distillation proceeds and the mixture in the distilling flask becomes more and more enriched in the higher-boiling component, to obtain pure distillate, more theoretical plates are needed. Unfortunately, the efficiency of a given fractionation column is fixed. The result is that as a distillation proceeds and as more theoretical plates are needed, a point is reached in which the column can no longer provide the same separation as in the beginning, resulting in distillate that is no longer as pure as at the start. This will become apparent when we look at a distillation curve.

Distillation Curve

The BP-Composition curves were discussed to illustrate the workings of a distillation. A more useful graph, one that you will produce in your experiment, is a distillation curve, which plots temperature *vs.* volume of distillate. The following is an example. As the distillation on an unknown mixture is carried out, the experimenter records the temperature of distillate as soon as the first drop is collected and every several drops thereafter. Once the data is collected, a graph such as the following is made.

**FRACTIONAL DISTILLATION OF A MIXTURE OF
TWO UNKNOWN LIQUIDS**

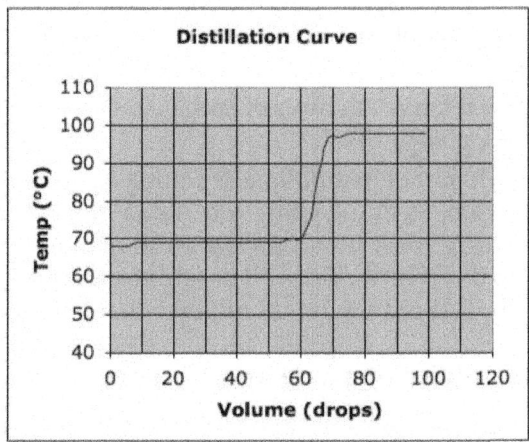

How to interpret this graph: note that distillation occurs at a relatively stable temperature of about 69° until about 60 drops has distilled. This temperature plateau represents the BP of the lower boiling component, in this case, about 69°. The relatively stable temperature between 0 and 60 drops shows that relatively pure material is distilling during this time. At 60 drops the temperature rises and reaches another plateau at about 98° and 70 drops. The distillation proceeds from 70 drops until distillation ends at 100 drops. The second BP plateau, at 98°, is that of the second,higher-boiling component. The ratio of pure components is 60 drops :30 drops or 2:1.

Note that between 60 and 70 drops the temperature continually rises. The 10 drops consist of a mixture of the two components. Why doesn't the temperature shoot directly up to the second plateau after the entire first component has distilled? Recall that as the distillation proceeds and the lower-boiling component is removed, more theoretical plates are required to attain the desired purity. However the fractionation column only has so many theoretical plates so at some point it can no longer handle the separation and the result is that a mixture will distill until the entire lower-boiling component has been removed.

At that point the pure second component begins distilling as evidenced by the second temperature plateau. Obtaining the impure intermediate fraction must be accepted. If a better separation were desired a better fractionation column would be required. To obtain the maximum efficiency in any distillation, the distillation rate must be kept constant and slow (for our apparatus a rate of about 1 drop per 20 – 30 seconds produces the best separation).Distillation rate is the single most important variable that contributes to an effective distillation.

Experimental Procedure

In the first part of this experiment, to practice the technique of distillation, you will separate a mixture of cyclohexane and toluene (tol-you-ene) by simple and fractional distillation and compare the two methods. In the second part of this experiment, you will distill an unknown mixture and

(1) determine the identity of the components by observing their boiling points (BPs), and

(2) determinethe relative amounts of each by measuring the amount of distillate collected at each BP.

The most difficult part of distillations is setting up the small apparatus properly and attaining a slow steady heating rate. Remember that a slow and steady heating rate is the single most important factor in obtaining a good distillation. Take your time to do it correctly. You will use this technique in later experiments. If a first attempt fails, just redo it. It will not take that long.

Simple Distillation of Cyclohexane and Toluene

Because the electrically heated sand bath takes time to reach operating temperature, turn it on to a setting of just under 40(NEVER HIGHER) as soon as you enter the lab (make sure the orange power light comes on). For a sketch of the apparatus, refer to the figure at the end of this procedure and also to the photos on the course website. Check the 5 mL round-bottomed (RB) flask from your kit and replace it if it is cracked. A flask that has even a small crack may break upon heating. In every distillation, always check the inside of the plastic connectors to see that the ring of plastic in the center is not frayed too badly.

If it is, the connection will leak and vapor will be lost. Samples of frayed and good connectors will be on display in the lab. Replace frayed connectors with new ones, which are available on the equipment replacement shelves. If you have a question about the condition of the connector, ask your TA. The connector should be checked before each distillation because it may have deteriorated during the previous distillation. Connectors are expensive though so do not dispose of them unless absolutely needed.

Set-up

To the 5 mL round-bottomed (RB) flask, add 2.0 mL of dry cyclohexane (measure in fume hood, using a small graduated cylinder), 2.0 mL of dry toluene (fume hood), and a couple of boiling chips. A clean transfer from the graduated cylinder to the flask may be accomplished by use of a pipet. Clamp the neck of the flask securely to a ring stand using a small three-pronged clamp. Using the black plastic connector, attach the distilling head to the RB flask. Make sure the clamp is placed as far as possible down on the neck of the flask so the black connector will fit tightly. Otherwise the connection may leak. To the distilling head, attach the thermometer adaptor and thermometer.

USE CAUTION when inserting the thermometer into the thermometer adaptor. Hold the thermometer close to the adaptor and push and twist gently away from you. Breakage could result in serious injury. If excessive force is needed, STOP and ask for help. Report broken thermometers to your TA.

To further avoid possible leaks, confirm that all glass pieces are seated firmly in the connectors and that the apparatus stands vertically. A common problem is to attach the three-pronged clamp in such a way that stress is placed on the connection, resulting in a leak. Clamp the apparatus carefully to prevent such leaks (remember to place the clamp as far down on the neck of the flask as possible). The collection vial should be pushed up well onto the distilling head at an angle of about 45° and held in place with a 6" piece of copper wire. One end of the wires hould be twisted around the lip of the vial and the other twisted around the vertical part of the distilling head.

The vial should be held in such a position that the outlet of the distilling head neither touches nor comes too close to the inside of the vial. During the distillation, drops dripping off the end of the distilling head need to be counted. If the vial is touching the head, instead of forming drops, the liquid will dribble down the side of the vial, making it impossible to count drops. The collection vial should be immersed in a 100 mL beaker filled with just enough ice/water to cool the collected distillate, to keep the distillate from evaporating.

The beaker can be held in place at an angle with your large three-pronged clamp attached to a second ring stand. For everything to fit, the sand bath must be pushed as far away from the collection side of the apparatus as possible. This also helps to ensure that the cold beaker will not touch the hot sand bath, which could cause the beaker to crack. Refer to the figure at the end of this handout. A sample set-up will also be on display in the lab. In all distillations remember to use a boiling chip to promote smooth boiling.

Heating

Using a spatula, make a small depression in the sand. To begin the distillation, lower the apparatus down into the depression so that the lower part of the RB flask is immersed in the hots and. Start by not pushing the flask too deeply into the sand. Otherwise overheating may occur and the distillation may proceed too rapidly. The depth can be adjusted as needed. Setting it down into a small depression from the start makes it easier to scrape more sand onto the flask later on if more heat is needed. The liquid will soon begin to boil. Boiling should be gentle enough so that the hot vapors move slowly up into the distilling head, eventually reaching the thermometer bulb. This will take several minutes.

The temperature reading will not change much until the vapors actually reach the bulb. Shortly thereafter liquid will begin to condense in the side-arm of the distilling head and begin to drip into the collection vial. Record the temperature at which the first drop falls into the vial and then at 4 drop intervals throughout the distillation. Afterward a plot of temperature vs # of drops will be made. On a micro scale, for an efficient separation, the rate of distillation should be about 1 drop per 20 - 30 seconds. Also, the most effective separation occurs when the collection rate is kept steady. The heating rate, and thus the distilling rate, is controlled by moving small amounts of hot sand onto or off of the distilling flask with a spatula.

Usually after the lower-boiling component has distilled, more heat is needed to distill the higher-boiling component. This usually results in a temporary drop in temperature and a slowed or stopped drip rate. To resume distillation,scrape more hot sand around the distilling flask. When little liquid remains in the distilling flask,stop the distillation by raising the apparatus out of the hot sand. Allow it to cool before disassembling it. A distillation should always be stopped before the distilling flask runs dry. In some cases, high boiling explosive compounds such as peroxides may be present. If the flask runs dry and the temperature rises too much, an explosion may result.

Even if the flask were allowed to run dry,some material would be left behind in the apparatus (HOLDUP of the apparatus). The holdup of material in the apparatus together with the material purposely left behind in the flask represents a loss of material. This is a necessary sacrifice and results in a decreased yield of distillate.

Fractional distillation of Cyclohexane and Toluene

The procedure is the same as that for the simple distillation except that a distilling column packed with copper sponge is placed between the distilling flask and the distilling head. The column should not be packed too tightly or too loosely. Samples of correct amounts of copper sponge and correctly filled columns will be available near the balance area for viewing. Some lockers may contain a glass tube that looks similar to the distillation column. Care must be taken to use the distilling column, **NOT** the similar but longer and slightly narrower chromatography column.

If the chromatography column is used by mistake, the connections will leak and material will be lost. It is not necessary to clean the apparatus after the simple distillation. Because the distilling path is longer, heat loss is greater in the fractional distillation, so it is helpful to wrap slightly-crinkled Al foil loosely around the column to insulate the column from drafts and to help keep the heat in (if the Al foil is tightly wrapped it will defeat the purpose by conducting heat away from the column. The trapped air in the crinkled, loosely wrapped Al foil is what serves as an insulator.) This helps to provide the heat necessary to distill the material through the apparatus at a steady rate. It will take longer for vapor to reach the thermometer in the fractional distillation. In the write-up, on the same graph, plot the temperature vs# of drops for both the simple and fractional distillation. Neither will look as "perfect" as the example in the above discussion.

Fractional Distillation of an Unknown Mixture

(Note: if you distill too rapidly and the entire sample distills over too quickly or if you fail to record the temperature or change vials, simply recombine the liquids and redistill. In most cases you will not need a new sample.) For this experiment it is not necessary to wash the apparatus or to change the copper sponge. These items are contaminated only with volatile (easily evaporated) compounds. Simply dry the apparatus in the fume hood by blowing air gently over and through the glassware to evaporate most of the remaining cyclohexane and toluene.

Certain items such as distilling columns are more effectively dried by connecting them to a source of vacuum and drawing air through them. A vacuum outlet exists in the fume hood. You will be given about 4 mL (measure it exactly with a graduated cylinder and record the amount) of a mixture of two of the compounds given in the following table. The two components will be present in the ratio of 1:2, 1:1, or 2:1. To minimize evaporation keep samples inclosed containers when not in use.

compound	boiling point (BP) (°C)
acetone	56
hexane	69
2-methyl-2-propanol	82
heptane	98
toluene	111
1-butanol	117

In the distillation of the unknown mixture, after the first component has distilled over its temperature plateau (± a few degrees) and the temperature begins to rise more steeply, the collection vial should be changed to prevent contamination of the first component with the second. The material collected between the BPs of the two substances is a mixture of the two and should be considered to be an impure intermediate fraction. Once the temperature reaches a second plateau, this should be considered to be the BP of the second substance and the collection vial should again be changed to collect pure component number two.

Thus, three vials of distillates will be collected- one containing pure lower-boiling component, one containing the intermediate mixture, and one containing pure higher-boiling component. Using the BPs and amounts of each pure fraction, determine the identities and ratios of the unknowns. As in all distillations, because a small amount of material is purposely left behind in the flask and because of holdup in the apparatus, the amount of the second component that is actually collected in the collection vial will be less than the amount in the original mixture.

To determine the ratio in the original mixture, use the total amount of original mixture (~4 mL) and subtract the amount of the first component collected to determine the amount of the second component (remember that because of hold-up not all of the second component will distill over). If pressed for time, as soon as the second BP plateau is clearly reached, the distillation may be stopped. Collecting separate fractions is not done in the distillations of cyclohexane and toluene because these experiments are simply done to learn the operation of distillation. In distillations in general however, once the first fraction has distilled, the vial must be changed to collect the intermediate, impure fraction, then, once the second pure component begins to distill, the vial must be changed again to collect that pure component separately. The goal is to end up with pure separated compounds.

SAVE THE COPPER WIRE AND LEAVE THE COPPER SPONGE IN THE COLUMN. THESE WILL BE USED IN A FUTURE EXPERIMENT

BEFORE LEAVING THE LAB: shut off the sand bath, place wastes in the proper containers, place all of your equipment into your locker and lock up, clean up your work areas, close the fume hood sash completely, and ask your TA for her or his signature. No other student in the world has a better organic lab in which to work as you have in the ISB. Let's try to keep it in great shape.

WASTE: Place all liquids into the ORGANIC LIQUID WASTE container.

Things to Watch Out For In Distillations:

(1) The thermometer is positioned incorrectly - this leads to observed temperatures which are incorrect. The top of the thermometer bulb must be even with the bottom of the side arm on the distilling head. It must not touch the copper packing or the glass apparatus.

(2) Distillation is too rapid due to excessive heating - this leads to a poor separation. Start over if this occurs. If this happens with your unknown, do not dispose of the liquids. Simply recombine all liquids and red is till the mixture.

(3) Not enough heating - this leads to reflux (a condition in which the vapors condense and return to the distilling flask) instead of distillation. Supply enough heat so that the distillation proceeds steadily at a rate of about one drop per 20 - 30 seconds.

(4) After the low-boiling fraction has distilled, the distillation may slow or stop and the temperature may fluctuate or drop - this is because more heat is required to distill the higher-boiling fraction. Pile more hot sand on the distilling flask so that the distillation proceeds at a rate of one drop per 20- 30 seconds. This is more likely to happen with the fractional distillation because of the longer path. It is advisable, especially if the lab is drafty, to insulate the distilling column in the fractional distillation with loosely-wrapped aluminum foil.

(5) The distilling column is packed too tightly with copper sponge - this leads to a situation known as flooding of the column, in which a plug of liquid collects in the column. Distillation will be severely hampered in such a case.

(6) The distilling column is packed too loosely - a poor separation results.

(7) With this equipment a measured BP may be considered to be reliable only after the temperature has leveled off and the distillation rate is about 1 drop per 20-30 seconds. Even then the measured BP may be a few degrees off.

(8) Liquid boils in the flask but none is collected in the vial. This may be due to a leaking connector. Be sure that the connector is not frayed and that the clamp is not pushing up on the connector, causing a bad connection.

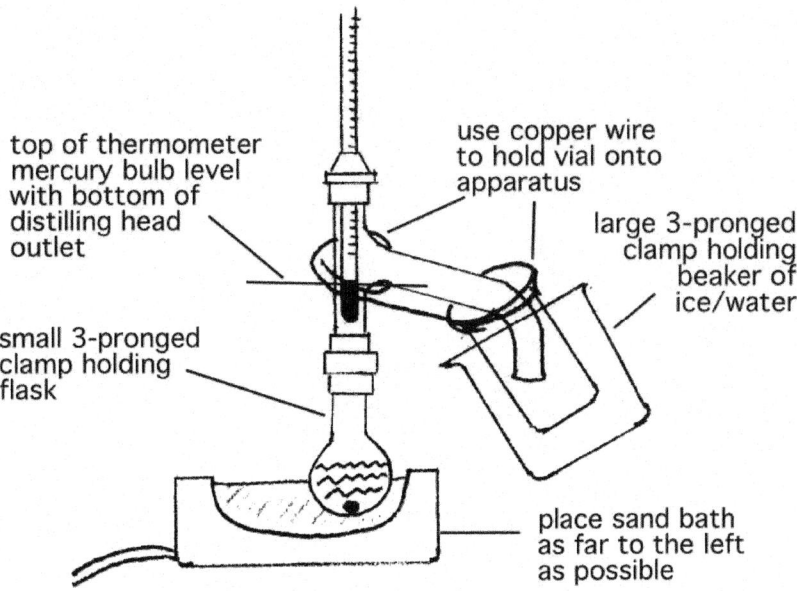

top of thermometer mercury bulb level with bottom of distilling head outlet

use copper wire to hold vial onto apparatus

large 3-pronged clamp holding beaker of ice/water

small 3-pronged clamp holding flask

place sand bath as far to the left as possible

Fig. Simple Distillation Apparatus.

Chapter 13

FRACTIONAL DISTILLATION

Fractional distillation is the separation of a mixture into its component parts, or fractions, such as in separating chemical compounds by their boiling point by heating them to a temperature at which one or more fractions of the compound will vaporize. It is a special type of distillation. Generally the component parts boil at less than 25°C from each other under a pressure of one atmosphere. If the difference in boiling points is greater than 25°C, a simple distillation is used.

Laboratory Setup

Fractional distillation in a laboratory makes use of common laboratory glass-ware and apparatuses, typically including a Bunsen burner, a round-bottomed flask and a condenser, as well as the single-purpose fractionating column.

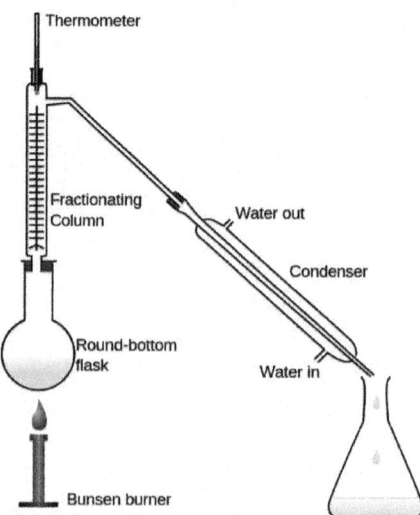

Fig. Fractional distillation. An Erlenmeyer flask is used as a receiving flask. Here the distillation head and fractionating column are combined in one piece.

Apparatus

- heat source, such as a hot plate with a bath, and ideally with a magnetic stirrer.
- distilling flask, typically a round-bottom flask
- receiving flask, often also a round-bottom flask
- fractionating column
- distillation head
- thermometer and adapter if needed
- condenser, such as a Liebig condenser, Graham condenser or Allihn condenser
- vacuum adapter (not used in image to the right)
- boiling chips, also known as anti-bumping granules
- Standard laboratory glassware with ground glass joints, *e.g.* quickfit apparatus.

Discussion

As an example consider the distillation of a mixture of water and ethanol. Ethanol boils at 78.4°C while water boils at 100°C. So, by heating the mixture, the most volatile component (ethanol) will concentrate to a greater degree in the vapor leaving the liquid. Some mixtures form azeotropes, where the mixture boils at a lower temperature than either component. In this example, a mixture of 96% ethanol and 4% water boils at 78.2°C; the mixture is more volatile than pure ethanol. For this reason, ethanol cannot be completely purified by direct fractional distillation of ethanol-water mixtures.

The apparatus is assembled as in the diagram. (The diagram represents a batch apparatus as opposed to a continuous apparatus.) The mixture is put into the round bottomed flask along with a few anti-bumping granules(or a Teflon coated magnetic stirrer bar if using magnetic stirring), and the fractionating column is fitted into the top. The fractional distillation column is set up with the heat source at the bottom on the still pot. As the distance from the stillpot increases, a temperature gradient is formed in the column; it is coolest at top and hottest at the bottom. As the mixed vapor ascends the temperature gradient, some of the vapor condenses and revaporizes along the temperature gradient.

Each time the vapor condenses and vaporizes, the composition of the more volatile component in the vapor increases. This distills the vapor along the length of the column, and eventually the vapor is composed solely of the more volatile component (or an azeotrope). The vapor condenses on the glass platforms, known as trays, inside the column, and runs back down into the liquid below, refluxing distillate. The efficiency in terms of the amount of heating and time required to get fractionation can be improved by insulating the outside of the column in an insulator such as wool, aluminium foil or preferably a vacuum jacket. The hottest tray is at the bottom and the coolest is at the top.

At steady state conditions, the vapor and liquid on each tray are at *equilibrium*. The most volatile component of the mixture exits as a gas at the top of the column. The vapor at the top of the column then passes into the condenser, which cools it down until it liquefies. The separation is more pure with the addition of more trays (to a practical limitation of heat, flow, *etc.*) Initially, the condensate will be close to the azeotropic composition, but when much of the ethanol has been drawn off, the condensate becomes gradually richer in water. The process continues until all the ethanol boils out of the mixture. This point can be recognized by the sharp rise in temperature shown on the thermometer.

The above explanation reflects the theoretical way fractionation works. Normal laboratory fractionation columns will be simple glass tubes (often vacuum-jacketed, and sometimes internally silvered) filled with a packing, often small glass helices of 4 to 7 mm diameter. Such a column can be calibrated by the distillation of a known mixture system to quantify the column in terms of number of theoretical trays. To improve fractionation the apparatus is set up to return condensate to the column by the use of some sort of reflux splitter (reflux wire, gago, Magnetic swinging bucket, *etc.*) - a typical careful fractionation would employ a reflux ratio of around 4:1(4 parts returned condensate to 1 part condensate take off).

In laboratory distillation, several types of condensers are commonly found. The Liebig condenser is simply a straight tube within a water jacket, and is the simplest (and relatively least expensive) form of condenser. The Graham condenser is a spiral tube within a water jacket, and the Allihn condenser has a series of large and small constrictions on the inside tube, each increasing the surface area upon which the vapor constituents may condense.

Alternate set-ups may utilize a multi–outlet distillation receiver flask (referred to as a "cow" or "pig") to connect three or four receiving flasks to the condenser. By turning the cow or pig, the distillates can be channeled into any chosen receiver. Because the receiver does not have to be removed and replaced during the distillation process, this type of apparatus is useful when distilling under an inert atmosphere for air-sensitive chemicals or at reduced pressure. A Perkin triangle is an alternative apparatus is often used in these situations because it allows isolation of the receiver from the rest of the system, but does require removing and reattaching a single receiver for each fraction.

Vacuum distillation systems operate at reduced pressure, thereby lowering the boiling points of the materials. Anti-bumping granules, however, become ineffective at reduced pressures.

Industrial Distillation

Fractional distillation is the most common form of separation technology used in petroleum refineries, petrochemical and chemical plants, natural gas processing and cryogenic air separation plants. In most cases, the distillation is operated at a continuous steady state. New feed is always being added to the distillation column and products are always being removed. Unless the process is disturbed due to changes in feed, heat, ambient temperature, or condensing, the amount of feed being added and the amount of product being removed are normally equal. This is known as continuous, steady-state fractional distillation.

Industrial distillation is typically performed in large, vertical cylindrical columns known as "distillation or fractionation towers" or "distillation columns" with diameters ranging from about 65 centimeters to 6 meters and heights ranging from about 6 meters to 60 meters or more. The distillation towers have liquid outlets at intervals up the column which allow for the withdrawal of different fractions or products having different boiling points or boiling ranges. By increasing the temperature of the product inside the columns, the different hydrocarbons are separated. The "lightest" products (those with the lowest boiling point) exit from the top of the columns and the "heaviest" products (those with the highest boiling point) exit from the bottom of the column.

For example, fractional distillation is used in oil refineries to separate crude oil into useful substances (or fractions) having different hydrocarbons of different boiling points. The crude oil fractions with higher boiling points:

- have more carbon atoms
- have higher molecular weights
- are more branched chain alkanes
- are darker in color
- are more viscous
- are more difficult to ignite and to burn

Fig. Typical industrial fractional distillation columns.

Large-scale industrial towers use reflux to achieve a more complete separation of products. Reflux refers to the portion of the condensed overhead liquid product from a distillation or fractionation tower that is returned to the upper part of the tower as shown in the schematic diagram of a typical, large-scale industrial distillation tower. Inside the tower, the reflux liquid flowing downwards provides the cooling needed to condense the vapors flowing upwards, thereby increasing the effectiveness of the distillation tower. The more reflux is provided for a given number of theoretical plates, the better the tower's separation of lower boiling materials from higher boiling materials. Alternatively, the more reflux provided for a given desired separation, the fewer theoretical plates are required.

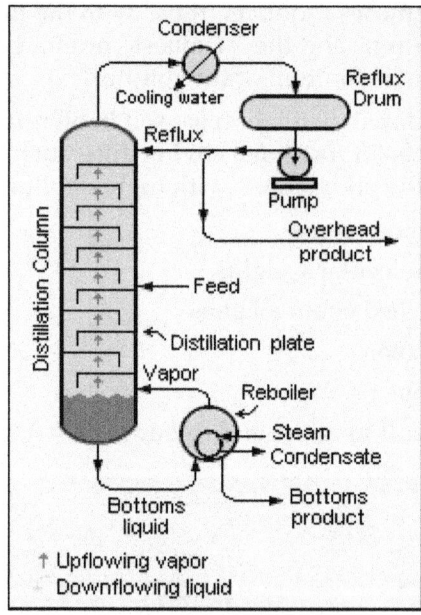

Fractional distillation is also used in air separation, producing liquid oxygen, liquid nitrogen, and highly concentrated argon. Distillation of chlorosilanes also enable the production of high-purity silicon for use as a semiconductor.

In industrial uses, sometimes a packing material is used in the column instead of trays, especially when low pressure drops across the column are required, as when operating under vacuum. This packing material can either be random dumped packing (1-3" wide) such as Raschig rings or structured sheet metal. Typical manufacturers are Koch, Sulzer and other companies. Liquids tend to wet the surface of the packing and the vapors pass across this wetted surface, where mass transfer takes place.

Unlike conventional tray distillation in which every tray represents a separate point of vapor liquid equilibrium the vapor liquid equilibrium curve in a packed column is continuous. However, when modeling packed columns it is useful to compute a number of "theoretical plates" to denote the separation efficiency of the packed column with respect to more traditional trays. Differently shaped packings have different surface areas and void space between packings. Both of these factors affect packing performance.

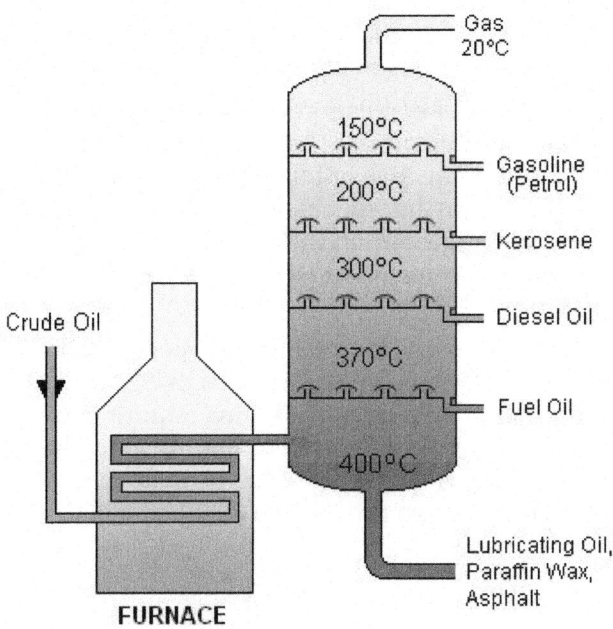

Design of Industrial Distillation Columns

Design and operation of a distillation column depends on the feed and desired products. Given a simple, binary component feed, analytical methods such as the McCabe–Thiele method or the Fenske equation can be used. For a multi-component feed, simulation models are used both for design and operation.

Moreover, the efficiencies of the vapor–liquid contact devices (referred to as *plates* or *trays*) used in distillation columns are typically lower than that of a theoretical 100% efficient equilibrium stage. Hence, a distillation column needs more plates than the number of theoretical vapor–liquid equilibrium stages.

Reflux refers to the portion of the condensed overhead product that is returned to the tower. The reflux flowing downwards provides the cooling required for condensing the vapours flowing upwards. The reflux ratio, which is the ratio of the (internal) reflux to the overhead product, is conversely related to the theoretical number of stages required for efficient separation of the distillation products. Fractional distillation towers or columns are designed to achieve the required separation efficiently.

The design of fractionation columns is normally made in two steps; a process design, followed by a mechanical design. The purpose of the process design is to calculate the number of required theoretical stages and stream flows including the reflux ratio, heat reflux and other heat duties. The purpose of the mechanical design, on the other hand, is to select the tower internals, column diameter and height. In most cases, the mechanical design of fractionation towers is not straight-forward. For the efficient selection of tower internals and the accurate calculation of column height and diameter, many factors must be taken into account. Some of the factors involved in design calculations include feed load size and properties and the type of distillation column utilized.

The two major types of distillation columns used are tray and packing columns. Packing columns are normally used for smaller towers and loads that are corrosive or temperature sensitive or for vacuum service where pressure drop is important. Tray columns, on the other hand, are used for larger columns with high liquid loads. They first appeared on the scene in the 1820s. In most oil refinery operations, tray columns are mainly used for the separation of petroleum fractions at different stages of oil refining.

In the oil refining industry, the design and operation of fractionation towers is still largely accomplished on an empirical basis. The calculations involved in the design of petroleum fractionation columns require in the usual practice the use of numerable charts, tables and complex empirical equations. In recent years, however, a considerable amount of work has been done to develop efficient and reliable computer-aided design procedures for fractional distillation.

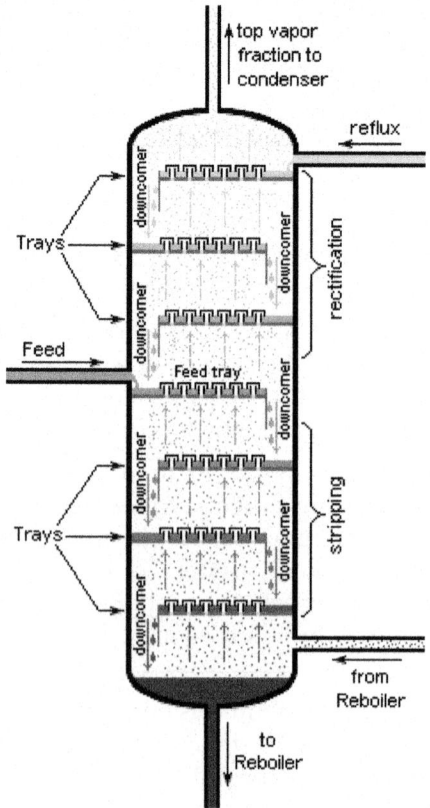

Fig. Chemical engineering schematic of typical bubble-cap trays in a distillation tower.

AZEOTROPIC DISTILLATION

In chemistry, azeotropicdistillation[1] is any of a range of techniques used to break an azeotrope in distillation. In chemical engineering, azeotropic distillation usually refers to the specific technique of adding another component to generate a new, lower-boiling azeotrope that is heterogeneous (*e.g.* producing two, immiscible liquid phases), such as the example below with the addition of benzene to water and ethanol. This practice of adding an entrainer which forms a separate phase is a specific sub-set of (industrial) azeotropic distillation methods, or combination thereof. In some senses, adding an entrainer is similar to extractive distillation.

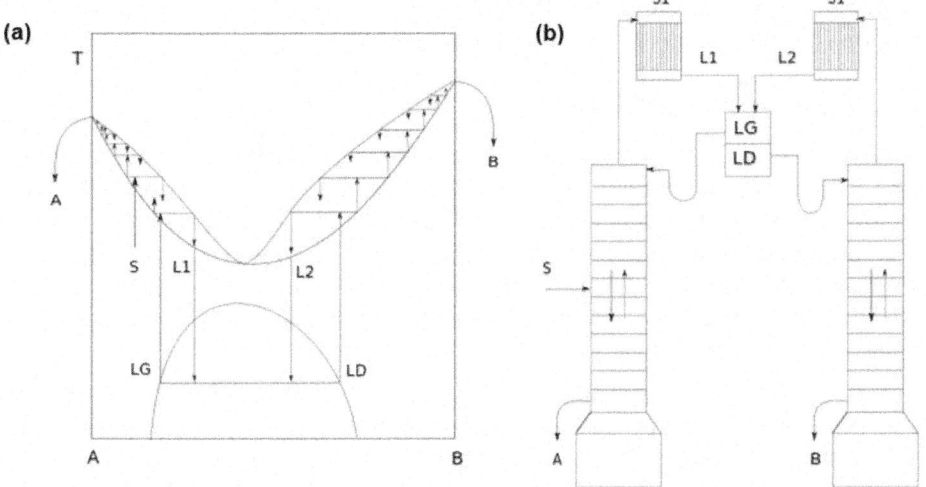

Fig. (a) Phase diagram and (b) process flow diagram of an apparatus for the azeotropic distillation with material separation agent. In this case the phase diagram includes a zone where components are not miscible, so following the condensation of the azeotrope, it is possible to separate the liquid components through decantation.

Material Separation Agent

The addition of a Material Separation Agent, such as benzene to an ethanol/water mixture, changes the molecular interactions and eliminates the azeotrope. Added in the liquid phase, the new component can alter the activity coefficient of various compounds in different ways thus altering a mixture's relative volatility. Greater deviations from Raoult's law make it easier to achieve significant changes in relative volatility with the addition of another component. In azeotropic distillation the volatility of the added component is the same as the mixture, and a new azeotrope is formed with one or more of the components based on differences in polarity. If the material separation agent is selected to form azeotropes with more than one component in the feed then it is referred to as an entrainer. The added entrainer should be recovered by distillation, decantation, or another separation method and returned near the top of the original column.

Distillation of Ethanol/Water

A common historical example of azeotropic distillation is its use in dehydrating ethanol and water mixtures. For this, a near azeotropic mixture is sent to the final column where azeotropic distillation takes place. Several entrainers can be used for this specific process: benzene, pentane, cyclohexane, hexane, heptane, isooctane, acetone, and diethyl ether are all options as the mixture. Of these benzene and cyclohexane have been used the most extensively. However, because benzene has been discovered to be a carcinogenic compound, its use has declined. While this method was the standard for dehydrating ethanol in the past, it has lost favor due to the high capital and energy costs associated with it. Another favorable method and less toxic than using benzene to break the azeotrope of the ethanol-water system is to use toluene instead.

Pressure-swing Distillation

Another method, pressure-swing distillation, relies on the fact that an azeotrope is pressure dependent. An azeotrope is not a range of concentrations that cannot be distilled, but the point at which the activity coefficients of the distillates are crossing one another. If the azeotrope can be "jumped over", distillation can continue, although because the activity coefficients have crossed, the water will boil out of the remaining ethanol, rather than the ethanol out of the water as at lower concentrations.

To "jump" the azeotrope, the azeotrope can be moved by altering the pressure. Typically, pressure will be set such that the azeotrope will be closer to 100% concentration. For ethanol, that may be 97%. Ethanol can now be distilled up to 97%. It will actually be distilled to something slightly less, like 96.5%. The 96.5% alcohol is then sent to a distillation column that is under a different pressure, one that pulls the azeotrope down, maybe to 96%. Since the mixture is already above the 96% azeotrope, the distillation will not get "stuck" at that point and the ethanol can be distilled to whatever concentration is needed.

Molecular Sieves

For the distillation of ethanol for gasoline addition, the most common means of breaking the azeotrope is the use of molecular sieves. Ethanol is distilled to 96%, then run over a molecular sieve which adsorbs water from the mixture. The concentration is now above 96% and can be further distilled. The sieve is heated to remove the water and reused.

BATCH DISTILLATION

Batch distillation refers to the use of distillation in batches, meaning that a mixture is distilled to separate it into its component fractions before the distillation still is again charged with more mixture and the process is repeated. This is in contrast with continuous distillation where the feedstock is added and the distillate drawn off without interruption. Batch distillation has always been an important part of the production of seasonal, or low capacity and high-purity chemicals. It is a very frequent separation process in the pharmaceutical industry.

Batch Rectifier

The simplest and most frequently used batch distillation configuration is the **batch rectifier**, including the alembic and pot still. The batch rectifier consists of a pot (or reboiler), rectifying column, a condenser, some means of splitting off a portion of the condensed vapour (distillate) as reflux, and one or more receivers.

The pot is filled with liquid mixture and heated. Vapour flows upwards in the rectifying column and condenses at the top. Usually, the entire condensate is initially returned to the column as reflux. This contacting of vapour and liquid considerably improves the separation. Generally, this step is named start-up. The first condensate is the *head*, and it contains undesirable components. The last condensate is the *feints* and it is also undesirable, although it adds flavor. In between is the *heart* and this forms the desired product.

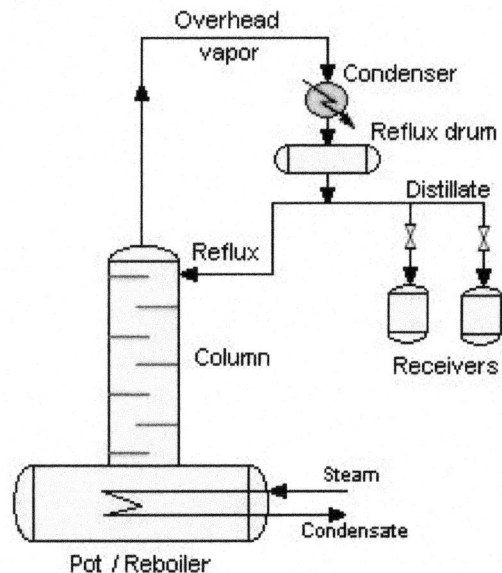

Fig. Diagram of a Batch Rectifier.

The head and feints may be thrown out, refluxed, or added to the next batch of mash/juice, according to the practice of the distiller. After some time, a part of the overhead condensate is withdrawn continuously as distillate and it is accumulated in the receivers, and the other part is recycled into the column as reflux.

Owing to the differing vapour pressures of the distillate, there will be a change in the overhead distillation with time, as early on in the batch distillation, the distillate will contain a high concentration of the component with the higher relative volatility. As the supply of the material is limited and lighter components are removed, the relative fraction of heavier components will increase as the distillation progresses.

Batch Stripper

The other simple batch distillation configuration is the **batch stripper**. The batch stripper consists of the same parts as the batch rectifier. However, in this case, the charge pot is located above the stripping column.

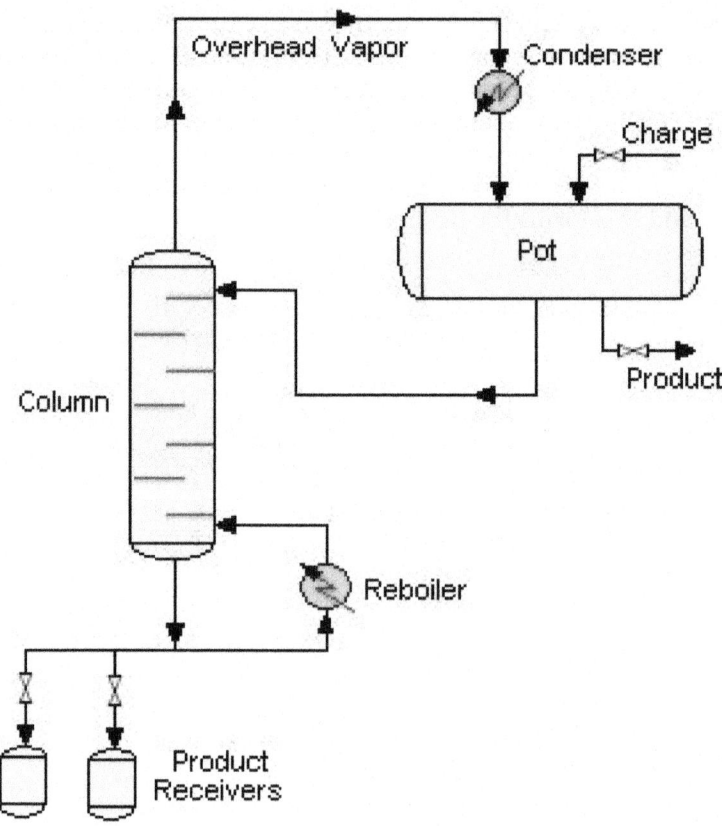

Fig. Diagram of a Batch Stripper.

During operation (after charging the pot and starting up the system) the high boiling constituents are primarily separated from the charge mixture. The liquid in the pot is depleted in the high boiling constituents, and enriched in low boiling ones. The high boiling product is routed into the bottom product receivers. The residual low boiling product is withdrawn from the charge pot. This mode of batch distillation is very seldom applied in industrial processes.

Middle Vessel Column

A third feasible batch column configuration is the **middle vessel column**. The middle vessel column consists of both a rectifying and a stripping section and the charge pot is located at the middle of the column.

Feasibility Studies

Generally, the feasibility studies of batch distillation are based on analyses of the following maps:

- Residue curve map
- still path map
- distillate path map
- different column profile maps

During the feasibility studies, the following basic simplifying assumptions are made:

- infinite number of equilibrium stages
- infinite reflux ratio
- negligible tray hold-up in the two column sections
- quasi-steady state in the column
- constant molar overflow

Bernot*et al.* used the batch distillation regions to determine the sequence of the fractions. According to Ewell and Welch, a batch distillation region gives the same fractions upon rectification of any mixture lying within it. Bernot *et al.* examined the still and distillate paths for the determination of the region boundaries under high number of stages and high reflux ratio, named maximal separation. Pham and Doherty in pioneering work described the structure and properties of residue curve maps for ternary heterogeneous azeotropic mixtures.

In their model, the possibility of the phase separation of the vapour condensed is not taken into consideration yet. The singular points of the residue curve maps determined by this method were used to assign batch distillation regions by Rodriguez-Donis *et al.* and Skouras *et al.*Modla *et al.* pointed out that this method may give misleading results for the minimal amount of entrainer. Lang and Modla extended the method of Pham and Doherty and suggested a new, general method for the calculation of residue curves and for the determination of batch distillation regions of heteroazeotropic distillation.

Lelkes*et al.* published a feasibility method for the separation of minimum boiling point azeotropes by continuously entrainer feeding batch distillation. This method has been applied for the use of a light entrainer in the batch rectifier and stripper by Lang *et al.*(1999) and it applied for maximum azeotropes by Lang *et al.* Modla*et al.* extended this method for batch heteroazeotropic distillation under continuous entrainer feeding.

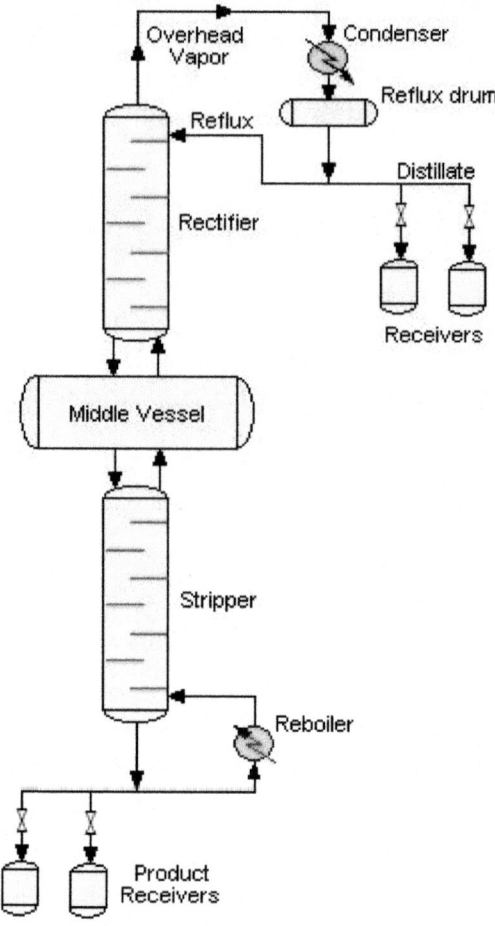

Fig. Diagram of a Middle Vessel Column.

EXTRACTIVE DISTILLATION

Extractive distillation is defined as distillation in the presence of a miscible, high boiling, relatively non-volatile component, the solvent, that forms no azeotrope with the other components in the mixture. The method is used for mixtures having a low value of relative volatility, nearing unity. Such mixtures cannot be separated by simple distillation, because the volatility of the two components in the mixture is nearly the same, causing them to evaporate at nearly the same temperature at a similar rate, making normal distillation impractical.

The method of extractive distillation uses a separation solvent, which is generally non-volatile, has a high boiling point and is miscible with the mixture, but doesn't form an azeotropic mixture. The solvent interacts differently with the components of the mixture thereby causing their relative volatilities to change.

This enables the new three-part mixture to be separated by normal distillation. The original component with the greatest volatility separates out as the top product. The bottom product consists of a mixture of the solvent and the other component, which can again be separated easily because the solvent does not form an azeotrope with it. The bottom product can be separated by any of the methods available.

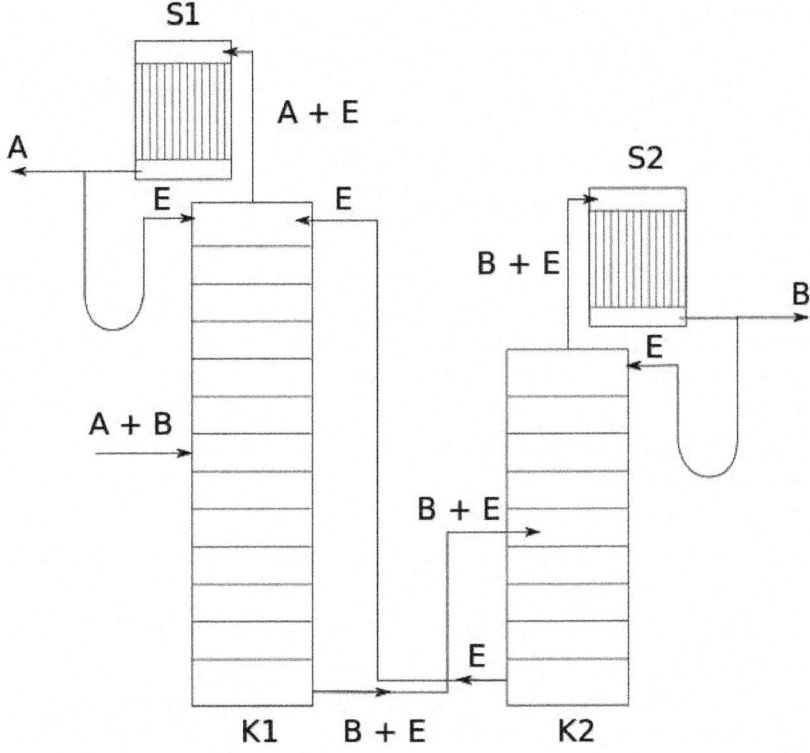

Fig. Process flow diagram showing an extractive distillation apparatus. In this case the mixture components A and B are separated in the first column through the solvent E (recovered in the second column).

FRACTIONAL FREEZING

Fractional freezing is a process used in process engineering and chemistry to separate substances with different melting points. It can be done by partial melting of a solid, for example in zone refining of silicon or metals, or by partial crystallization of a liquid, as in **freeze distillation**, also called **normal freezing** or **progressive freezing**.

Partial crystallization can also be achieved by adding a dilute solvent to the mixture, and cooling and concentrating the mixture by evaporating the solvent, a process called **solution crystallization**. Fractional freezing is generally used to produce ultra-pure solids, or to concentrate heat-sensitive liquids.

Freeze Distillation

Freeze distillation is a misnomer, since it is not distillation but rather a process of enriching a solution by partially freezing it and removing frozen material that is poorer in the dissolved material than is the liquid portion left behind. Such enrichment parallels enrichment by true distillation, where the evaporated and re-condensed portion is richer than the liquid portion left behind.

The detailed situation is the subject of thermodynamics, a subdivision of physics of importance to chemistry. Without resorting to mathematics, the following can be said for a mixture of water and alcohol:

- Freezing in this scenario begins at a temperature significantly below 0°C.
- The first material to freeze is not the water, but a dilute solution of alcohol in water.
- The liquid left behind is richer in alcohol, and as a consequence, further freezing would take place at progressively lower temperatures. The frozen material, while always poorer in alcohol than the (increasingly rich) liquid, becomes progressively richer in alcohol.
- Further stages of removing frozen material and waiting for more freezing will come to naught once the liquid uniformly cools to the temperature of whatever is cooling it.
- If progressively colder temperatures are available,
  o the frozen material will contain progressively larger concentrations of alcohol, and
  o the fraction of the original alcohol removed with the solid material will increase.
- In practice, unless the removal of solid material carries away liquid, the degree of concentration will depend on the final temperature rather than on the number of cycles of removing solid material and chilling.
- Thermodynamics gives fair assurance, even without more information about alcohol and water than that they freely dissolve in each other, that
  o even if temperatures *somewhat below* the freezing point of ethyl alcohol are achieved, there will still be alcohol and water mixed as a liquid, and
  o at some still lower temperature, the remaining alcohol-and-water solution will freeze without an alcohol-poor solid being separable. The best-known freeze-distilled beverages are applejack and ice beer. Ice wine is the result of a similar process, but in this case, the freezing happens *before* the fermentation, and thus it is sugar, not alcohol, that gets concentrated.

Purification of Solids

When a pure solid is desired, two possible situations can occur. If the contaminant is soluble in the desired solid, a multiple stage fractional freezing is required, analogous to multistage distillation. If, however, a eutectic system forms (analogous to an azeotrope in distillation), a very pure solid can be recovered, as long as the liquid is not at its eutectic composition (in which case a mixed solid forms, which can be hard to separate) or above its eutectic composition (in which case the undesired solid forms).

Concentration of Liquids

When the requirement is to concentrate a liquid phase, fractional freezing can be useful due to its simplicity. Fractional freezing is also used in the production of fruit juice concentrates and other heat-sensitive liquids, as it does not involve heating the liquid (as happens during evaporation).

Desalination

Fractional freezing can be used to desalinate sea water. In a process that naturally occurs with sea ice, frozen salt water, when partially melted, leaves behind ice that is of a much lower salt content. Because sodium chloride lowers the melting point of water, the salt in sea water tends to be forced out of pure water while freezing. Likewise, the frozen water with the highest concentration of salt melts first. Either method decreases the salinity of the frozen water left over, and with multiple runs can be drinkable.

Alcoholic Beverages

Fractional freezing can be used as a simple method to increase the alcohol concentration in fermented alcoholic beverages, a process sometimes called freeze distillation. Examples are applejack, made from hard cider, and ice beer. In practice, while not able to produce an alcohol concentration comparable to distillation, this technique can achieve some concentration with far less effort than any practical distillation apparatus would require. Freeze distillation of alcoholic beverages is illegal in some countries, including the United States.

The danger of freeze distillation of alcoholic beverages, is that unlike heat distillation, where the methanol and other impurities can be separated from the finished product, freeze distillation does not remove them. Thus the ratio of impurities may be increased compared to the total volume of the beverage. This concentration may cause side effects to the drinker, leading to intense hangovers and a condition known as "apple palsy" (although this term has also simply been used to refer to intoxication, especially from applejack.)

Alternative Fuels

Fractional freezing is commonly used as a simple method to reduce the gel point of biodiesel and other alternative diesel fuels, whereby esters of higher gel point are removed from esters of lower gel point through cold filtering, or other methods to reduce the subsequent alternative fuel gel point of the fuel blend. This process employs fuel stratification whereby components in the fuel blend develop a higher specific gravity as they approach their respective gel points and thus sink to the bottom of the container, where they can be removed.

STEAM DISTILLATION

Steam distillation is a special type of distillation(a separation process) for *temperature sensitive* materials like natural aromatic compounds. It once was a popular laboratory method for purification of organic compounds, but has become obsolete by vacuum distillation. Steam distillation remains important in certain industrial sectors.

Many organic compounds tend to decompose at high sustained temperatures. Separation by distillation at the normal (1 atmosphere) boiling points is not an option, so water or steam is introduced into the distillation apparatus. The water vapor carries small amounts of the vaporized compounds to the condensation flask, where the condensed liquids phase separate, allowing for easy collection. This process effectively allows for distillation at lower temperatures, reducing the deterioration of the desired products. If the substances to be distilled are very sensitive to heat, steam distillation may be applied under reduced pressure, thereby reducing the operating temperature further.

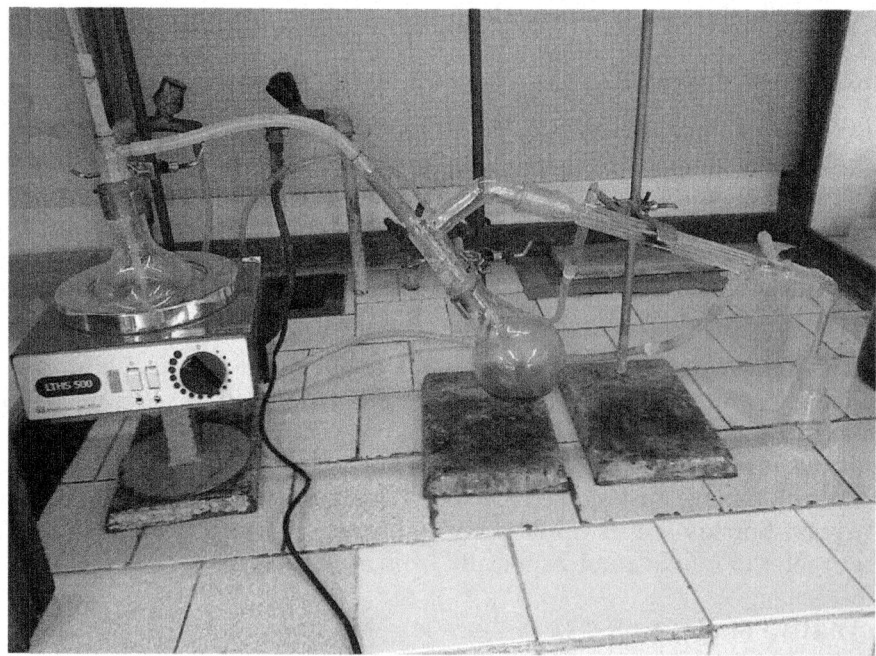

Fig. Steam distillation apparatus.

After distillation the vapors are condensed as appropriate. Usually the immediate product is a two-phase system of water and the organic distillate, allowing for separation of the components by decantation, partitioning or other suitable methods.

Principle

When a mixture of two practically immiscible liquids is heated while being agitated to expose the surface of one liquid to the vapor phase, each constituent independently exerts its own vapor pressure as a function of temperature as if the other constituent were not present. Consequently, the vapor pressure of the whole system increases. Boiling begins when the sum of the partial pressures of the two immiscible liquids just exceeds the atmospheric pressure(approximately 101kPa at sea level). In this way, many organic compounds insoluble in water can be purified at a temperature well below the point at which decomposition occurs. For example, the boiling point of bromobenzene is 156°C and the boiling point of water is 100°C, but a mixture of the two boils at 95°C. Thus, bromobenzene can be easily distilled at a temperature 61°C below its normal boiling point.

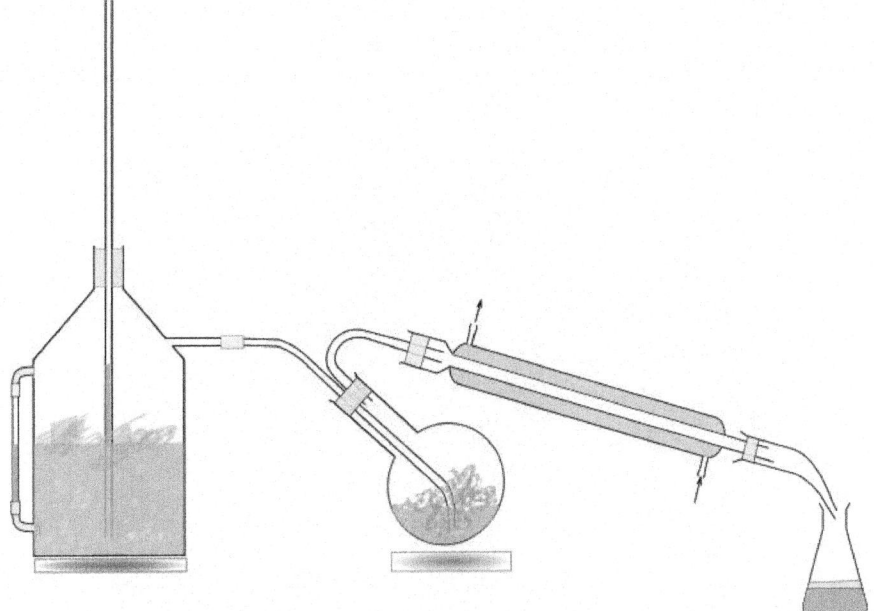

Fig. Steam distillation apparatus in a lab.

Applications

Steam distillation is employed in the isolation of essential oils, for use in perfumes, for example. In this method, steam is passed through the plant material containing the desired oils. Eucalyptus oil and orange oil are obtained by this method on the industrial scale. Steam distillation is also sometimes used to separate intermediate or final products during the synthesis of complex organic compounds.

Steam distillation is also widely used in petroleum refineries and petrochemical plants where it is commonly referred to as "steam stripping".

Steam distillation also is an important means of separating fatty acids from mixtures and for treating crude products such as tall oils to extract and separate fatty acids, soaps and other commercially valuable organic compounds.

Equipment

On a lab-scale steam distillations are carried out using steam generated outside the system and piped through macerated biomass or steam generation in-situ using a Clevenger-type apparatus.

VACUUM DISTILLATION

Vacuum distillation is a method of distillation whereby the pressure above the liquid mixture to be distilled is reduced to less than its vapor pressure(usually less than atmospheric pressure) causing evaporation of the most volatile liquid(s) (those with the lowest boiling points). This distillation method works on the principle that boiling occurs when the vapor pressure of a liquid exceeds the ambient pressure. Vacuum distillation is used with or without heating the mixture.

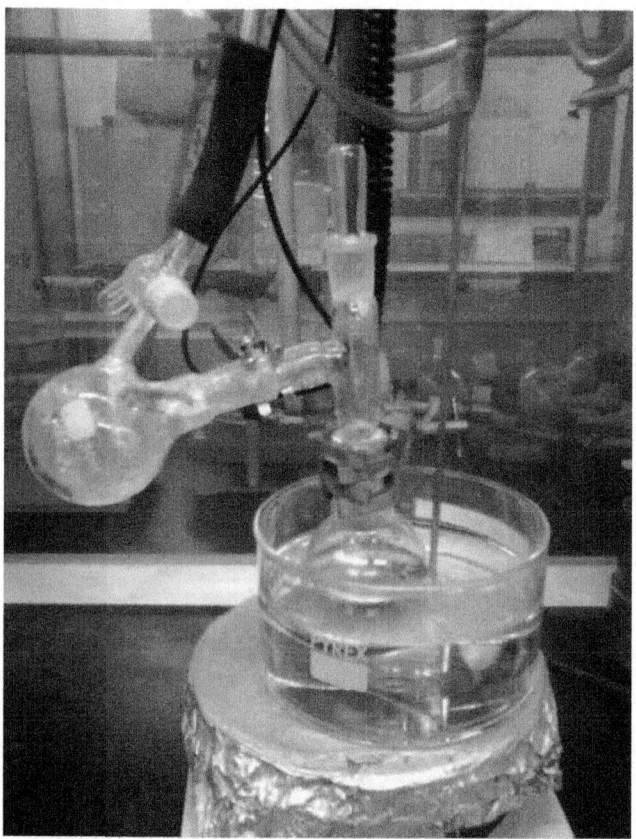

Fig. At atmospheric pressure, dimethyl sulfoxide boils at 189°C. Under a vacuum, it distills off into the connected receiver at only 70°C.

Laboratory-scale Applications

Laboratory-scale vacuum distillation is used when liquids to be distilled have high atmospheric boiling points or chemically change at temperatures near their atmospheric boiling points. Temperature sensitive materials (such as beta carotene) also require vacuum distillation to remove solvents from the mixture without damaging the product. Another reason vacuum distillation is used is that compared to steam distillation there is a lower level of residue build up. This is important in commercial applications where heat transfer is produced using heat exchangers.

There are many laboratory applications for vacuum distillation as well as many types of distillation set-ups and apparatuses.

Safety is an important consideration when using glassware as part of the set-up. All of the glass components should be carefully examined for scratches and cracks which could result in implosions when the vacuum is applied. Wrapping as much of the glassware with tape as is practical helps to prevent dangerous scattering of glass shards in the event of an implosion.

Rotary Evaporation

Rotary evaporation is a type of vacuum distillation apparatus used to remove bulk solvents from the liquid being distilled. It is also used by environmental regulatory agencies for determining the amount of solvents in paint, coatings and inks.

Rotary evaporation set-ups include an apparatus referred to as a *Rotovap* which rotates the distillation flask (sometimes called the *still pot*) to enhance the distillation. Rotating the flask throws up liquid on the walls of the flask and thus increases the surface area for evaporation.

Heat is often applied to the rotating distillation flask by partially immersing it in a heated bath of water or oil. Typically, the vacuum in such systems is generated by a water aspirator or a vacuum pump of some type.

Distillation of High-boiling and/or Air Sensitive Materials

Some compounds have high boiling point temperatures as well as being air sensitive. A simple laboratory vacuum distillation glassware set-up can be used, in which the vacuum can be replaced with an inert gas after the distillation is complete.

However, this is not a completely satisfactory system if it is desired to collect fractions under a reduced pressure.

For better results or for very air sensitive compounds, either a Perkin triangle distillation set-up or a short-path distillation set-up can be used.

Perkin Triangle Distillation Set-up

The Perkin triangle set-up uses a series of Teflon valves to allow the distilled fractions to be isolated from the distillation flask without the main body of the distillation set-up being removed from either the vacuum or the heat source, and thus can remain in a state of reflux.

To do this, the distillate receiver vessel is first isolated from the vacuum by means of the Teflon valves.

The vacuum over the sample is then replaced with an inert gas (such as nitrogen or argon) and the distillate receiver can then be stoppered and removed from the system.

Vacuum Distillation Set-up Using a Short-path Head

Vacuum distillation of moderately air/water-sensitive liquid can be done using standard Schlenk-line techniques. When assembling the set-up apparatus, all of the connecting lines are clamped so that they cannot pop off.

Once the apparatus is assembled, and the liquid to be distilled is in the still pot, the desired vacuum is established in the system by using the vacuum connection on the short-path distillation head. Care is taken to prevent potential "bumping" as the liquid in the still pot degases.

While establishing the vacuum, the flow of coolant is started through the short-path distillation head. Once the desired vacuum is established, heat is applied to the still pot.

If needed, the first portion of distillate can be discarded by purging with inert gas and changing out the distillate receiver.

When the distillation is complete: the heat is removed, the vacuum connection is closed, and inert gas is purged through the distillation head and the distillate receiver. While under the inert gas purge, remove the distillate receiver and cap it with an air-tight cap. The distillate receiver can be stored under vacuum or under inert gas by using the side-arm on the distillation flask.

Industrial-scale Applications

Industrial-scale vacuum distillation has several advantages. Close boiling mixtures may require many equilibrium stages to separate the key components. One tool to reduce the number of stages needed is to utilize vacuum distillation. Vacuum distillation columns typically used in oil refineries have diameters ranging up to about 14 meters (46 feet), heights ranging up to about 50 meters (164 feet), and feed rates ranging up to about 25,400 cubic meters per day (160,000 barrels per day).

Vacuum distillation increases the relative volatility of the key components in many applications. The higher the relative volatility, the more separable are the two components; this connotes fewer stages in a distillation column in order to effect the same separation between the overhead and bottoms products. Lower pressures increase relative volatilities in most systems.

A second advantage of vacuum distillation is the reduced temperature requirement at lower pressures. For many systems, the products degrade or polymerize at elevated temperatures.

Vacuum distillation can improve a separation by:

- Prevention of product degradation or polymer formation because of reduced pressure leading to lower tower bottoms temperatures,
- Reduction of product degradation or polymer formation because of reduced mean residence time especially in columns using packing rather than trays.
- Increasing capacity, yield, and purity.

Another advantage of vacuum distillation is the reduced capital cost, at the expense of slightly more operating cost. Utilizing vacuum distillation can reduce the height and diameter, and thus the capital cost of a distillation column.

Vacuum Distillation in Petroleum Refining

Petroleum crude oil is a complex mixture of hundreds of different hydrocarbon compounds generally having from 3 to 60 carbon atoms per molecule, although there may be small amounts of hydrocarbons outside that range. The refining of crude oil begins with distilling the incoming crude oil in a so-called *atmospheric distillation column* operating at pressures slightly above atmospheric pressure.

Vacuum distillation can also be referred as "low temperature distillation"

In distilling the crude oil, it is important not to subject the crude oil to temperatures above 370 to 380°C because the high molecular weight components in the crude oil will undergo thermal cracking and form petroleum coke at temperatures above that. Formation of coke would result in plugging the tubes in the furnace that heats the feed stream to the crude oil distillation column. Plugging would also occur in the piping from the furnace to the distillation column as well as in the column itself.

The constraint imposed by limiting the column inlet crude oil to a temperature of less than 370 to 380°C yields a residual oil from the bottom of the atmospheric distillation column consisting entirely of hydrocarbons that boil above 370 to 380°C.

To further distill the residual oil from the atmospheric distillation column, the distillation must be performed at absolute pressures as low as 10 to 40 mmHg(also referred to as Torr) so as to limit the operating temperature to less than 370 to 380°C.

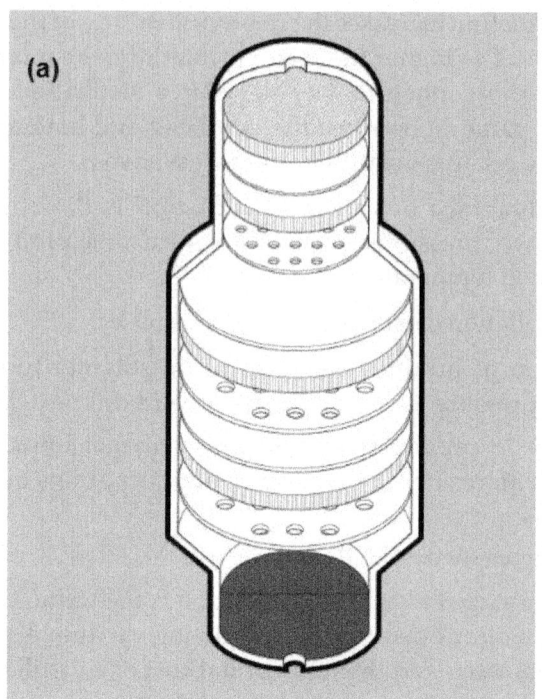

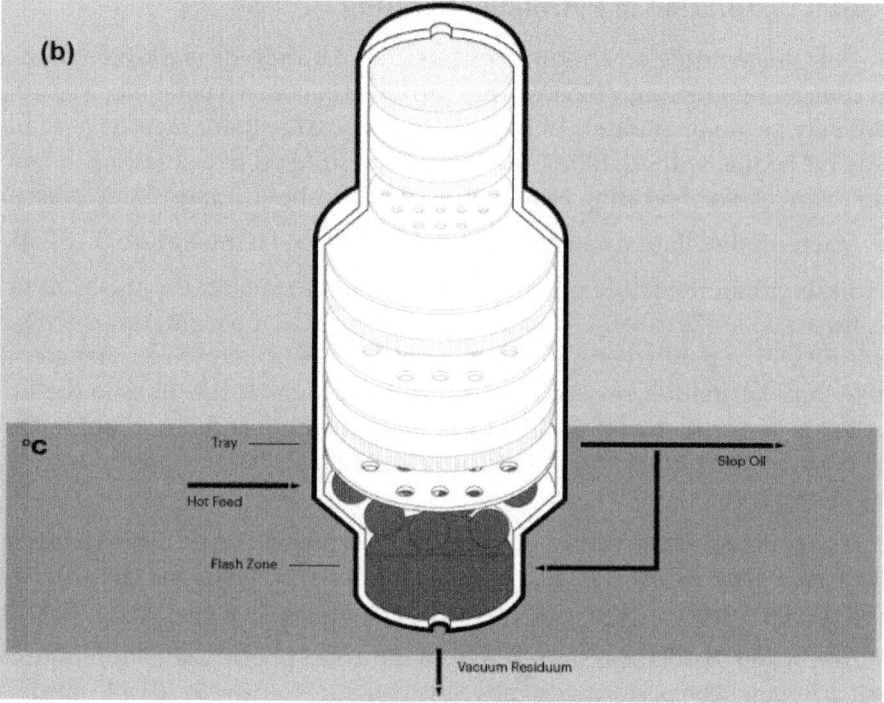

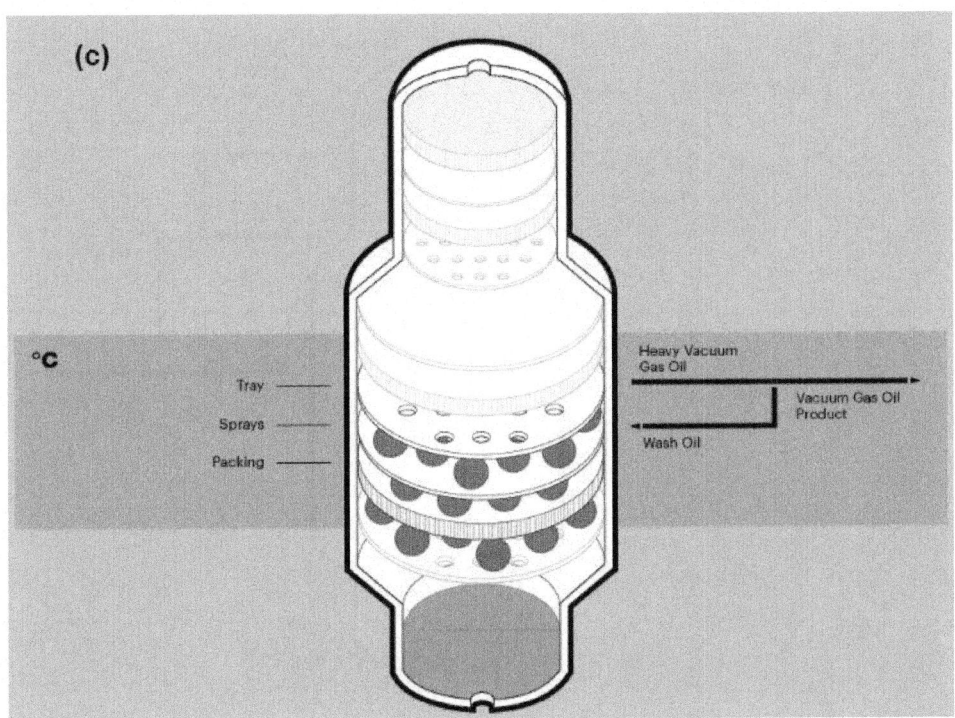

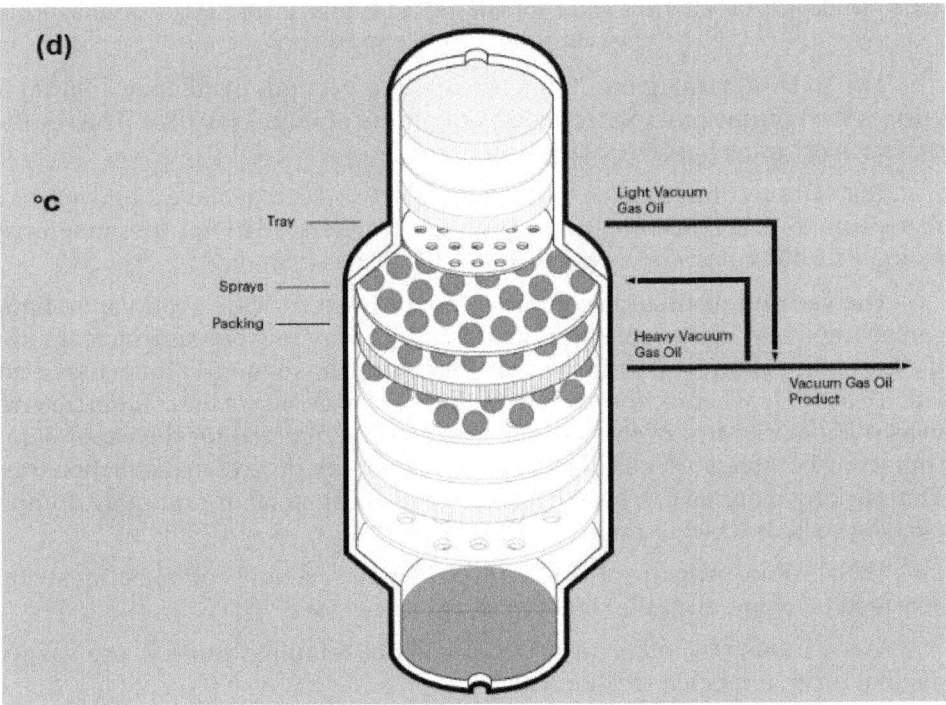

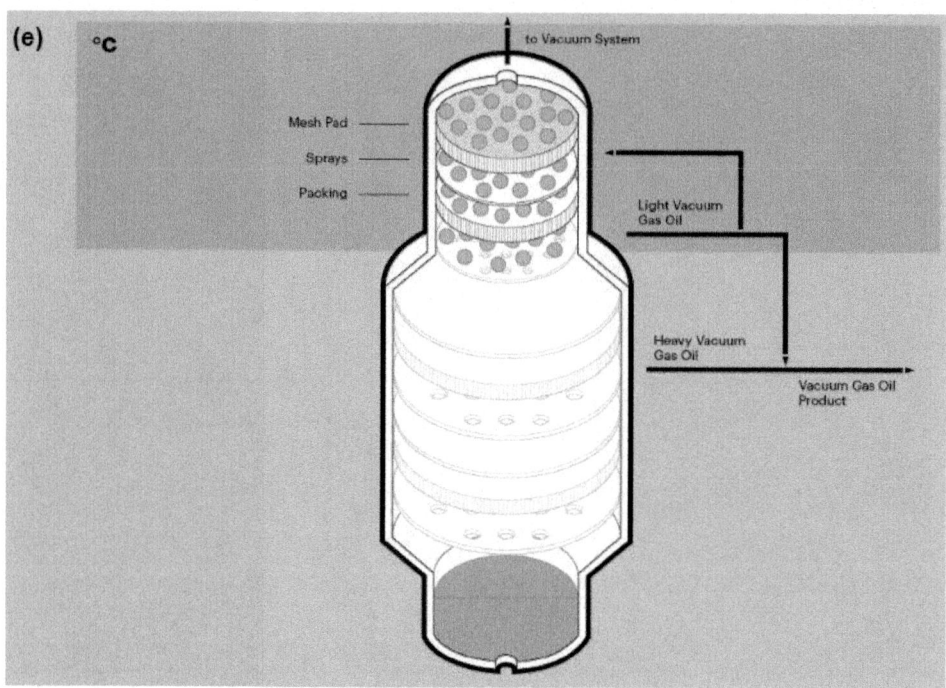

Fig. (a)-(e) is a simplified process diagram of a petroleum refinery vacuum distillation column that depicts the internals of the column and a photograph of a large vacuum distillation column in a petroleum refinery.

The 10 to 40 mmHg absolute pressure in a vacuum distillation column increases the volume of vapor formed per volume of liquid distilled. The result is that such columns have very large diameters.

Distillation columns such those in Images 1 and 2, may have diameters of 15 meters or more, heights ranging up to about 50 meters, and feed rates ranging up to about 25,400 cubic meters per day (160,000 barrels per day).

The vacuum distillation column internals must provide good vapor–liquid contacting while, at the same time, maintaining a very low pressure increase from the top of the column top to the bottom. Therefore, the vacuum column uses distillation trays only where withdrawing products from the side of the column (referred to as *side draws*). Most of the column uses packing material for the vapor–liquid contacting because such packing has a lower pressure drop than distillation trays. This packing material can be either structured sheet metal or randomly dumped packing such as Raschig rings.

The absolute pressure of 10 to 40 mmHg in the vacuum column is most often achieved by using multiple stages of steam jet ejectors.

Many industries, other than the petroleum refining industry, use vacuum distillation on a much a smaller scale.

Chapter 14

A CHEMICAL ENGINEERING PERSPECTIVE ON THE ORIGINS OF LIFE

Martha A. Grover*, Christine Y. He, Ming-Chien Hsieh and Sheng-Sheng Yu

School of Chemical & Biomolecular Engineering, Georgia Institute of Technology, 311 Ferst Dr. NW, Atlanta, GA 30032, USA; E-Mails: christine.he@chbe.gatech.edu (C.Y.H.); mhsieh34@gatech.edu (M.-C.H.); syu47@gatech.edu (S.-S.Y.)

*   Author to whom correspondence should be addressed; E-Mail: martha.grover@chbe.gatech.edu; Tel.: +1-404-894-2878 or +1-404-894-2866.
    Academic Editor: Michael Henson

ABSTRACT

Atoms and molecules assemble into materials, with the material structure determining the properties and ultimate function. Human-made materials and systems have achieved great complexity, such as the integrated circuit and the modern airplane. However, they still do not rival the adaptivity and robustness of biological systems. Understanding the reaction and assembly of molecules on the early Earth is a scientific grand challenge, and also can elucidate the design principles underlying biological materials and systems. This research requires understanding of chemical reactions, thermodynamics, fluid mechanics, heat and mass transfer, optimization, and control. Thus, the discipline of chemical engineering can play a central role in advancing the field. In this paper, an overview of research in the origins field is given, with particular emphasis on the origin of biopolymers and the role of chemical engineering phenomena. A case study is presented to highlight the importance of the environment and its coupling to the chemistry.

Keywords

Origins of life; complex system; evolution.

1. INTRODUCTION

How life evolved from chemicals on the early Earth is an open scientific question and an area of active research. If this process — the transition from non-life to life — can be understood, then the underlying design principles could also be applied to engineer materials and systems with a robustness and adaptability approaching that of biological systems. It is difficult to determine exactly what were the conditions on the early Earth, and certainly they would have been dynamic and spatially heterogeneous. However, constraints have been established by the scientific community, such as the period of time between the formation of the Earth (4.5 billion years ago) and the first fossil evidence of single-cellular life (3.5 billion years ago) [1]. Further constraints provide scientists with guidance on "plausibly prebiotic" conditions during this time, such as the composition of the atmosphere [2], the ocean [3], and appearance of dry land [4,5].

In modern biology, nucleic acid polymers store information, and this information is read by the ribosome to produce proteins; proteins then catalyze chemical reactions. Because the ribosome is highly conserved across all life forms, its emergence has even been identified as the "universal ancestor" of all life on Earth [6]. In addition to RNA, proteins are an integral component of the ribosome, suggesting a "chicken and egg" problem: which came first, proteins or nucleic acid polymers [7]? In modern biology, the polymerization processes for both polymer types are accomplished with complex enzyme machinery, but on the early Earth such protein catalysts would not have been available. However, the RNA World hypothesis provides a potential solution to the "chicken and egg" problem.

The idea of catalytic RNA (ribozymes) [8–10] and its subsequent discovery in 1982 [11] spawned the RNA World hypothesis, in which RNA served both functions (information storage and catalysis) in early life, with DNA and proteins appearing later in the evolutionary process [12,13]. Alternative pre-RNA world scenarios have also been proposed, where an even earlier genetic system eventually evolved into RNA [14–16]. However, robust and high-yield prebiotic routes to the non-enzymatic polymerization of RNA (or pre-RNA) continue to elude the origins of life community. The origin of life was most likely a complex system, not having a single component that evolved to become complex, but rather involving cooperation and co-evolution of multiple components [17], including small molecules [18,19], proteins [20,21] and lipids [22,23].

Amino acids, nucleotides, sugars, and lipids are the building blocks of life. These first three components react and polymerize to form proteins, the nucleic acid polymers of DNA and RNA, and polysaccharides. These biopolymers are then encapsulated in self-assembled lipid membranes to form cells. Although polysaccharides are central to living systems, they have not been studied much in a prebiotic context [24]. Sugars are described in this manuscript only in the context of ribose for DNA and RNA, but there remains much research to be done on the prebiotic origins of polysaccharides. The focus of this paper is on the formation of peptides (short proteins) and nucleic acids polymers (DNA and RNA). The chemical structures of the monomers, polymers, and their assemblies are given in Figures 1 and 2.

A possible source of protein monomers was elucidated by Miller and Urey in the 1950s [25], who established a path from simple prebiotic molecules to amino acids [26]. In their seminal experiment, published in 1953, an environment containing water, methane, ammonia, and hydrogen sulfide was heated and then subjected to an electric spark. Amino acids constituted 2% of the carbon, and sugars were also formed. Nucleobases were not identified in these reactions, but adenine was later found to be formed in concentrated solutions of hydrogen cyanide [27], and all four nucleobases of RNA have since been produced in various prebiotic environments, including in neat formamide with UV irradiation [28] and in aqueous aerosols [29]. Over the years, the scientific consensus on the atmosphere of the early Earth has changed, suggesting that Miller's conditions may have been too reducing. However, amino acids and nucleobases have since been produced under a wide range of "prebiotically plausible' atmospheric conditions [30].

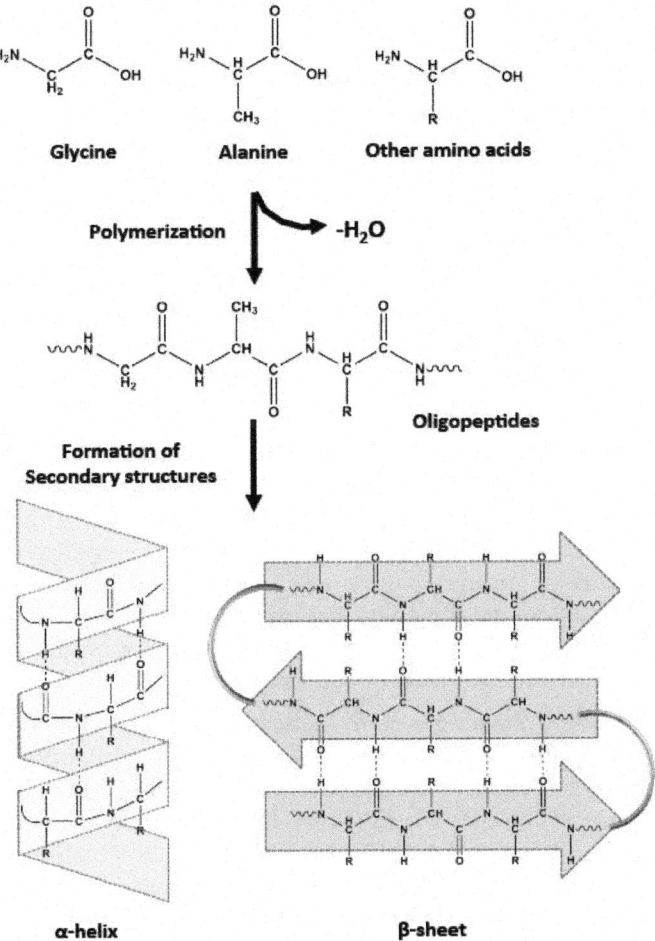

Figure 1. The components of proteins — amino acid monomers polymerize and then assemble. The assembled structures enable catalytic function.

The formation of the monomers does not imply the formation of their polymers, which has its own unique set of challenges in a prebiotic context. All three types of biopolymers are formed via condensation polymerization, producing a single water molecule per bond formed. Water is the solvent of life, and geologists predict that a significant ocean volume existed on the prebiotic Earth [3]. However, polymerization reactions in aqueous media drive the polymerization reaction toward the reactants, via hydrolysis, since water is a product of the reaction. In origins of life research, this is referred to as the "water problem" [31]. Dilution is another challenge for prebiotic polymerization, due the large volume of the ocean and the relatively small amount of monomer that would have been present.

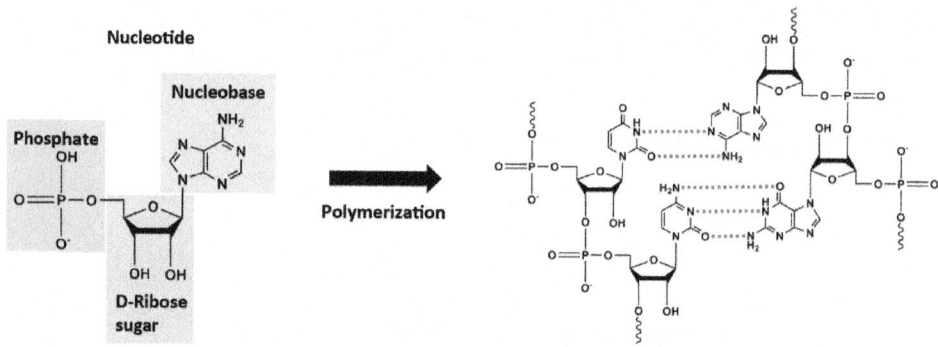

Figure 2. The components of ribonucleic acid (RNA). The nucleotide monomer is comprised of a nucleobase (A, G, C, U), the sugar ribose, and a phosphate group. The nucleotides polymerize, thereby storing information through the sequence of bases. Polymers strands then base pair with their complementary bases via hydrogen bonding. In DNA, uracil is replaced with thymine. Beyond hydrogen bonding, DNA further assembles into a double helix.

An additional hurdle to non-enzymatic polymerization is the fact that the thermodynamics of bond formation (amide bond in proteins, phosphodiester bond in nucleic acids) are energetically unfavorable, making high yields difficult to obtain at equilibrium conditions. Living systems are open systems, driven far from equilibrium by inputs of mass and energy, and biopolymers are held in non-equilibrium states by the energetic input of ATP (adenosine triphosphate). Similarly, the non-enzymatic prebiotic synthesis of biopolymers must be understood in this open non-equilibrium context [32,33].

To overcome the challenges of prebiotic biopolymer formation, some researchers have proposed alternative chemistries, yielding "protobiopolymers" that over time would have evolved into the biopolymers of modern life. For example, Miller proposed that peptide nucleic acid (PNA) could have been a potential precursor to the phosphodiester bond of RNA and DNA [34], which eliminates the need to form the high-energy phosphodiester bond. Orgel suggested that the ester bond may have preceded the amide bond in peptides and proteins [15]. Although the ester bond is less stable than the amide bond, this could actually have been beneficial in the early stages of life, preventing the irreversible formation of the cyclic dimer, a thermodynamic "dead end" [35].

Chemical reactions are central to the origins of life question, but chemistry alone cannot provide the answer. Whether or not the original polymers were "alternative" or not, the chemistry cannot be considered independent from the environment. To fully address the challenges of non-enzymatic polymerization, hydrolysis, and dilution, the chemistry and the environment must be considered together as a system. Non-aqueous solvents may have concentrated monomers while also excluding water. Mineral surfaces such as iron sulfide or clay may have served as catalysts, and also as concentrating agents. In fact, an inorganic start to life was proposed by Cairns-Smith [36], based on information storing minerals such as clay or mica. Dynamic environmental conditions would necessarily have been present on the prebiotic Earth [37]; periodicity driven by day/night, tidal, or seasonal cycles may have driven polymerization under non-equilibrium conditions, prior to the advent of the energy storing molecule adenosine triphosphate (ATP).

The concept of a prebiotic soup, suggested by Darwin as a "warm pond" [38], reflects the chemical complexity of the early Earth, but it does not explicitly highlight the spatial heterogeneity that would have also been present across the Earth. Phase separation is central to many origins of life studies. Even the early Miller-Urey experiments exhibited multiple phases, yielding a solid "tholin" product that has never been fully characterized. Phase separation has also been proposed as an early precursor to cellular compartmentalization [39]. The transport of fluid, mass, and heat in bulk and along surfaces drives many of the prebiotic environments proposed: water must be evaporated in wet/dry cycling environments [40], and thermal gradients drive replication via convective flow in porous media [41]. Such phenomena lie squarely in the field of chemical engineering.

Moreover, the concepts of optimization and feedback are central to the subsequent stages in the origins of life, and more broadly to chemical evolution. With an environment and chemistry that can support biopolymer (or protobiopolymer) formation, a combinatorial explosion of polymer sequences could be formed. Modern biology employs 20 amino acids and a four-letter nucleic alphabet, although these monomer sets may have been somewhat different during the early stages of evolution [42]. In fact, the current genetic code for amino acid incorporation into proteins appears to be near optimum for error minimization [43]. This is the ultimate "genetic algorithm," and the inspiration for directed evolution. During an evolutionary process in the prebiotic soup (actually a heterogeneous and multiphase "stew"), protein selection may have been driven by stability rather than catalytic function; since folded proteins are both more stable and more functional, over time proteins with catalytic function may have been amplified in the population [44].

Feedback is another concept central to many models of replication and amplification of nucleic acid polymers, or their earlier predecessors. Regulation through negative feedback naturally occurs as species are accumulated and subsequently bind to each other—inhibiting the formation of additional product. Modern biology uses feedback loops extensively to regulate conditions, creating robust function in the presence of an uncertain environment. As primitive life evolved, such regulatory loops would confer further advantage at the system (*e.g.*, cellular) level. In addition to negative feedback, Eigen proposed a model of autocatalysis, in which sequences with the highest replication rates will dominate the population through positive feedback [45]. However, the selection is only for replication, not for any beneficial function. While this model represents "survival of the fittest," with a single winner in the end, Eigen further proposed the evolution of a hypercycle, in which multiple sequences can coexist in a cross-catalytic network, to achieve higher-level function at the system level. Experimentally, cross-catalytic networks have been demonstrated in RNA [46,47] and peptide [48] systems. Although Eigen does not identify what is the chemical nature of his replicator, his model is most consistent with RNA and the RNA World hypothesis, since RNA has the capability to both store information and to catalyze reactions (as a ribozyme). Subsequent modeling of autocatalysis has made an explicit connection to the RNA World [17,49]. However, the origin of life was undoubtedly a more complicated system, not consistent with a reductionist perspective and a "silver bullet" single answer.

A transition in evolution occurs when a collection of simpler components begin to act as a system, so that selection moves up to the system level. Examples outlined by Maynard-Smith and Szathmary [50] also include the transition from single-cellular to multicellular life, and the evolution of human language. Similarly, cooperation among diverse biopolymer types was required in the emergence of the first cell. While proteins and nucleic acids might originally have evolved separately, as RNA and Protein Worlds, the point in time at which they began to cooperate as the ribosome has been described as the origin of life on Earth [6].

Over the years, the scientific community has attempted to define "life" [51], and a modern commonly used working definition is "a self-sustaining chemical system capable of Darwinian evolution [52]." However, with only one example of life in the universe available, ambiguity remains [53]. Despite the debate about the exact point of transition from chemistry to biology, research into the evolution of chemicals on the prebiotic Earth is an active area of scientific research — and the topic of this paper.

A number of comprehensive review papers have been written on origins of life chemistry over the years, for example, [7,17,20,54-57]. The purpose of this paper is to provide an introduction to the field for chemical engineers, with an overview of key concepts and findings, while particularly highlighting the chemical *engineering* phenomena — not only chemical kinetics and thermodynamics, but also transport, phase separation, optimization, and control. While the origins of life field is broad and includes substantial research from *e.g.*, geologists and physicists, this contribution is focused around the *chemical origins of biopolymers* on the early Earth. Work by the authors and collaborators is highlighted to some extent, as it represents a more chemical engineering approach, but the intent is to cover the key papers and discoveries in the field. The background section is divided into three sections: monomers, polymers, and assemblies, as illustrated in Figure 3.

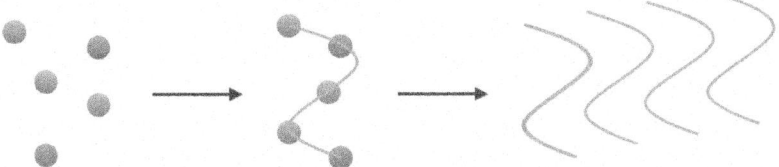

Figure 3. Progression from monomers to polymers to assemblies.

2. BACKGROUND

2.1. Monomers

The first step toward uncovering the chemical origins of life is to understand the source of the monomers that polymerize into proteins and nucleic acid polymers. The exact conditions on the prebiotic Earth are not known, but geologists have identified likely chemical species and environments, based on constraints from mineral evidence and geochemical principles. Hydro thermal vents deep in the ocean are one possible environment [58,59], as well as air-water interfaces on the ocean surface [60] or the surface of aerosol particles [29]. Exposed dry land emerged before 4.3 billion years ago [4,5], prior to the origin of life, providing dry environments that could more efficiently promote condensation reactions. The ocean-land interface may also have provided a conducive environment, in which tidal pools cycle between hydrated and dehydrated states, alternatively providing mixing, concentration, and dehydration. UV irradiation is also expected to have played a significant role in prebiotic chemistry. The ozone layer was not yet developed on the prebiotic Earth, so significant UV irradiation may have contributed to

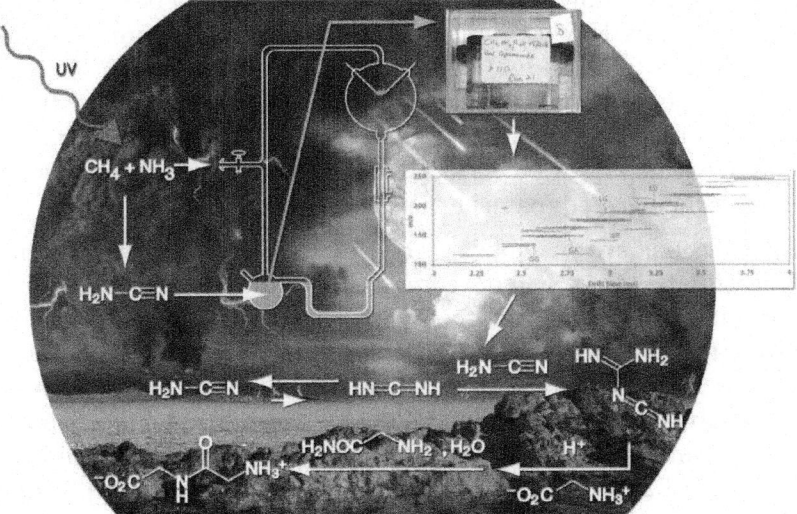

Figure 4. Schematic of the Miller-Urey apparatus, reproduced from the cover of Ref. [66]. Reprinted with permission from Wiley, Hoboken, NJ, USA, 2014.

photocatalysis [28,61] as well as photodegradation [3]. Mineral surfaces may have also catalyzed early reactions [62] prior to the emergence of protein catalysts. In addition to thermal energy, reactions could have been driven forward (and backward) by external energy sources such as electrical discharges [25] and cosmic rays [63]. Lipids can also be generated under similar environments [64,65], providing a pathway toward early cellular compartmentalization.

The star·ting material on the prebiotic Earth is typically delineated beginning with simple stable molecules foduiantedd btiohm rooeucuglehpooulyt etrhizaetoun niveerresaerly, ianrtchlcuonddiintiognwlnathteei r, moe- ecular hydrogen, molecular nitrogen, carbon dioxyiadnaem, dceainrbon monoxFerne, ammonia, anow methe dane. tion oese molecules may then react to form slightly more compyiecatepdolmmeroiz laeticonuledse,r ssimuuclahte apsrehbiyotdic reongvireonnmcenytsa. n(B iadcke- (HCN), formaldehyde (HCHO), and glycine (NH$_2$CH$_2$COOH). During the original Miller-Urey experiments in the 1950's, amino acids (including glycine) were generated using water, molecular hydrogen, methane, and ammonia [25]. See Figure 4. However, methane and ammonia are unstable under high UV irradiation, so this particular environment may not be plausible. More recent research focuses on less reducing environments, such as a two-phase system with water, molecular nitrogen, and carbon dioxide over liquid water [60]. The nature of the chemical starting material also feeds back to the thermodynamics and kinetics of all aqueous-phase reactions, by altering the pH of the ocean (*e.g.*, amount of dissolved carbon dioxide); the pH of the prebiotic ocean is thought to have been slightly acidic [3].

How did these simple chemicals react to form the small-molecule building blocks of life? — amino acids, nucleobases, sugars, and lipids. While the exact pathways may never be known, many pathways can certainly be excluded based on prebiotic constraints, leaving a smaller number of plausibly prebiotic candidates. The scientific question is thus "How could life have emerged?" To answer this question, the chemicals and the environment must be considered together as a system, which may also include a periodically varying environment that drives the chemical reactions. The kinetics of these reactions must be understood, as well as the long-term thermodynamic behavior.

Molecules that are unstable may be problematic for prebiotic chemistry, although the situation is really more complicated. Although the Miller-Urey experiments operated on a closed system, living systems are open, driven away from equilibrium by mass and energy fluxes. Thus, the transition to living systems on the early Earth should also be considered with an open system. Even molecules that are unstable can be accumulated if the net rate of production is positive [58]. The apparent instability of amino acids under UV exposure could have been mitigated by UV absorbing molecules in the prebiotic ocean [3]. However, the instability of ribose has led researchers to consider alternative, more stable sugars in proto-nucleic acid polymers [67]. Stability arguments have also been employed to suggest which nucleobase pair came first: because of the instability of cytosine, the AU pair in RNA may have preceded CG [68].

The specificity of reactions is another important consideration. In general, a large number of products were produced in the Miller-Urey experiments and in most subsequent experiments from mixtures of a few simple chemicals. The formose reaction produces ribose from formaldehyde, but it also produces a large number of other products [69]. Catalysts are needed for the formose reaction, and the nature of the catalyst influences the distribution of products [70]. Mineral [62] as well as amino acid catalysts [70] have been used. However, none create a high selectivity toward ribose, and this remains a challenge in understanding the origin of nucleic acid polymers.

The yield is also an important factor in any prebiotic reaction pathway. If a molecule is made in only a trace amount, it may not provide a reliable source for subsequent polymerization. Miller and Urey analyzed their samples with paper chromotography [25], which limited the number of species that could be detected. However, modern analytical chemistry techniques enable many more species to be identified, if only in trace amounts. Recent analysis of previously unanalyzed Miller-Urey samples identified a dozen amino acids, as well as numerous dipeptides [66]. This lack of specificity may in fact be important to generate the diversity needed for subsequent evolution — if the species are all present in significant quantity. Other species are presumably also present, such as alternative prebiotic versions of amino acids or nucleobases. In a complex mixture, it is impractical to identify all species, and but rather one searches for compounds of particular interest in the sample. Alpha hydroxy acids have been suggested as an early version of amino acids [15], and were found in later Miller-Urey experiments [71]. Urazole is a heterocyclic compound that reacts readily with ribose and can be formed under plausibly prebiotic conditions [72]. Although modern analytical chemistry is a powerful tool for prebiotic chemistry, still many challenges remain in fully characterizing any complex mixture representing the "prebiotic soup".

Compared to amino acids (protein monomers), the synthesis of nucleotides is much more complicated. Nucleotides consist of a nucleobase, a sugar, and a phosphate group, and are polymerized to form RNA and DNA. The glycosidic bond connecting the RNA nucleobases and sugar is difficult to form directly, but can be accomplished for some nucleobases using dry-down reactions that remove the water product [73]. To overcome this challenge, Miller and co-workers [72], and later Hud and co-workers [74], proposed alternative nucleobases that are more reactive. Powner and co-workers presented an alternative route to pyrimidine nucleoside formation, using reactants that are chemically distinct from nucleobases or sugars [75]. Powner and co-workers also demonstrated a one-pot system that generates both nucleosides and lipids [76]. The source of the phosphorus in nucleotides has been an open question in the field, since most phosphorus would have been stored in minerals such as apatite that are unreactive. However, the mineral schreibersite could have been provided by meteorites and has been demonstrated to phosphorylate the hydroxyl group of glycerol [77].

It is also possible that the monomers of life were not generated on Earth, but rather were extraterrestrial in origin [78]. Meteorite samples show significant amounts of amino acids, and interstellar ice can generate lipids [61]. The late heavy bombardment period occurred around 4 billion years ago and delivered a significant total mass to the Earth [79]. In fact, monomers were generated on Earth and in space – the better question is what was their more significant source.

2.2. Polymers

The polymerization of amino acids and nucleotides into proteins and nucleic acid polymers (DNA, RNA) is considered to be an early and necessary step in the origins of life. These biopolymers store genetic information (our design) and catalyze chemical reactions (how we operate). It is difficult to envision a living system without biopolymers, although alternative crystal-based living systems have also been proposed [36]. Achieving biopolymers of significant length is a critical step in chemical evolution; without a sufficient length, nucleic acids cannot form the double helix that is needed for replication (6–7 nucleotides), and peptides cannot fold into catalytically functional units (≈20 amino acids). Recent work shows that assemblies of multiple shorter peptides [80] or RNA oligomers [81] can also catalyze chemical reactions, but even achieving this length is challenging under prebiotic conditions, due to unfavorable kinetics and thermodynamics [82,83]. Amide and phosphodiester bonds are typically formed through condensation reactions (*i.e.*, generating water), which is thermodynamically unfavorable in an aqueous solution such as the ocean. Catalysts are needed to increase the forward rate of polymerization, and even then the monomers often require additional chemical modification, or "activation," to achieve significant length and yield [84]. For these reasons, alternative monomers and their polymers have also been proposed as pathways toward modern biopolymers.

As in the case of monomer generation, the coupling between the chemicals and the environment is critical to understanding prebiotic polymerization of biopolymers. Deep sea hydrothermal vents have been extensively studied as a site for peptide and nucleic acid polymerization. At their high temperatures and pressures, non-enzymatic oligomerization of amino acids [85] and nucleic acids [86] can proceed. However, yields are low, and much of the amino acid becomes trapped in the cyclic dimer, known as diketopiperazine (DKP), providing a thermodynamic dead end [87]. Stability is an additional challenge at these high temperatures. The peptide bond is resistant to hydrolysis at moderate temperatures and neutral pH, but in a high temperature aqueous solution it undergoes rapid hydrolysis. In addition, the amino acid monomers themselves are unstable, undergoing decarboxylation. An open system with inlet and outlet mass flow and monomer generation (*i.e.*, a hydrothermal vent) might be able to accumulate peptide faster than it degrades, although this concept and design may not be consistent with heating timescales expected in hydrothermal vents [88].

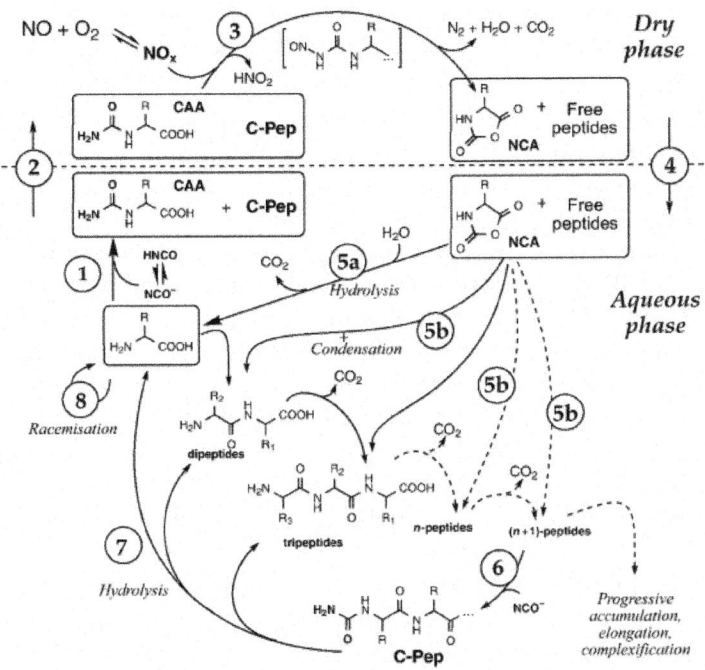

Figure 5. Proposed chemical pathway for peptide polymerization, using nitrogen oxide and a cyclic environment. From Ref. [89]. Reprinted with permission from Wiley, Hoboken, NJ, USA, 2002.

The air-water interface is another potential environment for polymerization. This interface could facilitate polymerization by aligning charged and zwitterionic amino acids, as well as excluding some water [90]. However, to date this approach has only been demonstrated for amino acid ester monomers, which do not generate water during their polymerization. Another environment that minimizes water contact is the hydration/dehydration cycle, which could occur at the interface between water and land. Periodic environments have been used to form peptides from amino acids, although without chemical modification ("activation") of the amino acids the yields are very low [40]. Chemical activation of amino acids has been achieved using, for example, carbonyl imidazole [91], COS [92], and NO [89], as well as alumina [93]. A schematic from Ref. [89] is shown in Figure 5. The prebiotic availability of carbonyl imidazole is questionable due to its high reactivity, while COS might be available from volcanic reactions [92]. However, because these activating agents are consumed in the reaction (unlike a catalyst), finding a prebiotic source in high enough quantities is problematic. Activation has also been a key component of nucleic acid polymerization: organic bases have been used as activiting agents [94] and cyclic nucleotides have been used as activated monomers [95]. Activated RNA was also used in conjunction with a cold environment (*i.e.*, ice) [96]. Although water is still present, at low temperatures the hydrolysis of the amide bond is reduced, and RNA up to 17-mers were achieved at 90% yield. As with amino acid polymerization, a prebiotic and robust source of activated nucleotide has not yet been established. In fact, in contrast to amino acids, a prebiotic and robust source of unactivated nucleotides remains elusive.

In addition to chemical activation of the monomers, catalysts are also frequently used to facilitate prebiotic polymerization. In modern biology, enzymes (*i.e.*, protein catalysts) provide efficient and specific catalysts for linking amino acids and nucleotides into polymers. Prior to the emergence of catalytically functional proteins, mineral surfaces may have served this role [97]. By aligning amino acids on the surface [98] the DKP trap can be overcome to form much longer peptides (*e.g.*, 55-mers) [91]. Mineral surfaces were similarly used to elongate nucleic acids. In both cases, the monomers were chemically activated, suggesting that the mineral surface alone is not sufficient to facilitate the polymerization. Another possible disadvantage of mineral surfaces is that the polymers may be difficult to release from the surface, especially as the length of the polymer grows [99]. Salt addition has also been used to promote peptide formation under cycling conditions, with and without mineral surfaces [100]. However, the yield and peptide length are still low.

Lipids are another potential catalyst and environment for amino acid and nucleotide polymerization. The polar headgroup of the lipid could align the monomers, while water would be excluded from this hydrophobic environment. Cationic lipids were used to facilitate peptide polymerization, although fatty acids are a more likely prebiotic candidate [101]. RNA polymerization of up to 50-mers was reported in 2008 using fatty acids and unactivated polymers, although the yield and robustness of this reaction must be addressed [102]. Non-aqueous small molecule solvents have also been proposed recently, as another route to exclude water from the vicinity of the reaction. Prebiotically plausible small molecules such as urea and formamide provide the components for an ionic liquid or deep eutectic solvent. Preliminary work suggests that such environments can promote phosphorylation, in the presence of phosphorus-containing minerals [103].

Due to the challenges associated with the robust prebiotic polymerization of amino acids and nucleotides, researchers have also investigated the polymerization of alternative "protobiopolymers." For example, the ester bond has been suggested as a prebiotic version of the peptide bond [15]. The monomers could be α-hydroxy acids, which are found in Miller-Urey type experiments [71] and are chemically similar to amino acids (substitution of a hydroxyl group for the amine). Previous studies have shown that the ribosome is capable of forming ester bonds between activated hydroxy acids in addition to forming amide bonds between amino acids, which has led to the long-standing hypothesis that polyesters could have come before polypeptides in the early stages life [104]. Although the ester bond is less stable than the amide bond, its formation is thermodynamically favorable (unlike the amide bond), and it does not form the cyclic dimer trap associated with amino acids. Polyester oligomers were formed from unactivated monomers in a drying reaction [105] and more recently in a cyclic hydration/dehydration environment [35].

Prebiotic alternatives to RNA and DNA have been extensively pursued due to the many challenges associated with the prebiotic generation of either RNA or DNA. The peptide nucleic acid (PNA) replaces the phosphodiester bond with an amide bond, and still forms a double helix [34]. Although more prebiotically accessible than the phosphodiester bond, it still inherits all the challenges associated with peptide bond formation; to date only formation with chemical activation has been reported [106]. The glyoxylate linkage for nucleic acids is another proposed linkage, which unlike the peptide and phosphodiester bond is thermodynamically favored [107]. Nucleotide dimers were successfully linked in a drying experiment, albeit at low yield.

The previous discussion of polymerization focuses on early Earth environments, but delivery of material from outer space is an alternative hypothesis. While the extraterrestrial supply of monomers to the early Earth has scientific support, the supply of polymers by meteorites is less viable. Short oligomers of amino acids and hydroxy acids have been found in meteorites, and thus could have been delivered to the early Earth [108]. However, no evidence suggests that long (*i.e.*, functional) biopolymers would have been delivered in significant quantity.

2.3. Assemblies

Polymeric assemblies are essential for performing the fundamental tasks of life—storing and transferring information via the nucleic acid double helix, and catalyzing chemical reactions with folded proteins. In both cases, non-covalent interactions (*e.g.*, hydrogen bonding between the side chains) create more complicated topology and structure, enabling the performance of practical tasks. However, if the polymers are not long enough, these assemblies are not stable. Nucleotide monomers of present day RNA do not base pair in water under ambient conditions. The dipeptide of phenylalanine does form fibrils in water due to the hydrophobic phenyl rings on the side chains [109], but for most peptides a longer length is also required for assembly and catalysis. Understanding the assembly of biopolymers under early Earth conditions is essential for uncovering the route to chemical evolution and the subsequent transition to living systems.

2.3.1. Selection

The specific characteristics of the monomer side chains determine the assembled structure of both proteins and nucleic acid polymers. There are twenty natural amino acids and four nucleic acid monomer types; within each class, the monomer type is distinguished based on its side chain. Although random polymer sequences might have formed through chemical reactions on the early Earth, not all sequences would assemble. It is this assembly that could have driven selection for functional sequences. New and Pohorille describe a selection process for proteins, in which the proteins that fold are both more stable to hydrolysis (by excluding water) and more catalytic [44]. Over time, a higher concentration of catalytic proteins would emerge from a random sequence pool, due to this apparent coincidence in which stability and catalysis are correlated through folding.

Just as hydrolysis provides an early selection pressure for peptides, nucleic acid polymers assembled through base pairing might also provide protection against hydrolysis. However, the emergence of a genetic code requires the copying of sequences, such that information can be propagated from generation to generation. In modern biology, this is achieved through templated replication—the assembly of DNA into a double-stranded helix, with copying achieved via complicated enzyme machinery. Achieving this templated replication in a prebiotic, non-enzymatic context presents many challenges, as articulated recently by Szostak [110]. An early RNA or DNA replicator could not copy itself with high fidelity [111], producing instead a group of similar sequences with random mutations that are collectively called a quasispecies [45].

Experimental demonstration of non-enzymatic, template-directed replication remains difficult, with successful results demonstrated for fairly short, often highly specific sequences (*e.g.*, palindromic [112,113] or self-complementary [114,115]). Challenges include the relative rates of duplex re-assembly versus copying, the separation of mother and daughter strands after copying has occurred, and the fidelity of the copying process [110]. To copy a nucleic acid duplex, the duplex strands must be separated into single strands, each of which serves as a template for copying. However, even moderately short nucleic acid duplexes are quite stable owing to Watson Crick base pair stacking interactions [116,117], so that the result of copying is a template-copy duplex that is difficult to separate—a problem known as strand inhibition [118,119]. Many researchers have approached these problems by focusing on chemical modification of the modern nucleic acid structure. redIn one approach, template or fragment strands are chemically modified at specific sites so that once ligation occurs, the resulting template-copy duplex is destabilized, resulting in enhanced turnover numbers [120–122]. However, it is debatable whether the proposed alternative backbones or base replacements are prebiotically plausible.

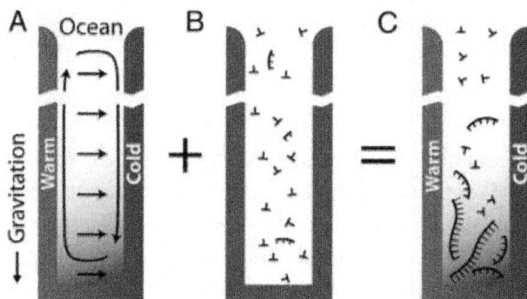

Figure 6. Accumulation of long RNA oligomers in a pore, driven by a thermal gradient. From Ref. [41].

Several studies focused on the role that physical and environmental factors, rather than chemical alternatives, could have played in non-enzymatic replication. In one approach to strand inhibition, template strands are immobilized on a solid particle support while undergoing cycles where complementary fragments are fed, ligated, washed, and product strands re-immobilized; this process resulted in desirable "exponential" replication (doubling of the template population with every round) [123]. In a similar system, additional binding of downstream "micro-helper" oligomers to the template strand allowed for efficient copying of all four RNA nucleobases [124]. Though not a self-sustained system, these studies emphasize the potential importance of solid-phase chemistry and periodic replacement of monomers/fragments in driving replication processes. Mast et al. proposed that replication could have been driven by thermal gradients in a tall, thin pore that is cold on one side and hot on the other [41], as shown in Figure 6. In this scenario, nucleic acid oligomers of different sizes undergo two transport phenomena: thermal convection caused by expansion of water as it heats up and thermophoresis of the oligonucleotides from the hot to cold wall. These combined effects result in an accumulation of longer oligomers at a rate exponentially dependent on length, which would be impossible in equilibrium reaction conditions. A recent study further demonstrates polymerization [125].

Heterogeneity of monomers is another key issue in the assembly of both peptides and nucleic acid polymers. In a complex prebiotic soup, many chemical variants could be present [57]. For example, formation of nucleosides from bases and ribose form a mix of 2'-5' and 3'-5' ribose linkages. Although this disorder has been viewed as a major challenge for understanding the origin of the modern 3'-5' linkage, recent research shows that polymers with mixed linkages can also base pair and provide functional structures [126]. Moreover, the resulting decrease in stability might have helpful in overcoming the strand inhibition problem. However, a loss of stability might also lead to higher mutation rates. Too much mutation results in the complete loss of information over multiple rounds of replication, and this critical mutation rate is known as the "error threshold" [127].

Another major source of heterogeneity is the chirality of the monomers. A major feature of modern biopolymers is the predominance of one chirality over the other, as seen in the L-amino acids and D-sugars of life today. In order to achieve assembly of modern biopolymers, homochirality is required. The transition to homochirality likely began with mirror symmetry breaking — the generation of a slight excess of one enantiomer over the other in a set of molecules — that was followed by enantiomeric amplification [128,129]. The initial enantiomeric excess may have come from outer space, as enantiomeric excesses of L-amino acids have been found on the Murchison meteorite [130]. Photolysis by circularly polarized light in the interstellar medium has been proposed as a means of generating this initial chiral asymmetry [131]. Others have proposed asymmetric adsorption of organic molecules onto mineral or crystal surfaces with handedness, such as calcite [132], as a geochemical route for the origin of homochirality. Enantiomeric enrichment from an initial imbalance can occur through physical [133,134] or chemical [135] amplification processes. Predominantly L-handed amino-oxazolines — precursors to RNA — are formed from the reaction of amino-oxazole with glyceraldehyde at a 1% excess of the L-enantiomer in the presence of chiral amino acids [136]. After the emergence of L-RNA, L-amino acids could then have determined the handedness of the D-sugars we see in life today [137].

Hud and coworkers recently proposed an alternative scenario for nucleic acid polymerization and assembly [138]. Since the modern nucleotides do not base pair through hydrogen bonding in water, it is unclear how base pairing in nucleic acid polymers would have emerged. Almost 25 years ago, the groups of [139] and Whitesides [140] reported that molecules similar to modern nucleobases, including barbituric acid, triaminopyrimidine, cyanuric acid and melamine, will assemble in organic solvents through specific hydrogen bonding. Investigating these systems in water, the Hud laboratory demonstrated the formation of linear assemblies containing thousands of complementary monomers tagged with these nucleobase analogs[141]. This ordering through assembly could then provide a catalyst for polymerization of protonucleic acid polymers, by aligning their functional groups. Additionally, these molecules are attractive protonucleic acid nucleobase candidates, as triaminopyrimidine forms glycosidic linkages with ribose to produce nucleosides in water, and barbituric acid, cyanuric acid, and melamine have been found in model prebiotic reactions [142,143].

2.3.2. Evolution

Evolution is a process of optimization, in which a system improves its ability to function within its current environment. Selection is one necessary component of evolution, and, as described in the previous section, selection of proteins and nucleic acids may have occurred through both attrition (hydrolysis) and amplification (replication). Mutation is another necessary component of evolution. While mutations to the genetic code of individual members are more or less random, the environment acts on this random population, removing some members and allowing the remaining ones to utilize environmental resources and propagate. The early evolution of RNA, proteins, and their predecessors can be framed in terms of optimization on fitness landscapes, which relate monomer sequence to Darwinian fitness [144]. The modern form of RNA is understood to be optimal, with its moderate base pairing strength being "essential for the evolution of a rich diversity of nucleic acid-related biological functions" [56].

The process of evolution confers robustness at the overall population level, under a dynamic and uncertain environment. Transitions in the evolutionary process occur when the selection moves to a higher level of organization [145]. Unlike engineering optimization, in which the goal is to find the global maximum, evolutionary optimization is operating on a time-varying fitness landscape defined by the environment. Just as biological organisms are optimized under evolution, molecules on the early Earth would also be evolving under a selective pressure.

As articulated by New and Porohille [44], selection for folded catalytic proteins should occur in an aqueous environment, due to the pressure of hydrolysis. However, at this point the catalysis would simply be an accident, for no particular purpose. A key transition in early chemical evolution is the formation of catalytic networks, such that selection moves to the higher level of the network, as opposed to acting on each individual polymer. Eigen and Schuster presented the concept of a hypercycle [45], which is a cross-catalytic chemical network, and formalized this concept mathematically using mass-action kinetics. More chemically detailed models have been proposed [146,147]. However, several practical questions remain, such as (1) How would the hypercycle form? Must all components emerge simultaneously, and what is the probability of this event? (2) What would be the functions in the first cross-catalytic network? (3) How can an individual polymer benefit from the catalytic "fruits of its labor," prior to the emergence of cell walls? In a well-mixed system, all polymers should benefit equally.

Walker et al. recently presented a scenario and simulation to address these three questions, based on mass-action kinetics and Fickian diffusion [148]. In this proposal, the polymers replicate by copying their own sequence (similar to DNA replication), and the catalytic function of the polymer is to produce more monomer. The first functional sequence produces more monomer of Type A, while a later emerging functional sequence produces more monomer of Type B. Thus, the hypercycle can be built up step by step. The first functional sequence does benefit from its own function (out-competes other sequences), but the greater benefit is realized — for both sequences — when both monomer types are catalyzed (cooperation between sequences). Although no cell compartmentalization is included in the model, the reactions occur on a surface and with limited diffusion. A region of parameter space is identified in which surface diffusion limits the mobility of the monomer, such that a functional sequence can benefit most from its product, with all sequences benefitting to a lesser extent. Although non-functional sequences are often described as "parasites" in the origins field [149], in this scenario these sequences may later evolve new functions, enabling more complex reaction networks to evolve step by step.

Although surfaces may have served an early role in localization of species, compartmentalization likely emerged early in chemical evolution. Compartmentalization plays several fundamental roles: concentrating organic molecules relative to the large volume of the ocean, providing a semi-permeable barrier with the environment, generating physical and chemical gradients which drive biological processes, and eventually coupling genotype with phenotype once compartmental survival/reproduction became functionally tied to the genetic material contained within. Oparin was the first to propose that systems of membrane-bound colloidal particles, which he called coacervates, were the first self-replicating entities [150]. Since then, researchers have studied how spatial organization could have arisen through formation of both membranes [151] and microenvironments not encapsulated by membranes.

Aqueous two-phase systems have been proposed as early non-membrane systems — a mixture of polyethylene glycol and dextran in water will segregate into microdroplets rich in one organic molecule over the other. Local RNA concentration is increased by up to three orders of magnitude in this system and is accompanied by a significant increase in the RNA cleavage rate by a hammerhead ribozyme [39]. Coupled mononucleotide-cationic amino acid species have also been found to form suspended microdroplets in water that sequester various organics [152]. Another microenvironment that was likely prebiotically abundant is the eutectic phase of water and ice, which is formed when water with salts/metal-ions is cooled below its depressed freezing point but above its eutectic point (at which the whole system freezes). This system has been found to promote the elongation of an RNA primer on a template via single nucleotide addition [153].

Research into the formation of membranes has generally focused on short-chain fatty acids, which can be generated under prebiotic conditions (see Section 2.1), rather than the phospholipids that compose cellular membranes. It has been found that encapsulation of RNA in fatty acid vesicles generates an osmotic pressure, due to the counterions associated with the RNA, that drives the uptake of additional fatty acid material at the expense of other vesicles without RNA [154]. Thus, coupling the presence of nucleic acid within a vesicle-to-vesicle growth confers a system-level advantage to the membrane-nucleic acid system, allowing it to compete with other vesicles for material. Similarly, efflux of fatty acids from a vesicle can be reduced by incorporation of higher phospholipid content; thus, the selective advantage of retaining membrane material could have led to the appearance of the less permeable phospholipid membranes that we find in life today [155].

3. CASE STUDY

The coupling between the chemistry and the environment is critical to understanding the origin of functional biopolymers. Here we consider three different environmental scenarios associated with a cycling environment. The chemical system studied is the polymerization of malic acid, the α-hydroxy acid previously studied by one of the authors [35]. This polyester was shown to form under alternating hot/dry and cold/wet conditions, yielding a potential proto-protein. However, the environment considered was highly idealized, with sudden switches between two different conditions. Under the hot condition of 85 °C, the relative humidity of the air was extremely dry, since it was heated from air at ambient temperature. Prior to the addition of water, the system was cooled to room temperature and capped, and then was heated back to 60 °C, a temperature at which hydrolysis is low. Monomer conversion of 60% could then be achieved [35], as shown in Figure 7. The purpose of this case study is to investigate polymerization under more realistic scenarios, in which the temperature varies gradually and the water mass transfer is coupled to the temperature.

3.1. Modeling

The simulation developed in Ref. [35] includes polymerization kinetics and mass transfer of water:

$$P_n + P_m \underset{k_{-1}}{\overset{k_1}{\rightleftharpoons}} P_{n+m} + H_2O \tag{1}$$

$$\frac{dz_i}{dt} = \frac{1}{V}\left(k_1 \sum_{j=1}^{i-1} z_{i-j}z_j + 2k_{-1}W \sum_{j=i+1}^{\infty} z_j - 2k_1 z_i \sum_{j=1}^{\infty} z_j - k_{-1}W(i-1)z_i \right) \tag{2}$$

$$\frac{dW}{dt} = k_1 \sum_{i=1}^{\infty} z_i \sum_{j=1}^{\infty} z_j - k_{-1}W \sum_{i=1}^{\infty}(i-1)z_i - R_{v,w} \tag{3}$$

$$R_{v,w} = K_{y,w}\left(\frac{P_w^* x_w}{P_t}\right) = K_P x_w \tag{4}$$

$$k_1 = (3.0 \times 10^{15}) \exp\left(\frac{-1.5 \times 10^4}{T}\right) \tag{5}$$

$$k_{-1} = (1.9 \times 10^8) \exp\left(\frac{-8.6 \times 10^3}{T}\right) \tag{6}$$

where P_n is a polymer containing n monomers, k_1 [L/mol-h] is the forward rate constant for polymerization, and k_{-1} [L/mol-h] is the reverse reaction. Because the liquid volume V is changing during drying and rehydration, Equations (2) and (3) are formulated in terms of total moles, where z_i is the total number of polymers of length i and W the number of moles of water in the liquid phase. The time is t [h], and the rate of water evaporation is $R_{v,w}$ [mol/h]. Water is also generated via this condensation reaction process. The driving force for evaporation is calculated using Raoult's Law, with $K_{y,w}$ providing the mass transfer coefficient. This coefficient was combined with the saturation pressure P_w^* and the total pressure P_t to yield K_P. The air was assumed to be sufficiently dry that its effect on mass transfer could be neglected, and at 85 °C, K_P was estimated from data as 0.022 mol/h. The prefactors and activation energies in eqns. (5) and (6) were also estimated from data, to determine the dependence of the reaction kinetics on temperature T [K].

The model structure used in Ref. [35] was adapted from Ref. [156], which also included the effect of water content in the air:

$$R_{v,w} = K_{y,w}(y_w^* - y_w) = K_{y,w}\left(\frac{P_w^* x_w}{P_t} - y_w\right) \tag{7}$$

where y_w and y_w^* are the mole fractions of water in the air and the amount at saturation conditions. In the present paper this effect is included in the model, assuming that the temperature dependence of $K_{y,w}$ can be neglected over 60–85 °C, and modeling the saturation pressure of water according to Antoine's Law as

$$P_w^* = A + \frac{B}{C+T} \tag{8}$$

with A = 8.07131, B = 1730.63, and C = 233.426 [157]. The units on T here are in °C, with P_w^* in [mmHg]. Using Antoine's equation at 85 °C and K_P = 0.0022 mol/h [35], a value of $K_{y,w}$ = 0.0038 mol/h is obtained.

3.2. Environmental Scenarios

Three distinct dynamic environments are considered here:

1. The case of sudden temperature switches as implemented in Ref [35]. At a high temperature of 85 °C, the air is assumed to be completely dry, and at the low temperature of 60 °C the system is capped, allowing no transfer of water. The hot period lasts 18 h and the cold period is 6 h. Eight cycles are performed, corresponding to eight days. (Note: the day length on the early Earth was actually closer to 12 h, but that effect should not change these results substantially.)

2. The temperature is varied sinusoidally with a 24 h period over 8 days. The temperature varies between 60 and 85 °C, the same levels as in Case 1. The system is open to mass transfer at all times, and the water content (mole fraction) in the air is constant, at the saturation level for 60 °C. This water level is motivated by an environmental scenario in which the air becomes saturated at night and dries out during the dry as the temperature heats up.

3. The temperature is cycled as in the previous case, but the system is now closed to the mass transfer of water. The system does contain gas in the head space, and transfer between the liquid and gas phases can occur. However, the total water content is fixed. This case is reiminscent of reaction in the pore in a rock, another possible origins of life scenario. In this case, the amount of polymerization depends strongly on the amount of gas in the headspace. The total pressure therefore increases as the temperature rises.

3.3. Results

The system behavior associated with the three scenarios is compared in Figure 7. In all scenarios, the liquid phase is initialized with 5.55 µmol of water and 2.5×10^{-3} µmol of malic acid monomer. Case 1 is identical to the most successful case studied in Ref. [35]. At the beginning of each cold phase, 5.55 µmol of water is added to rehydrate the system, corresponding to rain, dew, or an incoming tide. After eight days, 60% of the monomer has been incorporated into oligomer. During the cold segments the water content is high and free monomer is released, as shown in Figure 7b,c. Nevertheless, the monomer concentration continues to ratchet downward during the first few cycles, and then stabilizes into cyclic steady state. Because the system is capped during the cold period, the water content in the gas phase (Figure 7d) is not modeled for this case.

In Case 2, no water is directly added during the process, but water is still generated from the reaction and can transfer between the gas and liquid phases. The monomer conversion is about half that of Case 1, as seen in Figure 7b. However, it is still possible to form a significant number of ester bonds, despite the presence of water in the liquid phase at higher temperatures. With a different sinusoidal temperature profile, one could match the same mean temperature and deviation as in Case 1, and possibly achieve enhanced conversion. (However, the model was not validated for a wider temperature range, so this case is not presented.) As shown in Figures 7c, the amount of water in the liquid phase is small relative to Case 1, indicating that the water that is present initially and that is generated by the reaction can be removed efficiently through mass transfer.

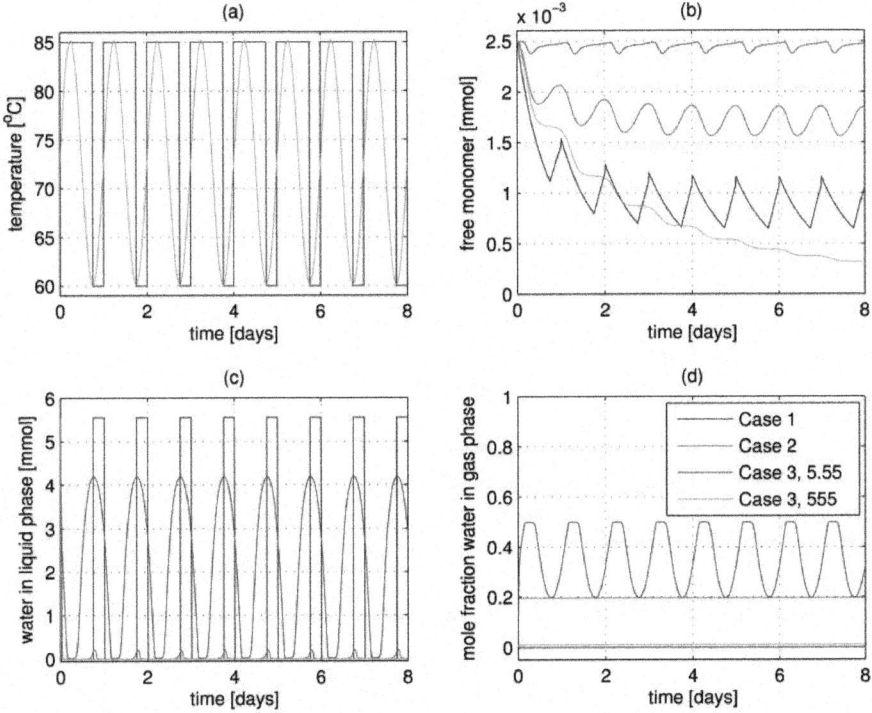

Figure 7. Monomer comsumption versus time for the three scenarios. (a) temperature versustime; (b) free monomer versus time; (c) water in liquid phase versus time; (d) water in gas phase versus time. For Case 3, results with small (5.55 μmol) and large (555 μmol) headspace are shown.

Two different versions of Case 3 are shown in Figure 7. In the first, the amount of gas (in moles) is equal to the amount of water initially present. In the second version, the amount of gas is one hundred times that amount. As shown in Figure 7, virtually no ester bond formation is achieved when the headspace is small, due to the high water content in the liquid at elevated temperature. However, a conversion greater than Case 1 is achieved for the larger headspace. As the gas phase water content is lowered in Case 2, it approaches the behavior of Case 3 for the larger headspace, since their temperature profiles are the same.

These results demonstrate that the reaction is very sensitive to the details of the environment, and that according to the model significant ester polymerization can be achieved under these more realistic scenarios. However, many simplifying assumptions made in the modeling should be considered more closely, such as ideal gas and ideal solution behavior. The reaction kinetics in the model are not dependent on viscosity or chain length, and the mass transfer is limited by the liquid side only. From these simplified models, it is clear that phase equilibria and mass transfer are critical effects to be included in origins of life kinetics modeling, and can be used to design new experiments and systems.

4. FUTURE OUTLOOK

The lessons learned from origins of life research can be applied to modern engineering problems as well. Selection has recently been applied to an exhaustive tripeptide library to identify sequences that assemble to form a hydrogel, with application to medicine and consumer products [158]. Because the relationship between peptide sequence and assembly structure is not well understood, selection can be a valuable approach for materials discovery, and can help to elucidate design principles in peptide assembly.

More broadly, Lehn pioneered the concept of dynamic combinatorial chemistry [139]. By moving beyond the chemistry of covalent bonding, supramolecular chemistry with reversible interactions can be used to create a more life-like library of diverse molecular assemblies, from simpler and continuously exchanging components. An assembly that binds to a particular target can then be selectively removed from the library, while the remaining library components continuously re-equilibrate. As a result, the best binder is continually produced by the library, at least until one of its constituents is completely depleted. The original motivation for this idea was catalysis [139] but template-triggered amplification also provides a new pathway for drug design and synthesis [159].

Lehn specifically articulated the connection between dynamic combinatorial chemistry and evolution [160] — as an chemical embodiment of selection on a diverse pool of molecules, and responding to variations in external factors such as temperature or pH. Otto proposed self-replicating systems, in simulations and also based on modified peptides and disulfide exchange. The term "systems chemistry" has been coined by chemists to describe and embrace the complexity associated with dynamic combinatorial chemistry [161,162]. Engineering approaches can potentially aid in managing such complexity [163].

Ultimately, by understanding the chemical origins of life on Earth, we can better understand the design principles associated with biological materials and biological systems. It does not appear that the chemical origins of life can be understood from a reductionist perspective, but rather that a complex reaction network would have been present. Tools from chemical engineering may be ideally suited to tackling these grand challenge research questions.

Acknowledgments

This work was jointly supported by a McDonnell Foundation 21st Century Science Initiative Grant on Studying Complex Systems No. 220020271, and by NSF and the NASA Astrobiology Program, under the NSF/NASA Center for Chemical Evolution, CHE-1004570. The authors appreciate helpful suggestions from numerous members of the Center for Chemical Evolution.

Author Contributions

The paper was written primary by Martha Grover. The literature search was led by Ming-Chien Hsieh (Section 2.1), Sheng-Sheng Yu (Section 2.2) and Christine He (Section 2.3). Ming-Chien Hsieh, Sheng-Sheng Yu and Christine He also assisted with all aspects of the paper.

Conflicts of Interest

The authors declare no conflict of interest.

REFERENCES

1. Schopf, J.W. Cradle of Life: The Discovery of the Earth's Earliest Fossils; Princeton University Press: Princeton, NJ, USA, 1999.

2. Kasting, J.F. Earth's early atmosphere. Science 1993, 259, 920–926.

3. Cleaves, H.J.; Miller, S.L. Oceanic protection of prebiotic organic compounds from UV radiation. Proc. Natl. Acad. Sci. USA 1998, 95, 7260–7263.

4. Valley, J.W.; Cavosie, A.J.; Ushikubo, T.; Reinhard, D.A.; Larson, D.J.; Clifton, P.H.; Kelly, T.F.; Wilde, S.A.; Moser, D.E.; Spicuzza, M.J. Hadean age for a post-magma-ocean zircon confirmed by atom-probe tomography. Nat. Geosci. 2014, 7, 219–223.

5. Roth, A.S.G.; Bourdon, B.; Mojzsis, S.J.; Rudge, J.F.; Guitreau, M.; Blichert-Toft, J. Combined Sm-147, Sm-146-Nd-143,Nd-142 constraints on the longevity and residence time of early terrestrial crust. Geochem. Geophys. Geosyst. 2014, 15, 2329–2345.

6. Woese, C. The universal ancestor. Proc. Natl. Acad. Sci. USA 1998, 95, 6854–6859.

7. Orgel, L.E. Prebiotic chemistry and the origin of the RNA World. Crit. Rev. Biochem. Mol. Biol. 2004, 39, 99–123.

8. Woese, C. The Genetic Code; Harper & Row: New York, NY, USA, 1967; pp. 179–195.

9. Orgel, L.E. Evolution of the genetic apparatus. J. Mol. Biol. 1968, 38, 381–393.

10. Crick, F.H.C. The origin of the genetic code. J. Mol. Biol. 1968, 38, 367–379.

11. Kruger, K.; Grabowski, P.J.; Zaug, A.J.; Sands, J.; Gottschling, D.E.; Cech, T.R. Self-splicing RNA: Autoexcision and autocyclization of the ribosomal RNA intervening sequence of tetrahymena. Cell 1982, 31, 147–157.

12. Gilbert, W. Origin of life—The RNA World. Nature 1986, 319, 618–618.

13. Joyce, G.F. The antiquity of RNA-based evolution. Nature 2002, 418, 214–221.

14. Joyce, G.F.; Schwartz, A.W.; Miller, S.L.; Orgel, L.E. The case for an ancestral genetic system involving simple analogs of the nucleotides. Proc. Natl. Acad. Sci. USA 1987, 84, 4398–4402.

15. Orgel, L.E. Some consequences of the RNA world hypothesis. Orig. Life Evol. Biosph. 2003, 33, 211–218.

16. Orgel, L. Origin of life—A simpler nucleic acid. Science 2000, 290, 1306–1307.

17. Higgs, P.G.; Lehman, N. The RNA World: Molecular cooperation at the origins of life. Nat. Genet. Rev. 2015, 16, 7–17.

18. Orgel, L.E. Self-organizing biochemical cycles. Proc. Natl. Acad. Sci. USA 2000, 97, 12503–12507.

19. Wachtershauser, G. Evolution of the first metabolic cycles. Proc. Natl. Acad. Sci. USA 1990, 87, 200–204.

20. Danger, G.; Plasson, R.; Pascal, R. Pathways for the formation and evolution of peptides in prebiotic environments. Chem. Soc. Rev. 2012, 41, 5416–5429.

21. Rode, B.; Son, H.; Suwannachot, Y.; Bujdak, J. The combination of salt induced peptide formation reaction and clay catalysis: A way to higher peptides under primitive Earth conditions. Orig. Life Evol. Biosph. 1999, 29, 273–286.

22. Segre, D.; Ben-Eli, D.; Deamer, D.W.; Lancet, D. The lipid word. Orig. Life Evol. Biosph. 2001, 31, 119–145.

23. Deamer, D.W. The first living systems: A bioenergetic perspective. Microbiol. Mol. Biol. Rev. 1997, 61, 239–261.

24. Stern, R.; Jadrzejas, M.J. Carbohydrate polymers at the center of life's origins: The importance of molecular processivity. Chem. Rev. 2008, 108, 5061–5085.

25. Miller, S.L. A production of amino acids under possible primitive earth conditions. Science 1953, 117, 528–529.

26. Bada, J.L.; Lazcano, A. Prebiotic Soup–Revisiting the Miller Experiment. Science 2003, 300, 745–746.

27. Oro, J. Mechanism of synthesis of adenine from hydrogen cyanide under possible primitive Earth conditions. Nature 1961, 191, 1193–1194.

28. Barks, H.L.; Buckley, R.; Grieves, G.A.; Di Mauro, E.; Hud, N.V.; Orlando, T.M. Guanine, adenine, and hypoxanthine production in UV-irradiated formamide solutions: Relaxation of the requirements for prebiotic purine nucleobase formation. Chembiochem 2010, 11, 1240–1243.

29. Ruiz-Bermejo, M.; Menor-Salvan, C.; Osuna-Esteban, S.; Veintemillas-Verdaguer, S. Prebiotic microreactors: A synthesis of purines and dihydroxy compounds in aqueous aerosol. Orig. Life Evol. Biosph. 2007, 37, 123–142.

30. Cleaves, H.J.; Chalmers, J.H.; Lazcano, A.; Miller, S.L.; Bada, J.L. A reassessment of prebiotic organic synthesis in neutral planetary atmospheres. Orig. Life Evol. Biosph. 2008, 38, 105–115.

31. Benner, S.A.; Kim, H.J.; Carrigan, M.A. Asphalt, water, and the prebiotic synthesis of ribose, ribonucleosides, and RNA. Acc. Chem. Res. 2012, 45, 2025–2034.

32. Pross, A. Seeking the chemical roots of Darwinism: Bridging between chemistry and biology. Chem.: Eur. J. 2009, 15, 8374–8381.

33. Goldenfeld, N.; Woese, C.W. Life is Physics: Evolutionas a Collective Phenomenon Far From Equilibrium. Annu. Rev. Condens. Matter Phys. 2011, 2, 375–399.

34. Nelson, K.E.; Levy, M.; Miller, S.L. Peptide nucleic acids rather than RNA may have been the first genetic molecule. Proc. Natl. Acad. Sci. USA 2000, 97, 3868–3871.

35. Mamajanov, I.; MacDonald, P.J.; Ying, J.; Duncanson, D.M.; Dowdy, G.R.; Walker, C.A.; Engelhart, A.E.; FernÃa_ndez, F.M.; Grover, M.A.; Hud, N.V.; et al. Ester formation and hydrolysis during wet-dry cycles: Generation of far-from-equilibrium polymers in a model prebiotic reaction. Macromolecules 2014, 47, 1334–1343.

36. Cairns-Smith, A.G. The chemistry of materials for artificial Darwinian systems. Int. Rev. Phys. Chem. 1988, 7, 209–250.

37. Damer, B.; Deamer, B. Coupled Phases and Combinatorial Selection in Fluctuating Hydrothermal Pools: A Scenario to Guide Experimental Approaches to the Origin of Cellular Life. Life 2015, 5, 872–887.

38. Darwin, F. The Life and Letters of Charles Darwin, Including an Autobiographical Chapter 3; John Murray: London, UK, 1887; p. 18.

39. Strulson, C.A.; Molden, R.C.; Keating, C.D.; Bevilacqua, P.C. RNA catalysis through compartmentalization. Nat. Chem. 2012, 4, 941–946.

40. Lahav, N.; White, D.; Chang, S. Peptide formation in the prebiotic era: Thermal condensation of glycine in fluctuating clay environments. Science 1978, 201, 67–69.

41. Mast, C.B.; Schink, S.; Gerland, U.; Braun, D. Escalation of polymerization in a thermal gradient. Proc. Natl. Acad. Sci. USA 2013, 110, 8030–8035.

42. Philip, G.K.; Freeland, S.J. Did evolution select a nonrandom "alphabet" of amino acids? Astrobiology 2011, 11, 235–240.

43. Freeland, S.J.; Knight, R.D.; Landweber, L.F.; Hurst, L.D. Early fixation of an optimal genetic code. Mol. Biol. Evol. 2000, 17, 511–518.

44. New, M.H.; Pohorille, A. An inherited efficiences model of non-genomic evolution. Simul. Pract. Theory 2000, 8, 99–108.

45. Eigen, M. The Hypercycle: A Principle of Natural Self-Organization; Springer: Berlin, Germany, 1979.

46. Lincoln, T.A.; Joyce, G.F. Self-Sustained Replication of an RNA Enzyme. Science 2009, 323, 1229–1232.

47. Sczepanski, J.T.; Joyce, G.F. A cross-chiral RNA polymerase ribozyme. Nature 2014, 515, 440–442.

48. Ashkenasy, G.; Jagasia, R.; Yadav, M.; Ghadiri, M.R. Design of a directed molecular network. Proc. Natl. Acad. Sci. USA 2004, 101, 10872–10877.

49. Wu, M.; Higgs, P. Origin of self-replicating biopolymers: Autocatalytic feedback can jumpstart the RNA world. J. Mol. Evol. 2009, 69, 541–554.

50. Szathmary, E.; Maynard Smith, J. The major evolutionary transitions. Nature 1995, 374, 227–232.

51. Schroedinger, E. What is Life? In Based on Lectures Delivered under the Auspices of the Dublin; Institute for Advanced Studies at Trinity College: Dublin, Germany, 1944.

52. Defining Life. Astrobiology Magazine, 19 June 2002.

53. Cleland, C.E.; Chyba, C.F. Defining "Life". Orig. Life Evol. Biosph. 2002, 32, 387–393.

54. Joyce, G.F. RNA evolution and the origins of life. Nature 1989, 338, 217–224.

55. Ruiz-Mirazo, K.; Briones, C.; de la Escosura, A. Prebiotic Systems Chemistry: New Perspectives for the Origins of Life. Chem. Rev. 2014, 114, 285–366.

56. Eschenmoser, A. Chemical Etiology of Nucleic Acid Structure. Science 1999, 284, 2188–2124.

57. Krishnamurthy, R. On the emergence of RNA. Isr. J. Chem. 2015, doi:10.1002/ijch.201400180.

58. Hennet, R.J.C.; Holm, N.G.; Engel, M.H. Abiotic synthesis of amino-acids under hydrothermal conditions and the origin of life—A perpetual phenomenon. Naturwissenschaften 1992, 79, 361–365.

59. Martin, W.; Baross, J.; Kelley, D.; Russell, M.J. Hydrothermal vents and the origin of life. Nat. Rev. Microbiol. 2008, 6, 805–814.

60. Plankensteiner, K.; Reiner, H.; Schranz, B.; Rode, B.M. Prebiotic formation of amino acids in a neutral atmosphere by electric discharge. Angew. Chem.-Int. Ed. 2004, 43, 1886–1888.

61. Dworkin, J.P.; Deamer, D.W.; Sandford, S.A.; Allamandola, L.J. Self-assembling amphiphilic molecules: Synthesis in simulated interstellar/precometary ices. Proc. Natl. Acad. Sci. USA 2001, 98, 815–819.

62. Hayatsu, R.; Studier, M.H.; Anders, E. Origin of organic matter in early solar system. 4. Amino acids—Confirmation of catalytic synthesis by mass spectrometry. Geochim. Cosmochim. Acta 1971, 35, 939.

63. Kobayashi, K.; Tsuchiya, M.; Oshima, T.; Yanagawa, H. Abiotic synthesis of amino-acids and imidazole by proton irradiation of simulated primitive earth atmospheres. Orig. Life Evol. Biosph. 1990, 20, 99–109.

64. McCollom, T.M.; Ritter, G.; Simoneit, B.R.T. Lipid synthesis under hydrothermal conditions by Fischer-Tropsch-type reactions. Orig. Life Evol. Biosph. 1999, 29, 153–166.

65. Nooner, D.W.; Gibert, J.M.; Gelpi, E.; Oro, J. Closed system Fischer-Tropsch synthesis over meteoritic iron, iron-ore and nickel-iron alloy. Geochim. Cosmochim. Acta 1976, 40, 915–924.

66. Parker, E.T.; Zhou, M.S.; Burton, A.S.; Glavin, D.P.; Dworkin, J.P.; Krishnamurthy, R.; Fernandez, F.M.; Bada, J.L. A plausible simultaneous synthesis of amino acids and simple peptides on the primordial earth. Angew. Chem.–Int. Ed. 2014, 53, 8132–8136.

67. Larralde, R.; Robertson, M.P.; Miller, S.L. Rate of decomposition of ribose and other sugars — Implications for chemical evolution. Proc. Natl. Acad. Sci. USA 1995, 92, 8158–8160.

68. Levy, M.; Miller, S.L. The stability of the RNA bases: Implications for the origin of life. Proc. Natl. Acad. Sci. USA 1998, 95, 7933–7938.

69. Simonov, A.N.; Pestunova, O.P.; Matvienko, L.G.; Parmon, V.N. The nature of autocatalysis in the Butlerov reaction. Kinet. Catal. 2007, 48, 245–254.

70. Weber, A.L. The sugar model: Catalysis by amines and amino acid products. Orig. Life Evol. Biosph. 2001, 31, 71–86.

71. Miller, S.L.; Urey, H.C. Organic compound synthesis on the primitive Earth. Science 1959, 130, 245–251.

72. Kolb, V.M.; Dworkin, S.L.; Miller, S.L. Alternative bases in the RNA World: The prebiotic synthesis of urazole and its ribosides. J. Mol. Evol. 1994, 38, 549–557.

73. Fuller, W.D.; Orgel, L.E.; Sanchez, R.A. Studies in prebiotic synthesis .7. Solid-state synthesis of purine nucleosides. J. Mol. Evol. 1972, 1, 249.

74. Bean, H.D.; Sheng, Y.H.; Collins, J.P.; Anet, F.A.L.; Leszczynski, J.; Hud, N.V. Formation of a beta-pyrimidine nucleoside by a free pyrimidine base and ribose in a plausible prebiotic reaction. J. Am. Chem. Soc. 2007, 129, 9556–9557.

75. Powner, M.W.; Gerland, B.; Sutherland, J.D. Synthesis of activated pyrimidine ribonucleotides in prebiotically plausible conditions. Nature 2009, 459, 239–242.

76. Powner, M.W.; Sutherland, J.D. Prebiotic chemistry: A new modus operandi. Philos. Trans. R. Soc. B-Biol. Sci. 2011, 366, 2870–2877.

77. Pasek, M.A.; Harnmeijer, J.P.; Buick, R.; Gull, M.; Atlas, Z. Evidence for reactive reduced phosphorus species in the early Archean ocean. Proc. Natl. Acad. Sci. USA 2013, 110, 10089–10094.

78. Pizzarello, S. The chemistry of life's origin: A carbonaceous meteorite perspective. Acc. Chem. Res. 2006, 39, 231–237.

79. Abramov, O.; Mojzsis, S.M. Microbial habitability of the Hadean Earth during the late heavy bombardment. Nature 2009, 459, 419–422.

80. Rufo, C.M.; Moroz, Y.S.; Moroz, O.V.; Stohr, J.; Smith, T.A.; Hu, X.Z.; DeGrado, W.F.; Korendovych, I.V. Short peptides self-assemble to produce catalytic amyloids. Nat. Chem. 2014, 6, 303–309.

81. Adamala, K.; Engelhart, A.E.; Szostak, J.W. Generation of functional RNAs from inactive oligonucleotide complexes by non-enzymatic primer extension. J. Am. Chem. Soc. 2015, 137, 483–489.

82. Martin, R.B. Free energies and equilibria of peptide bond hydrolysis and formation. Biopolymers 1998, 45, 351–353.

83. Dickson, K.S.; Burns, C.M.; Richardson, J.P. Determination of the free-energy change for repair of a DNA phosphodiester bond. J. Biol. Chem. 2000, 275, 15828–15831.

84. Liu, Z.; Beaufils, D.; Rossi, J.C.; Pascal, R. Evolutionary Importance of the Intramolecular Pathways of Hydrolysis of Phosphate Ester Mixed Anhydrides with Amino Acids and Peptides. Sci. Rep. 2014, 4, 7440.

85. Imai, E.I.; Honda, H.; Hatori, K.; Brack, A.; Matsuno, K. Elongation of oligopeptides in a simulated submarine hydrothermal system. Science 1999, 283, 831–833.

86. Ogasawara, H.; Yoshida, A.; Imai, E.I.; Honda, H.; Hatori, K.; Matsuno, K. Synthesizing oligomers from monomeric nucleotides in simulated hydrothermal environments. Orig. Life Evol. Biosph. 2000, 30, 519–526.

87. Orgel, L. The origin of polynucleotide-directed protein synthesis. J. Mol. Evol. 1989, 29, 465–474.

88. Cleaves, H.J.; Aubrey, A.D.; Bada, J.L. An evaluation of the critical parameters for abiotic peptide synthesis in submarine hydrothermal systems. Orig. Life Evol. Biosph. 2009, 39, 109–126.

89. Commeyras, A.; Collet, H.; Boiteau, L.; Taillades, J.; Vandenabeele-Trambouze, O.; Cottet, H.; Biron, J.P.; Plasson, R.; Mion, L.; Lagrille, O.; et al. Prebiotic synthesis of sequential peptides on the Hadean beach by a molecular engine working with nitrogen oxides as energy sources. Polym. Int. 2002, 51, 661–665.

90. Griffith, E.C.; Vaida, V. In situ observation of peptide bond formation at the water-air interface. Proc. Natl. Acad. Sci. USA 2012, 109, 15697–15701.

91. Ferris, J.P.; Hill, A.R.; Liu, R.; Orgel, L.E. Synthesis of long prebiotic oligomers on mineral surfaces. Nature 1996, 381, 59–61.

92. Leman, L.; Orgel, L.; Ghadiri, M.R. Carbonyl sulfide-mediated prebiotic formation of peptides. Science 2004, 306, 283–286.

93. Budjak, J.; Rode, B.M. Peptide bond formation on the surface of activated alumina: Peptide chain elongation. Catal. Lett. 2003, 91, 149–154.

94. Huang, W.; Ferris, J.P. One-step, regioselective synthesis of up to 50-mers of RNA oligomers by montmorillonite catalysis. J. Am. Chem. Soc. 2006, 128, 8914–8919.

95. Costanzo, G.; Pino, S.; Ciciriello, F.; Di Mauro, E. Generation of long RNA chains in water. J. Biol. Chem. 2009, 284, 33206–33216.

96. Monnard, P.A.; Kanavarioti, A.; Deamer, D.W. Eutectic phase polymerization of activated ribonucleotide mixtures yields quasi-equimolar incorporation of purine and pyrimidine nucleobases. J. Am. Chem. Soc. 2003, 125, 13734–13740.

97. Milner-White, E.J. The relevance of peptides that bind FeS clusters, phosphate groups, cations or anions for prebiotic evolution. In Origins of Life: The Primal Self-Organization; Springer: Berlin, Germany, 2011; pp. 155–166.

98. Lambert, J.F. Adsorption and polymerization of amino acids on mineral surfaces: A review. Orig. Life Evol. Biosph. 2008, 38, 211–242.

99. Hud, N.V. Mineral surfaces: A mixed blessing for the RNA world? Astrobiology 2009, 9, 253–255.

100. Plankensteiner, K.; Reiner, H.; Rode, B.M. Stereoselective differentiation in the salt-induced peptide formation reaction and its relevance for the origin of life. Peptides 2005, 26, 535–541.

101. Zepik, H.H.; Rajamani, S.; Maurel, M.C.; Deamer, D. Oligomerization of thioglutamic acid: Encapsulated reactions and lipid catalysis. Orig. Life Evol. Biosph. 2007, 37, 495–505.

102. Rajamani, S.; Vlassov, A.; Benner, S.; Coombs, A.; Olasagasti, F.; Deamer, D. Lipid-assisted synthesis of RNA-like polymers from mononucleotides. Orig. Life Evol. Biosph. 2008, 38, 57–74.

103. Gull, M.; Zhou, M.; Fernandez, F.M.; Pasek, M.A. Prebiotic phosphate ester syntheses in a deep eutectic solvent. J. Mol. Evol. 2014, 78, 109–117.

104. Fahnestock, S.; Rich, A. Ribosome-catalyzed polyester formation. Science 1971, 173, 340–343.

105. Weber, A.L. Thermal synthesis and hydrolysis of polyglyceric acid. Orig. Life Evol. Biosph. 1989, 19, 7–19.

106. Nielsen, P. Peptide nucleic acids and the origin of life. Chem. Biodivers. 2007, 4, 1996–2002.

107. Bean, H.D.; Anet, F.A.L.; Gould, I.R.; Hud, N.V. Glyoxylate as a backbone linkage for a prebiotic ancestor of RNA. Orig. Life Evol. Biosph. 2006, 36, 39–63.

108. Pizarello, S.; Davidowski, S.K.; Holland, G.P.; Williams, L.B. Processing of meteoritic organic materials as a possible analog of early molecular evolution in planetary environments. Proc. Natl. Acad. Sci. USA 2013, 110, 15614–15619.

109. Reches, M.; Gazit, E. Designed aromatic homo-dipeptides: formation of ordered nanostructures and potential nanotechnological applications. Phys. Biol. 2006, 3, S10–S19.

110. Szostak, J.W. The eightfold path to non-enzymatic RNA replication. J. Syst. Chem. 2012, 3, 1–14.

111. Hagenbuch, P.; Kervio, E.; Hochgesand, A.; Plutowski, U.; Richert, C. Chemical primer extension: Efficiently determining single nucleotides in DNA. Angew. Chem.-Int. Ed. 2005, 44, 6588–6592.

112. Von Kiedrowski, G. A Self-Replicating Hexadeoxynucleotide. Angew. Chem. Int. Ed. 1986, 25, 932–935.

113. Li, T.; Nicolaou, K. Chemical self-replication of palindromic duplex DNA. Nature 1994, 369, 218–221.

114. Zielinski, W.S.; Orgel, L.E. Autocatalytic synthesis of a tetranucleotide analogue. Nature 1987, 327, 346–347.

115. Sievers, D.; von Kiedrowski, G. Self-replication of complementary nuceotide-based oligomers. Nature 1994, 369, 221–224.

116. Freier, S.M.; Kerzek, R.; Jaeger, J.A.; Sugimoto, N.; Caruthers, M.H.; Neilson, T.; Turner, D.H. Improved free-energy parameters for predictions of RNA duplex stability. Proc. Natl. Acad. Sci. USA 1986, 83, 9373–9377.

117. SantaLucia, J.; Hicks, D. The thermodynamics of DNA structural motifs. Annu. Rev. Biophys. Biomol. Struct. 2004, 33, 415–440.

118. Grossmann, T.N.; Strohback, A.; Seitz, O. Achieving Turnover in DNA-Templated Reactions. ChemBioChem 2008, 9, 2185–2192.

119. Fernando, C.; von Kiedrowski, G.; Szathmary, E. A Stochastic Model of Nonenzymatic Nucleic Acid Replication: "Elongators" Sequester Replicators. J. Mol. Evol. 2007, 64, 572–585.

120. Zhan, Z.Y.; Lynn, D.G. Chemical Amplification through Template-Directed Synthesis. J. Am. Chem. Soc. 1997, 119, 12420–12421.

121. Kausar, A.; McKay, R.D.; Lam, J.; Bhogal, R.S.; Tang, A.Y.; Gibbs-Davis, J.M. Tuning DNA stability to achieve turnover in template for an enzymatic ligation reaction. Angew. Chem. Int. Ed. 2011, 50, 8922–8926.

122. Dose, C.; Ficht, S.; Seitz, O. Reducing product inhibition in DNA-template-controlled ligation reactions. Angew. Chem. Int. Ed. 2006, 45, 5369–5373.

123. Luther, A.; Brandsch, R.; Kiedrowski, G.V. Surface-promoted replication and exponential amplification of DNA analogues. Nature 1998, 396, 245–248.

124. Deck, C.; Jauker, M.; Richert, C. Efficient enzyme-free copying of all four nucleobases templated by immobilized RNA. Nat. Chem. 2011, 3, 603–608.

125. Kreysing, M.; Keil, L.; Lanzmich, S.; Braun, D. Heat flux across an open pore enables the continuous replication and selection of oligonucleotides toward increasing length. Nat. Chem. 2015, 7, 203–208.

126. Engelhart, A.E.; Powner, M.W.; Szostak, J.W. Functional RNAs exhibit tolerance for non-heritable 2′-5′ versus 3′-5′ backbone heterogeneity. Nat. Chem. 2013, 5, 390–394.

127. Eigen, M. Self-organization of matter and evolution of biological macromolecules. Naturwissenschaften 1971, 58, 465–523.

128. Podlech, J. Origin of organic molecules and biomolecular homochirality. Cell. Mol. Life Sci. 2001, 58, 44–60.

129. Hein, J.E.; Blackmond, D.G. On the origin of single chirality of amino acids and sugars in biogenesis. Acc. Chem. Res. 2012, 45, 2045–2054.

130. Cronin, J.R.; Pizzarello, S. Enantiomeric excesses in meteoritic amino acids. Science 1997, 275, 951–955.

131. Bailey, J.; Chrysostomou, A.; Hough, J.H.; Gledhill, T.M.; McCall, A.; Clark, S.; Menard, F.; Tamura, M. Circular polarization in star-formation regions: Implications for biomolecular homochirality. Science 1998, 281, 672–674.

132. Hazen, R.M.; Filley, T.R.; Goodfriend, G.A. Selective adsorption of L- and D-amino acids on calcite: Implications for biochemical homochirality. Proc. Natl. Acad. Sci. USA 2001, 98, 5487–5490.

133. Glavin, D.P.; Dworkin, J.P. Enrichment of the amino acid L-isovaline by aqueous alteration on CI and CM meteorite parent bodies. Proc. Natl. Acad. Sci. USA 2009, 106, 5487–5492.

134. Goldberg, S.I. Enantiomeric Enrichment on the Prebiotic Earth. Orig. Life Evol. Biosph. 2007, 37, 55–60.

135. Blair, N.E.; Bonner, W.A. A model for the enantiomeric enrichment of polypeptides on the primitive Earth. Orig. Life 1981, 11, 331–335.

136. Hein, J.E.; Tse, E.; Blackmond, D.G. A route to enantiopure RNA precursors from nearly racemic starting materials. Nat. Chem. 2011, 3, 704–706.

137. Breslow, R.; Cheng, Z.L. L-amino acids catalyze the formation of an excess of D-glyceraldehyde, and thus of other D sugars, under credible prebiotic conditions. Proc. Natl. Acad. Sci. USA 2010, 107, 5723–5725.

138. Hud, N.V.; Cafferty, B.J.; Krishnamurthy, R.; Williams, L.D. The origin of RNA and 'my grandfather's axe'. Chem. Biol. 2013, 20, 466–474.

139. Lehn, J.M. Perspectives in supramolecular chemistry — From molecular recognition towards molecular information processing and self-organization. Angew. Chem. Int. Ed. 1990, 29, 1304–1319.

140. Whitesides, G.M.; Mathias, J.P.; Seto, C.T. Molecular self-assembly and nanochemistry: A chemical strategy for the synthesis of nanostructures. Science 1991, 254, 1312–1319.

141. Cafferty, B.J.; Gallego, I.; Chen, M.C.; Farley, K.I.; Eritja, R.; Hud, N.V. Efficient self-assembly in water of long noncovalent polymers by nucleobase analogs. J. Am. Chem. Soc. 2013, 135, 2447–2450.

142. Menor-Salvan, C.; Ruiz-Bermejo, M.; Guzman, M.I.; Osuna-Esteban, S.; Veintemillas-Verdaguer, S. Synthesis of pyrimidines and triazines in ice: Implications for the prebiotic chemistry of nucleobases. Chem.-A Eur. J. 2009, 15, 4411–4418.

143. Chen, M.C.; Cafferty, B.J.; Mamajanov, I.; Gallego, I.; Khanam, J.; Krishnamurthy, R.; Hud, N.V. Spontaneous prebiotic formation of a β-ribofuranoside that self-assembles with a complementary heterocycle. J. Am. Chem. Soc. 2014, 136, 5640–5646.

144. Athavale, S.S.; Spicer, B.; Chen, I.A. Experimental fitness landscapes to understand the molecular evolution of RNA-based life. Curr. Opin. Chem. Biol. 2014, 22, 35–39.

145. Maynard Smith, J.; Szathmary, E. The Major Transitions in Evolution; Oxford: New York, NY, USA, 1995.

146. Wattis, J.A.D.; Coveney, P.V. The origin of the RNA world: A kinetic model. J. Phys. Chem. B 1999, 103, 4231–4250.

147. Coveney, P.V.; Swadling, J.B.; Wattis, J.A.D.; Greenwell, H.C. Theory, modelling and simulation in origins of life studies. Chem. Soc. Rev. 2012, 41, 5430–5446.

148. Walker, S.I.; Grover, M.A.; Hud, N.V. Universal sequence replication, reversible polymerization and early functional biopolymers: A model for the initiation of prebiotic sequence evolution. PLoS ONE 2012, 7, doi:10.1371/journal.pone.0034166.

149. Boerlijst, M.; Hogeweg, P. Spiral wave structure in pre-biotic evolution: Hypercycles stable against parasites. Phys. D 1991, 48, 17–28.

150. Novak, V.J.A. Present state of the coacervate in coacervate theory — Origin and evolution of the cell structure. Orig. Life Evol. Biosph. 1984, 14, 513–522.

151. Monnard, P.A.; Deamer, D.W. Membrane self-assembly processes: Steps toward the first cellular life. Anat. Rec. 2002, 268, 196–207.

152. Koga, S.; Williams, D.S.; Perriman, A.W.; Mann, S. Peptide-nucleotide microdroplets as a step towards a membrane-free protocell model. Nat. Chem. 2011, 3, 720–724.

153. Monnard, P.A.; Szostak, J.W. Metal-ion catalyzed polymerization in the eutectic phase in water-ice: A possible approach to template-directed RNA polymerization. J. Inorg. Biochem. 2008, 102, 1104–1111.

154. Chen, I.A.; Roberts, R.W.; Szostak, J.W. The emergence of competition between model protocells. Science 2004, 305, 1474–1476.

155. Budin, I.; Szostak, J.W. Physical effects underlying the transition from primitive to modern cell membranes. Proc. Natl. Acad. Sci. USA 2011, 108, 5249–5254.

156. Harshe, Y.M.; Storti, G.; Morbidelli, M.; Gelosa, S.; Moscatelli, D. Polycondensation kinetics of lactic acid. Macromol. React. Eng. 2007, 1, 611–621.

157. Smith, J.M.; Van Ness, H.C.; Abbott, M.M. Introduction to Chemical Engineering Thermodynamics, 7th ed.; McGraw-Hill: New York, NY, USA, 2005; p. 682.

158. Frederix, P.W.J.M.; Scott, G.G.; Abul-Haija, Y.M.; Kalafatovic, D.; Pappas, C.G.; Javid, N.; Hunt, N.T.; Ulijn, R.W.; Tuttle, T. Exploring the sequence space for (tri-)peptide self-assembly to design and discover new hydrogels. Nat. Chem. 2015, 7, 30–37.

159. Otto, S. Dynamic combinatorial chemistry: A new method for selection and preparation of synthetic receptors. Curr. Opin. Drug Discov. Dev. 2003, 6, 509–520.

160. Lehn, J.M. Toward self-organization and complex matter. Science 2002, 295, 2400–2403.

161. Ludlow, R.F.; Otto, S. Systems Chemistry. Chem. Soc. Rev. 2008, 37, 101–108.

162. Otto, S. Dynamic molecular networks: From synthetic receptors to self-replicators. *Acc. Chem. Res.* 2012, *45*, 2200-2210.

163. Hernandez, A.F.; Wagner, M.J.; Grover, M.A. Model identification of a template-directed peptide network for optimization in a continuous reactor. *Chem. Commun.* 2014, *50*, 3849-3851.

INDEX